International Congress on Energy Fluxes and Radiation Effects

(EFRE 2014)

AA002778

Journal of Physics: Conference Series Volume 552

Tomsk, Russia
21-26 September 2014

ISBN: 978-1-63439-596-0
ISSN: 1742-6588

Printed from e-media with permission by:

Curran Associates, Inc.
57 Morehouse Lane
Red Hook, NY 12571

CURRAN ASSOCIATES INC.
proceedings
.com

Some format issues inherent in the e-media version may also appear in this print version.

Copyright© by the Institute of Physics
All rights reserved.

Printed by Curran Associates, Inc. (2015)

For permission requests, please contact the Institute of Physics
at the address below.

Institute of Physics
Dirac House, Temple Back
Bristol BS1 6BE UK

Phone: 44 1 17 929 7481
Fax: 44 1 17 920 0979

techtracking@iop.org

Additional copies of this publication are available from:

Curran Associates, Inc.
57 Morehouse Lane
Red Hook, NY 12571 USA
Phone: 845-758-0400
Fax: 845-758-2634
Email: curran@proceedings.com
Web: www.proceedings.com

TABLE OF CONTENTS

12TH INTERNATIONAL CONFERENCE ON MODIFICATION OF MATERIALS WITH PARTICLE BEAMS AND PLASMA FLOWS

012001 Investigation of Parameters and Uninterrupted Service of a Generator of Gas-discharge Plasma Based on a Nonsustained Hot-cathode Arc Discharge in Gas 1
D P Borisov, A D Korotaev, V M Kuznetsov, E V Chulkov

012002 Numerical Modelling of Physical Processes and Structural Changes in Metals Under Intensive Irradiation with Use of CRS Code: Dislocations, Twinning, Evaporation and Stress Waves 6
A E Mayer, E N Borodin, V S Krasnikov, P N Mayer

012003 Transport Charge of Gallium Arsenide Films Synthesized on Polycrystalline Silicon by Ion Ablation 13
A V Kabyshev, F V Konusov, G E Remnev, S K Pavlov

012004 on the Focused Beam Parameters of an Electron Gun with a Plasma Emitter 18
S Kornilov, N Rempe, A Beniyash, N Murray

012005 Discharge in the Saline Solutions in a Vicinity of the Threshold Voltages 24
Y D Korolev, I A Shemyakin, R V Ivashov, V S Kasyanov, N V Landl, Y H Sun, T Shao, Y Gao

012006 Study of Hydrogen Accumulation in Palladium and Silver- Palladium Alloy 31
N N Nikitenkov, Yu I Tyurin, V S Sypchenko, I V Dushkin, A N Nikitenkov, O V Vilkhivskaya

012007 Unbalance Magnetron Plasma Source for Ion Mass-separator 36
V L Paperny, V I Krasov, N V Astrakchantsev, N V Lebedev

012008 Boron Carbide Nanopowder Synthesized Using Electrical Discharge Plasma 42
I A Rakhmatullin, A A Sivkov, A F Makarova

012009 Ion Trajectories Calculation for Negatively Biased Needle Cathode in Volume Discharge Plasma 46
A G Remnev, K Uemura, A V Kozyrev, V V Lopatin

012010 Bias-assisted Magnetron Sputtering of Yttria-stabilised Zirconia Thin Films 52
A A Solovyev, S V Rabotkin, I V Ionov, A V Shipilova, A N Kovalchuk, A O Borduleva

INTERNATIONAL SYMPOSIUM ON HIGH-CURRENT ELECTRONICS

012011 Experimental Study of the Neck Formation in an X Pinch 58
A P Artyomov, S A Chaikovsky, A V Fedunin, N A Labetskaya, A G Rousskikh, A S Zhigalin, V I Oreshkin

012012 Electrical Discharge Phenomena Application for Solid Fossil Fuels In-situ Conversion 64
A A Bukharkin, V V Lopatin, S M Martemyanov, I A Koryashov

012013 Dynamics of Large-scale Instabilities in Conductors Electrically Exploded in Strong Magnetic Fields 68
I M Datsko, S A Chaikovsky, N A Labetskaya, V I Oreshkin, N A Ratakhin

012014 Effect of Electron Extraction from a Grid Plasma Cathode on the Generation of Emission Plasma 73
V N Devyatkov, N N Koval

012015 Effect of Laser Radiation Wavelength on Explosives Initiation Thresholds 79
R Akhmetshin, A Razin, V Ovchinnikov, A Skripin, V Tsipilev, V Oleshko, V Zarko, A Yakovlev

012016 Optimization of the Vacuum Insulator Stack of the MIG Pulsed Power Generator 83
G Khamzakhan, S A Chaikovsky

012017 Features of Pulsed Electron Beam Propagation with Nanosecond Duration in the Air Under the Low (Reduced) Pressure 89
G Kholodnaya, R Sazonov, D Ponomarev, G Remnev, A Vikanov

012018 Annular Electron Beam with Virtual Cathode in a Coaxial Diode with Magnetic Insulation 95
A A Grishkov, I K Kurkan, I V Pegel

012019 Current Interruption in a Low-pressure High-current Pulsed Discharge with Hollow Cathode 101
Y D Korolev, N V Landl, V G Geyman, O B Frants, A V Bolotov, V O Nekhoroshev, V S Kasyanov

012020 Diffusion of Fast Rising Strong Magnetic Fields Into Conductors 108
N A Labetskaya, V I Oreshkin, S A Chaikovsky, I M Datsko, N I Kuskova, A D Rud

012021 Numerical Simulation and Analysis of Energy Loss in a Nanosecond Spark Gap Switch 114
I V Lavrinovich, V I Oreshkin

012022 Oversized Interference Switches of Active Resonant Microwave Compressors 120
V Avgustinovich, S Artemenko, V Kaminsky, S Novikov, Yu Yushkov, V Osipov

012023 An Avalanche of Runaway Electrons Formed in an Air Discharge 125
E V Oreshkin, S A Barengolts, S A Chaikovsky, V I Oreshkin

012024 Study on Nanosecond Pulsed Electron Beam Generation 130
D Ponomarev, G Kholodnaya, G Remnev, M Kaikanov, R Sazonov

012025 Features of Silicon-containing Coatings Deposition from Ablation Plasma Formed by a Powerful Ion Beam 136
R Sazonov, G Kholodnaya, D Ponomarev, G Remnev, I Khailov

012026 Assessment of Thermodynamic Parameters of Plasma Shock Wave 143
O V Vasileva, Yu N Isaev, A A Budko, A I Filkov

012027 Experimental Research of the Fine Foil Explosion Dynamics 148
A S Zhigalin, A G Rousskikh, V I Oreshkin, S A Chaikovsky, N A Ratakhin, V V Kuznetsov

012028 Multi-borehole Electro-blast Method for Concrete Monolith Splitting Off 154
A S Yudin, N S Kuznetsova, V V Lopatin, N V Voitenko

012029 Effect of Electro-discharge Circuit Parameters on the Destructive Action of Plasma Channel in Solid Media 159
N S Kuznetsova, V V Lopatin, A S Yudin

012030 Measurements of High Voltage Pulses with Subnanosecond Rise Time 166
N M Bykov, I K Kurkan, A S Stepchenko

012031 Foil-less Plasma-filled Diode for HPM Generator 172
A A Eltchaninov, B M Kovalchuk, I K Kurkan, A A Zherlitsyn

INTERNATIONAL CONFERENCE ON RADIATION PHYSICS AND CHEMISTRY OF CONDENSED MATTER

012032 A Study Into Light Scattering and Absorption by Aluminum Nanoparticles in PETN 177
S B Aduev, D Nurmukhametov, A Zvekov, A Nikitin

012033 Theoretical Approach to Modelling the Low-barrier Chemical Reactions Initiated by Pulsed Electron Beam 184
N E Aktaev

012034 Portable Device for the Detection of Nitro-explosives Based on Optical Properties of Sensor's Material 189
A A Baranova, K O Khokhlov

012035 On Channeling Potential for Relativistic Electrons in Crossed Laser Field 194
E N Frolov, A V Dik, S B Dabagov, K P Artyomov

012036 Radiation Optical Effects in Commercial SiO_2:Ge Fibers 199
S A V Ishchenko, B V Shulgin, A S Vokhmintsev, I A Weinstein, L N Shalimov, N G Manko, A N Shtykov, G V Shestakov, E K Chistyakova, A A Shonokhova

012037 Modern Microcenter Heat Explosion Model 206
A V Kalenskii, V G Kriger, I Yu Zykov, M V Anan'Eva

012038 Pulsed Cathodoluminescence of LiF-Fe_2O_3 Crystals 212
V I Korepanov, P V Petikar, A A Kamrikova

012039 Modern Trends and Development in High-dose Luminescent Measurements 218
V Kortov

012040 Kinetics of the Cathodoluminescence Flash in Crystals with Competing Trapping Centres 225
V Lisitsyn, L Lisitsyna, Z Karipbaev, D Valiev

012041 Substitutional Impurities Influence on Hydrogen Sorption Properties of Titanium and Zirconium 230
T I Spiridonova, E V Tuch, A V Bakulin, S E Kulkova

012042 The Role of Deep Centers in Formation of Dosimetric Properties of Wide-gap Materials 235
S V Nikiforov, V S Kortov

012043 Photoluminescence of Nanosized Zn_2SiO_4:Mn Depending Upon Preparation Method 241
K A Petrovykh, V S Kortov, A A Rempel

012044 Initiation of Furazanotetrazinedioxide and Mixes on Its Basis by High-current Electron Beam 246
V I Oleshko, V E Zarko, V V Lysyk, V P Tsipilev, P I Kalmykov

012045 Inhomogeneous Change of Temperature of Ionic Crystals Under the Action of a Pulsed Electron Beam 251
S A Stepanov, V F Shtan'Ko, E P Chinkov

012046 Electrical and Photoelectric Properties of Polycrystalline Diamond Films Deposited from an Abnormal Glow Discharge 256
F V Konusov, A V Kabyshev, S A Linnik, A V Gaydaychuk, G E Remnev
012047 Calculation of the Spatial Distribution of Defects and Cascade- Probability Functions in the Materials 261
A I Kupchishin, A A Kupchishin, E V Shmygalev, T A Shmygaleva, K B Tlebaev
012048 Study of the Fluorescence Blinking Behavior of Single F_2 Color Centers in Lif Crystal 268
S V Boichenko, K Koenig, S A Zilov, V P Dresvianskiy, A P Voitovich, A L Rakevich, A V Kuznetsov, A V Bartul, E F Martynovich
012049 Formation of Neutron Fields for Radiation Technologies 272
V A Varlachev, E G Emets, S I Kuznetsov, A M Bogdan, N V Varlacheva
012050 Electric Pulse Breakdown and Rock Fracture in a Coupled Environment of Increased Pressure and Temperature 276
V F Vazhov, S Yu Datskvich, M Yu Zhurkov, V M Muratov
Author Index

Preface

The International Congress on Energy Fluxes and Radiation Effects 2014 (EFRE 2014) was held in Tomsk, Russia, on September 21-26, 2014. The organizers of the Congress were the Institute of High Current Electronics SB RAS and Tomsk Polytechnic University.

EFRE 2014 combines three international conferences which are regularly held in Tomsk, Russia: the 18th International Symposium on High-Current Electronics (18th SHCE), the 12th International Conference on Modification of Materials with Particle Beams and Plasma Flows (12th CMM) and the 16th International Conference on Radiation Physics and Chemistry of Condensed Matter (16th RPC). The International Conference on Radiation Physics and Chemistry of Condensed Matter is a traditional representative forum devoted to the discussion of the fundamental problems of physical and chemical non-linear processes in condensed matter (mainly inorganic dielectrics) under the action of particle and photon beams of all types including pulsed power laser radiation. The International Symposium on High-Current Electronics is held biannually in Tomsk, Russia. The program of the conferences covers a wide range of scientific and technical areas including pulsed power technology, ion and electron beams, high-power microwaves, plasma and particle beam sources, modification of materials, and pulsed power applications in chemistry, biology and medicine. The 12th International Conference on Modification of Materials with Particle Beams and Plasma Flows is devoted to the discussion of the fundamental and applied issues in the field of modification of materials properties with particle beams and plasma flows.

The six-day Congress brought together more than 250 specialists and scientists from different countries and organizations and provided an excellent opportunity to exchange knowledge, make oral contributions and poster presentations, and initiate discussion on the topics of interest.

The proceedings were edited by Victor Lisitsyn, Vladimir Lopatin, and Anna Bogdan.

We appreciate the contribution of the invited speakers and all participants, as well as sponsors "Intech Analytics" and "MICROSPLAV" for making the Congress successful.

Content from this work may be used under the terms of the Creative Commons Attribution 3.0 licence. Any further distribution of this work must maintain attribution to the author(s) and the title of the work, journal citation and DOI.
Published under licence by IOP Publishing Ltd

Peer review statement

All papers published in this volume of *Journal of Physics: Conference Series* have been peer reviewed through processes administered by the proceedings Editors. Reviews were conducted by expert referees to the professional and scientific standards expected of a proceedings journal published by IOP Publishing.

Content from this work may be used under the terms of the Creative Commons Attribution 3.0 licence. Any further distribution of this work must maintain attribution to the author(s) and the title of the work, journal citation and DOI.
Published under licence by IOP Publishing Ltd

Investigation of parameters and uninterrupted service of a generator of gas-discharge plasma based on a nonsustained hot-cathode arc discharge in gas

D P Borisov*, A D Korotaev, V M Kuznetsov and E V Chulkov
Tomsk State University, Tomsk 634050, Russia

E-mail: borengin@mail.ru

Abstract. The design of an electrode system and the characteristics of a gas-discharge plasma source (generator) based on an arc discharge in gas sustained by thermionic emission are presented. The results of an experimental investigation of the service life of a thermionic (hot) cathode of the plasma generator are reported as a function of the discharge current and voltage and the cathode filament power. It is shown that an increase in the burning voltage from 28 to 60 V considerably reduces the service life of the hot cathode, while the discharge current and the cathode filament power exert a less critical effect on its lifetime. The conditions necessary for achieving long-term service life parameters and high efficiency of plasma generation are found out.

1. Introduction

The generator of gas-discharge plasma based on non-self-sustained hot-cathode arc discharge developed in early 1990-s [1] has been until now the most acceptable facility for surface cleaning, etching, hard surfacing (nitriding), doping and plasma-assisted coating deposition due to simplicity of its design and high efficiency of plasma production [2, 3]. On the other hand, the use of a thermionic (hot) cathode in this plasma generator in addition to its high advantage of producing high-concentration plasma (~ 10^{11} cm^{-3}) at low pressures (≤ 0.1 Pa) and a potential of controlling its parameters in a wide range of operating pressures (0.1÷1 Pa) is, however, limited by its service life (lifetime). Due to high operating temperatures (≥ 2500°C) and bombardment and sputtering by plasma ions, this cathode fails (blows off) within an uncertain period of time (from a few to several tens of hours) under different operating modes, which makes the service life (lifetime) of the hot cathode in particular technological conditions indeterminable. Thus, an investigation of the dependence of the hot cathode lifetime on the plasma generator parameters and operating modes and identification of the conditions for increasing the continuous operation of the plasma generator itself (until replacement of the hot cathode) is of critical importance for achieving an improvement of vacuum-plasma technologies.

2. Gas-discharge plasma generator design and operation

A photograph and a concept scheme of the hot-cathode gas-discharge plasma generator are presented in Fig. 1 (a) and (b), respectively. The main element of the electrode system is a hybrid cathode consisting of a stainless-steel cylindrical hollow cathode *1* (Figure 1) measuring 86 mm in diameter

Content from this work may be used under the terms of the Creative Commons Attribution 3.0 licence. Any further distribution of this work must maintain attribution to the author(s) and the title of the work, journal citation and DOI.
Published under licence by IOP Publishing Ltd

and 380 mm in length and a thermionic cathode made of tungsten wires (filaments) immersed into it *2*. This hybrid cathode is placed into 0.02T magnetic field generated by a solenoid coil *3* and is clamped to a water-cooled flange of the cathode unit *4*, which is electrically insulated from the water-cooled housing of the plasma generator *5* via insulator *6*. The plasma generator in its housing *5*, which is under the anode potential, is placed onto the vacuum chamber, a hollow anode of the discharge *7*, in order to generate gas-discharge plasma by igniting a gas discharge in this chamber, which occurs as follows. When the thermionic cathode is hot enough and a voltage is supplied from the gas-discharge power supply, the electrons emitted from the hot cathode move between the vacuum chamber (anode) and the hybrid cathode (electrically coupled hollow and hot cathodes) along the magnetic field lines towards the anode (inner walls of the vacuum chamber). Since the working gas *8* is fed into the cathode cavity of the plasma generator, a region of elevated pressure is formed near the hot cathode. The presence of a region with increased concentration of the working-gas molecules and a flow of accelerated electrons, whose trajectories are elongated by the magnetic field, makes the ionization processes easier. The electrons, which bounce back from the potential barrier generated by the cathode cavity, oscillate in the cavity and effectively ionize the working gas molecules, causing ignition and burning of a non-self-sustained low-pressure arc discharge. The discharge persists between the cathode, made up by both the hot cathode and the hollow cathode connected to it, and the anode represented by the inner surface of the vacuum chamber. The gas-discharge plasma *9*, generated by this discharge, fills the vacuum chamber cavity and serves as a working medium for surface modification of the pieces immersed into it.

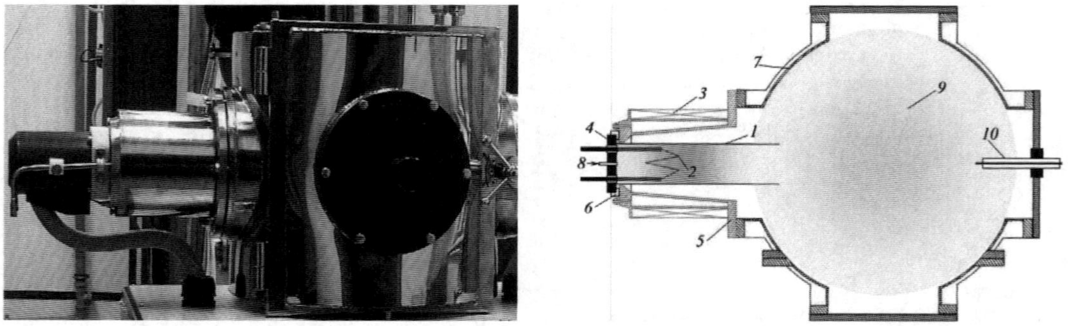

Figure 1. Photograph (a) and a concept scheme (b) of the gas-discharge plasma generator.

The characteristics of this plasma generator and the lifetime of its cathode were investigated during its operation with a cylindrical vacuum chamber 600 mm in diameter and 500 mm in height. The thermionic cathode used in all experiments was represented by a single VA-tungsten wire 0.8 mm in diameter and the length of the operating section 150 mm.

3. Dependence of the lifetime of a hot cathode on the plasma generator parameters

A characteristic feature and advantage of the hot-cathode gas-discharge plasma generator is the fact that its discharge current and hence the concentration of the resulting plasma can be adjusted by varying both the filament power and emissivity of the cathode (filament) and the value of the discharge electrical supply voltage. This is supported by the dependence of the plasma generator discharge current in argon at a pressure of 0.3 Pa on the filament power for different discharge voltage values - Figure 2.

A high discharge current can be achieved by increasing the number of filaments of the cathode, e.g., up to six. In the latter case, given their total filament power P_f=1920 W (320 W per every filament) at the discharge burning voltage U_d=60 V and pressure 0.3 Pa, the discharge current in argon would increase up to I_d=140 A. The flow of this current in the vacuum chamber generated plasma with a concentration n_i=2.4•10^{11} cm^{-3}, which was measured from the saturation ion current of the probe *10* (Figure 1).

The two available mechanisms of controlling the plasma generator discharge current also offer a possibility of sustaining the required current for a wide range of operating pressures. For instance, for the pressure varying from 0.67 to 0.05 Pa, by controlling the consumption of argon from 0.3 to 0.02 cm^3/s discharge current of I_d=22 A could be maintained within the entire pressure range due to an increase in the discharge pressure from 28 to 115 V at the same value of the filament power P_f=320 W. The same discharge current value I_d=22 A was maintained unchanged in the entire working gas pressure interval due to an increase in the filament power from 300 to 490 W for the same discharge voltage U_d=72 V.

The experiment on investigating the influence of the discharge voltage U_d on the hot-cathode lifetime τ was performed for the case of discharge burning in argon at the argon pressure p_{Ar}=0.3 Pa at the voltage values 28, 60 and 85 V and the average filament power P_f for these voltage values within the experimental time 620, 400 and 330 W, respectively, ensured the same discharge current I_d=30 A for every filament voltage used. The dependence under study is plotted in Figure 3.

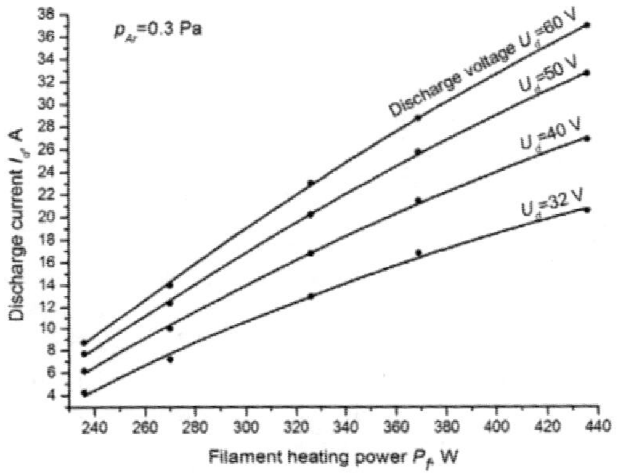

Figure 2. Plasma-generator discharge current versus the hot-cathode filament power for different values of discharge voltage.

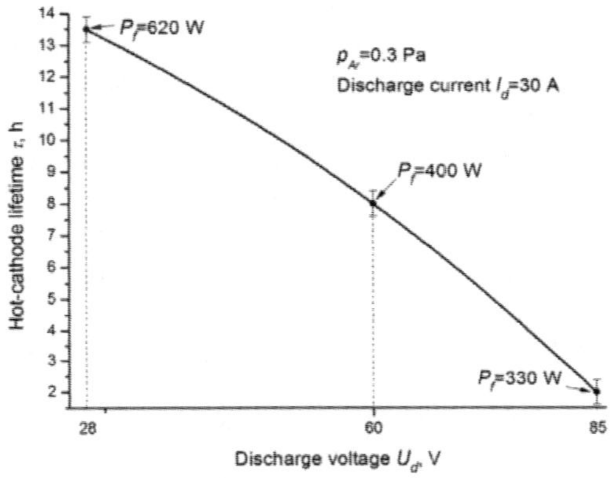

Figure 3. Hot-cathode lifetime as a function of the plasma generator discharge burning voltage.

It is evident from the plot that it is the discharge burning voltage rather than the hot-cathode filament power, which extremely shortens the hot-cathode lifetime. For instance, for a relatively low average filament power P_f=330 W, ruling out excessive overheating and accelerated evaporation of the filament material and a comparatively high discharge burning voltage U_d=85 V, the filament lifetime τ for the discharge current I_d=30 A was found to be 2 hours only. For the same discharge current, the filament lifetime of this plasma generator hot cathode operated in a low-burning-voltage mode (28 V) was found to be 13.5 hours, despite the fact that the average filament power for that mode was quite high P_f=620 W. Thus, it becomes evident that it is the wear of the filament by ion bombardment controlled by the burning voltage rather that the filament wear due to its overheating, which critically affects the cathode lifetime, shortening its service life and that of the plasma generator on the whole. For the discharge burning voltage U_d=60 V and the average (for the mode in question) filament power P_f=400 W and the same discharge current as in the above-mentioned examples (I_d=30 A), the lifetime τ of the hot-cathode filament was found to be 8 hours.

Relying on the results obtained in this experiment, we can formulate a guideline of how to extend the lifetime of tungsten wires in the filament-type hot cathodes operated in gas-discharge plasma generators – in order to do so we have to maintain their discharge burning voltage as low as possible. On the other hand, there is an assumption that a decrease in the discharge burning voltage of the plasma generator would "shorten" the propagation path of the plasma generated in the vacuum chamber, the chamber would not be completely filled with the plasma, and its concentration inside the vacuum chamber far from the generator would be low. This assumption relies on the fact that a decrease in the discharge burning voltage controlling the energy of primary electrons would shorten the path length of the latter in the space of the vacuum chamber due to a faster energy loss in their collisions with plasma particles. In view of this circumstance, the plasma produced by the plasma generator at low burning voltages would exert a significantly lower influence on the substrates located comparatively far from the plasma generator. In order to obtain a quantitative assessment of the effect of decreased plasma concentration far from the generator due to a decrease in the discharge burning voltage, we measured the plasma density at the distance l=330 mm from the outlet of the hollow cathode of the plasma generator using a probe *10* (Figure. 1). The resulting data for the same argon pressure as in the previous experiment are presented in Figure 4.

An analysis of the plot (Figure 4) of the saturation ion current density of the probe far from the plasma for the burning voltages (28, 60 and 85 V) used in the generator in the course of the lifetime tests demonstrated the following:

- the differences in the ion current picked off by the probe and hence the work piece being processed at a comparatively large distance from the plasma generator (l=330 mm) as a function of the discharge burning voltage are negligibly small;

- for the discharge current used in the life-time testing experiments equal to I_d=30 A (dotted line in Figure 4), the difference in the ion current densities obtained for the most differing burning voltages – 85 and 28 V – is as small as 0.0465 mA/cm^2 or ~11.5 % of the maximum ion current density, 0.408 mA/cm^2, which was obtained under these conditions at the burning voltage value 85 V.

Given the fact that the losses in the ion current density for thee burning voltage decreasing from 85 to 28 V at the same discharge current are negligible (~11.5 %), we can conclude that there is no need to considerably increase the discharge burning voltage, which substantially reduces the lifetime of the thermionic cathode of the plasma generator during technological plasma modification of product properties. It should also be noted that considering the burning-voltage data obtained as a result of lifetime tests – Figure 3, by operating the plasma generator at lower burning voltages we could considerably increase the lifetime of its hot cathode, e.g., by a factor of 6.75 in reducing the voltage from 85 to 28 V. On the other hand, a decrease in the ion current density on the work piece being modified, which is associated with the decreasing discharge burning voltage, could be compensated for by a small in value increase in the discharge current due to a higher filament power of the hot cathode of the generator. Specifically, Figure 4 implies that in order to make up for the ion-current loss resulting from the burning voltage decreasing from 85 to 28 V, we have to increase the plasma generator discharge current from 30 to only 33.5 A.

A possible application of the reserve in the discharge current made available by increasing the filament power of the hot cathode, rather than by increasing the filament voltage, without markedly decreasing the cathode lifetime could be determined via investigating the effects of the discharge current on the lifetime of the hot cathode.

Figure 5 depicts the lifetime values obtained as a result of the above investigations for the same hot-cathode filament at the discharge current values I_d equal to 10, 20 and 30 A and the respective filament power values P_f equal to 240, 300 and 400 W for every lifetime test, which ensure the same burning voltage U_d=60 V under all operating conditions under study.

It is evident from Figure 5 that the lifetime of hot cathode decreases nearly linearly with an increase in the discharge current: for the discharge current I_d=10 A it is τ=20 hours, for I_d=20 A – τ=13.4 hours, and for I_d=30 A – τ=8 hours. A threefold increase in the discharge current from 10 to 30 A is observed to result in a factor of 2.5 decrease (from 20 to 8 hours) in the lifetime of the plasma-generator hot cathode. In contrast to this observation, a threefold increase in the discharge burning voltage from 28 to 85 V achieved in the cathode filament lifetime tests using the plasma generator (Figure 3) resulted in a 6.75-fold decrease in the hot-cathode lifetime.

By comparing the cumulative results of all lifetime tests – plots in Figures 3 and 5, we can roughly (towards the lower bound) estimate that that lifetime τ of the generator cathode for the discharge current I_d=10 A at the burning voltage U_d=28 V would have a quite acceptable value for technological applications – no shorter than 33.75 hours. The lifetime τ of the generator hot cathode for the discharge current I_d=20 A and the burning voltage as low as U_d=28 V is also likely to be technologically acceptable – no shorter than 22.6 hours.

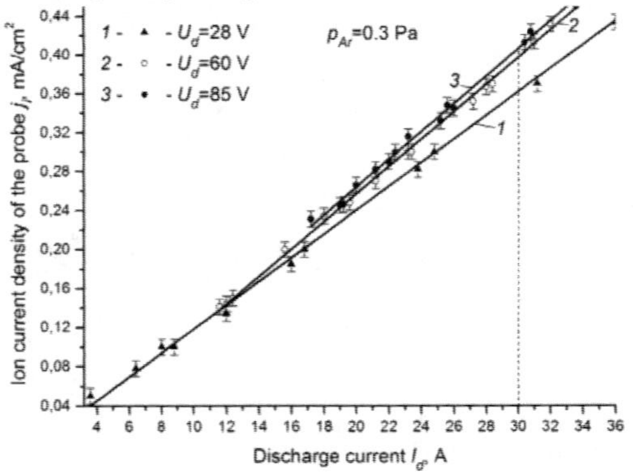

Figure 4. Behavior of the saturation ion current density of the probe far from the plasma generator for the discharge burning voltages 28, 60 and 85 V.

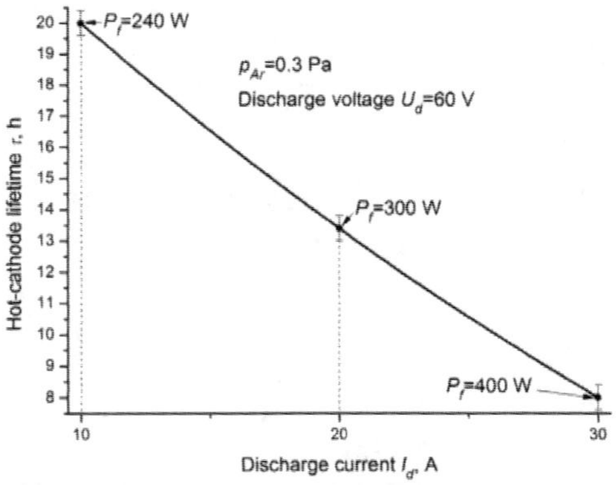

Figure 5. Dependence of the hot-cathode lifetime on the plasma generator discharge current.

4. Conclusion

The investigations of operation modes of a gas-discharge plasma generator relying on a non-self-sustained arc discharge, using a hot cathode made of a tungsten wire measuring 0.8 mm in diameter and 150 mm in the working section, have demonstrated that the lifetime of this cathode and the uninterrupted service of the plasma generator itself could be quite acceptable (more than 20 hours) for technological applications under the following conditions. The plasma generator has to be operated at the hot-cathode filament power providing low pressure of discharge burning (not higher than 28 V). The discharge current has to be not higher than 20 A; with a discharge current of 10 A ensuring 33 hours of an uninterrupted service of the hot cathode.

In order to achieve plasma of elevated concentration in the vacuum chamber space, which is dictated by the requirements of certain technological processes, an increase in the discharge current without any excessive load on the cathode filament reducing its lifetime could be achieved via using several single-type filaments.

Acknowledgement

Results were obtained by support of the Ministry of Education and Science of the Russian Federation (state task No.11.1655.2014/K).

References

[1] Borisov D P, Koval N N and Schanin P M 1994 *Rus.Phys. J.* **3** 295
[2] Borisov D P, Koval N N, Korotaev A D, Kuznetsov V M, Romanov V Ya, Terekhov P A and Chulkov E V 2013 *IEEE Trans. Plasma Sci.* **41** 2183
[3] Ivanov Yu F, Koval N N, Krysina O V et. al. 2010 *Rus.Phys. J.* **3/2** 119

Numerical modelling of physical processes and structural changes in metals under intensive irradiation with use of CRS code: dislocations, twinning, evaporation and stress waves

A E Mayer[1,*], E N Borodin[1,2], V S Krasnikov[1,3] and P N Mayer[1]
[1] Department of Physics, Chelyabinsk State University, 129 Bratyev Kashirinikh Street, Chelyabinsk 454001, Russia
[2] Institute of Problems of Mechanical Engineering of the Russian Academy of Sciences, 61 V.O. Bolshoj Avenue, St. Petersburg 199178, Russia
[3] Department of Physics, South-Ural State University, 76 Lenina Avenue, Chelyabinsk 454080, Russia

E-mail: mayer@csu.ru

Abstract. CRS computer program is presented, which calculates the dynamical deformation of metals under irradiation by high-current electron and powerful ion beams. The incorporated mathematical models allow one to calculate stresses, deformations and structural changes induced by the irradiation. The CRS code numerically solves the equations system, which consists of continua mechanics equations, supplemented by equations of dynamics and kinetics of structural defects: dislocations, grain boundaries, twins, micro-cracks and vapour bubbles. The dislocation plasticity model, the grain boundary sliding model, the mechanical twinning model, the spall fracture model and the non-equilibrium evaporation model are incorporated in the code. The energy release function for electron beam can be calculated within the code, while it can be exported from over programs for ion beam. The CRS code can be a useful tool in theoretical estimation and interpretation of experiments in the field of materials modification by intensive energy fluxes. Restricted rate of plastic deformation provides high values of shear stresses and action of several competitive plasticity mechanisms. Non-equilibrium evaporation of metal in the energy release zone leads to a metastable state of overheated melt, which results in formation of tensile wave following the stress wave in the solid part of the irradiated metal.

1. Introduction

Numerical simulation is an important instrument to reveal conditions and physical processes in the irradiated material, to analyse the obtained experimental data and even to predict possible results of the electron-beam and ion-beam treatment. The BETAIN computer code [1] was developed by A. P. Yalovets especially for the problem of intensive irradiation of metal targets by the beams of accelerated charged particles. It combines numerical solution of the continuum mechanics equations and the transport of fast particles. This code was used for evaluation of stress fields inside the irradiated material [2] and thermodynamic conditions on its surface [3]. In spite of universality and embedded solution of transport problem, the simplest model of perfectly plastic body is used for shear stresses in the BETAIN code, as well as equilibrium approximation for the liquid-vapour transition.

The presented here CRS computer program incorporates more complex plasticity model, which allows one to calculate stresses and deformations in the material of irradiated target together with the

Content from this work may be used under the terms of the Creative Commons Attribution 3.0 licence. Any further distribution of this work must maintain attribution to the author(s) and the title of the work, journal citation and DOI.
Published under licence by IOP Publishing Ltd

structural changes induced by the stresses. The dislocation plasticity model [4], the grain boundary sliding model [5] and the mechanical twinning model [6] are used for determination of shear stresses and density of lattice defects. Various combinations of these three plasticity mechanisms can reveal themselves depending on the material and the irradiation conditions. Thus, the possibility of accounting of all these mechanisms widens the domain of applicability of CRS. The spall fracture model [7] and the nonequilibrium evaporation model [8] describe the metal fracture near the irradiated [8] and back [9] surfaces. Accounting the non-equilibrium evaporation and the possible metastable state of expanded (overheated) liquid in the energy release zone makes the calculations of stresses more precise in comparison with the approximation of equilibrium liquid-vapour transition [1]. In addition, the non-equilibrium evaporation model allows one to estimate the size of melt drops in the ablated material. The fracture model [7] is necessary for prediction of the spall fracture in the solid material near the back free surface of target.

All mentioned features make the considered program more adequate for simulation of the material behaviour under intensive irradiation in spite of the corresponding complication of underlying models in comparison with [1]. The incorporated models had been previously verified by comparison with experimental results on high-speed impact, for example; and now they are applied to the problems of high-current electron and intensive ion irradiation of metal targets. The CRS makes calculations in 1D approximation, which is sufficient for the most of the problems concerning the metal irradiation.

2. Mathematical model

In general case there are three phases in the irradiated metal: solid, liquid and vapour. Solid and liquid are treated as a single condensed phase for simplicity; meanwhile, shear stresses are accounted for solid and not accounted for liquid. Liquid and solid states can be distinguished most accurately by a wide-range equation of state with explicit tracking of phase boundaries [10]; in the case of simpler equation of state [11] the controlling of temperature can be used. Accounting of separate vapour phase (voids) allows one to describe the dynamic fracture and non-equilibrium evaporation.

2.1. Substance dynamics

The irradiated metal is treated as a two-phase mixture of the carrying agent and the disperse phase. Condensed phase initially is the carrying agent, while pores and cracks form the disperse phase in those parts of metals where they appear. If the voids grow a lot and coalescence, the complete fracture occurs, the carrying agent and the disperse phase change over. The multiphase flow approach [12,13] is used, which is applicable if the characteristic size of phase heterogeneities is much smaller than the scale of flow. The one-velocity approximation is also used: both phases move with the same velocity, which is determined by stresses in the carrying agent [12]:

$$\rho\frac{d\upsilon}{dt} = -\frac{\partial P_c}{\partial z} + \frac{\partial S_{zz}}{\partial z}, \tag{1}$$

where υ is the substance velocity; z is the coordinate; P_c is the pressure in the carrying agent; S_{zz} and similar are the stress deviators, which are accounted only in the solid carrying agent; ρ is the average density of mixture. Subscript "c" identifies the carrying agent, while "d" identifies the disperse phase. Volume fractions of the carrying agent α_c and the disperse phase α_d:

$$\frac{d\alpha_c}{dt} = \alpha_d\frac{\partial\upsilon}{\partial z} - W, \qquad \frac{d\alpha_c}{dt} = -\alpha_d\frac{\partial\upsilon}{\partial z} + W, \tag{2}$$

where W is the growth rate of the dispersed phase volume per unit volume of mixture. True densities of phases:

$$\frac{d\rho_c}{dt} = \frac{\rho_c}{\alpha_c}\left(-\frac{\partial\upsilon}{\partial z} + W\right) - \frac{J}{\alpha_c}, \qquad \frac{d\rho_d}{dt} = -\frac{\rho_d}{\alpha_d}W + \frac{J}{\alpha_d}, \tag{3}$$

where J is the growth rate of the dispersed phase mass per unit volume of mixture. The average density in equation (1) is calculated as $\rho = \alpha_c \cdot \rho_c + \alpha_d \cdot \rho_d$. Equations for internal energies:

$$\alpha_c \rho_c \frac{dU_c}{dt} = -P_c \cdot \left(\frac{\partial \upsilon}{\partial z} - W \right) - Q + \rho_c \alpha_c \cdot D - J \cdot (U_{tr} - U_c) + \frac{\partial}{\partial z} \left(\kappa \frac{\partial T_c}{\partial z} \right) + Q_{pl},$$

$$\alpha_d \rho_d \frac{dU_d}{dt} = -P_c \cdot W + Q + \rho_d \alpha_d \cdot D + J \cdot (U_{tr} - U_d), \quad (4)$$

where Q is the power of heat exchange between phases per unit volume of mixture; T is the temperature; subscript "tr" identifies the phase which mass is increasing; Q_{pl} is additional heating rate due to the plastic deformation. The energy release function D for fast electrons is calculated by CRS with use of the method [14] for electron transport; for ions this function is exported from BETAIN [1]. Equations (2)–(4) follows from the conservation of volume, mass and energy in a mixture element; functions W, J and Q express the exchange rates between phases, including phase transition.

2.2. Dynamic fracture in liquid and solid states

Liquid is unstable against the phase transition into vapour at pressure lower than the saturation vapour pressure, particularly, at any negative pressure. It is typical for expanding heated layer of metal near the irradiated surface. In these conditions the cavitation occurs consisting in nucleation and growth of vapour cavities, it finally results in ablation of metal from the irradiated surface Nucleation of cavities demands the work against the surface tension; therefore, the melt can exist in a metastable state of expanded liquid (at negative pressure) during some time [15]. Nucleation of cavities due to the thermal fluctuations is dominant at high strain rates especially for the intensive irradiation. Spherical shape of cavities is preferable due to the surface tension; their growth is described by the Rayleigh-Plesset equation [16]. Equations for nucleation and growth of cavities [8] and the mass exchange due to the phase transition allow us to determine the exchange rates W, J and Q in this case.

A similar situation takes place in the expanded solid at the shock wave reflection from the back target surface. From the one hand, fracture in solid state is more complex because of complicated shape of voids (from spherical one to narrow cracks) and influence of shear stresses. From another hand, the mass and heat exchange rates, J and Q, can be supposed to be zero with high accuracy because the cavities are almost empty. The fracture model [7] is used in CRS, which considers nucleation (due to thermal fluctuations) and growth of cracks, it allows us to determine W; the model had been tested for wide range of strain rates–from 10^4 s^{-1} to 10^9 s^{-1}.

2.3. Plasticity

Total plastic strain of polycrystalline metal can be represented as a result of the combined action of three competing processes: i) the dislocation motion, ii) the mechanical twinning and iii) the sliding along the grain boundaries. According to this viewpoint, the plastic deformation tensor w_{ik} is represented by the following sum of three tensors: $w_{ik} = w_{ik}^{D} + w_{ik}^{tw} + w_{ik}^{gb}$, where w_{ik}^{D} is the part of plastic deformation caused by the dislocation motion, w_{ik}^{tw} is caused by the mechanical twinning and w_{ik}^{gb} is caused by the grain boundary sliding. Then the Hooke law [17] with accounting of the plastic strain w_{ik} is used for determination of the stress deviators:

$$S_{ik} = 2G\left[(2/3) u_{zz} \delta_{iz} \delta_{kz} - w_{ik} \right], \quad (5)$$

where G is the shear modulus; δ_{iz} is the bivalent mixed tensor; indexes *i* and *k* range over *x*, *y* and *z*. The geometrical deformation u_{zz} is induced by the macroscopic motion of substance:

$$\frac{du_{zz}}{dt} = \frac{\partial u}{\partial z}. \tag{6}$$

It should be noted that all components of the plastic strain and stress deviator tensors can be nonzero even in the case of considered 1D motion of substance [4].

2.3.1. Dislocation plasticity. Dislocations are characterized by scalar densities of mobile $\rho_{\mathrm{D}}^{\beta}$ and immobilized dislocations $\rho_{\mathrm{I}}^{\beta}$ and velocity of mobile dislocations V_{D}^{β} relative to substance; these quantities are determined for all possible slip systems. The slip system of dislocations is determined by the Burgers vector b_i^{β} and by the normal n_i^{β} and numerated by superscript β. The plastic strain w_{ik}^{D} produced by the dislocation plasticity can be found from the generalized Orowan equation [18]:

$$\frac{dw_{ik}^{\mathrm{D}}}{dt} = \sum_{\beta} \frac{1}{2}\left(b_i^{\beta} n_k^{\beta} + b_k^{\beta} n_i^{\beta}\right) V_{\mathrm{D}}^{\beta} \rho_{\mathrm{D}}^{\beta}. \tag{7}$$

The equation of dislocations motion is following:

$$m_0 \xi_{\beta}^3 \frac{\mathrm{d}V_{\mathrm{D}}^{\beta}}{\mathrm{d}t} = \left[\sum_i \sum_k S_{ik} b_i^{\beta} n_k^{\beta} \mp \frac{1}{2} bY\right] - B\xi_{\beta}^3 V_{\mathrm{D}}^{\beta}, \tag{8}$$

where $\xi_{\beta} = 1/\sqrt{1-\left(V_D^{\beta}/c_{\mathrm{t}}\right)^2}$ is the quasi-relativistic factor, which accounts that $\left|V_D^{\beta}\right| < c_{\mathrm{t}}$; $c_{\mathrm{t}} = \sqrt{G/\rho}$ is the transverse sound speed in the material; m_0 is the rest mass of dislocations; Y is the static yield strength; B is the phonon drag coefficient. Dislocations move only if the force of the shear stresses is higher than the resistance force $bY/2$; the sign "$\mp$" means that the resistance is always directed opposite to the dislocation motion. Kinetics equations for dislocation densities are described in [7] and reflect the balance between the generation of new dislocations, immobilization and annihilation of the existing one. Generation rate is determined through the energy dissipation rate at the plastic deformation–approximately 0.1 part of the dissipated energy is spent on formation of new defects [19].

2.3.2. Twinning. Twinning becomes an alternative plasticity mechanism at low temperatures and high strain rates [20,21]. Twins are supposed to be cylindrical and characterized by concentrations of mobile N_{TW}^{γ} and immobilized N_{IM}^{γ} twins (with fixed boundaries), their radii, R_{TW}^{γ} and R_{IM}^{γ}, and thicknesses, h_{TW}^{γ} and h_{IM}^{γ}; possible crystallographic orientations of twins are numerated by γ. The plastic strain caused by twinning is expressed through its volume fraction α^{γ}:

$$\frac{dw_{ik}^{\mathrm{TW}}}{dt} = \sum_{\gamma} \frac{d\alpha^{\gamma}}{dt}\left(\tau_i^{\gamma} n_k^{\gamma} + \tau_k^{\gamma} n_i^{\gamma}\right)\varepsilon_{\mathrm{TW}}, \tag{9}$$

where $\varepsilon_{\mathrm{TW}}$ is the deformation of twinned material, which is $\varepsilon_{\mathrm{TW}} = 1/\sqrt{2}$ in fcc metals [20]; unit vectors τ_i^{γ} and n_i^{γ} describe the crystallographic orientation of twins. The volume fraction is calculated from the concentration and size of twins, which are determined in accordance with [6].

2.3.3. Grain boundary sliding. Submodel of the grain boundary sliding [5] includes the equation for the plastic strain w_{ik}^{gb} based on the Maxwell model of highly viscous liquid [17]:

$$\frac{dw_{zz}^{\mathrm{gb}}}{dt} = \frac{S_{zz} \mp y_b}{G\mathrm{T}}, \tag{10}$$

where T is the relaxation time, which is proportional to the grain diameter [5]. A simple expression was obtained in [5] for the barrier resistance stress y_b –it is proportional to G.

3. Numerical implementation

The described above mathematical model is numerically realized in the CRS computer program written on FORTRAN. This program is designed to simulate the intensive actions on metals: the high-speed impact, intensive electron and ion irradiation. It is supplied by handy data output including the dynamic visualization. Various equations of state can be attached, including the interpolation one [11] and the tabulated one [10]. The bank of plasticity model constants includes aluminium, copper, nickel, iron and stainless steel at this point in time.

Method of separation by physical processes is used at the numerical solution. The substance dynamics is calculated by modification of the numerical method [22]; modification consists in eliminating of the artificial viscosity and accounting of the physical viscosity instead and allows one to obtain the stable solution by using of a fine enough computational grid [23]. Equation (8) for the dislocation velocity is solved with use of the approximate analytical solution. Other equations are solved by Euler method with varied time step. All the described in previous section physical processes can be accounted simultaneously in CRS, or part of them can be excluded for simplicity of analysis.

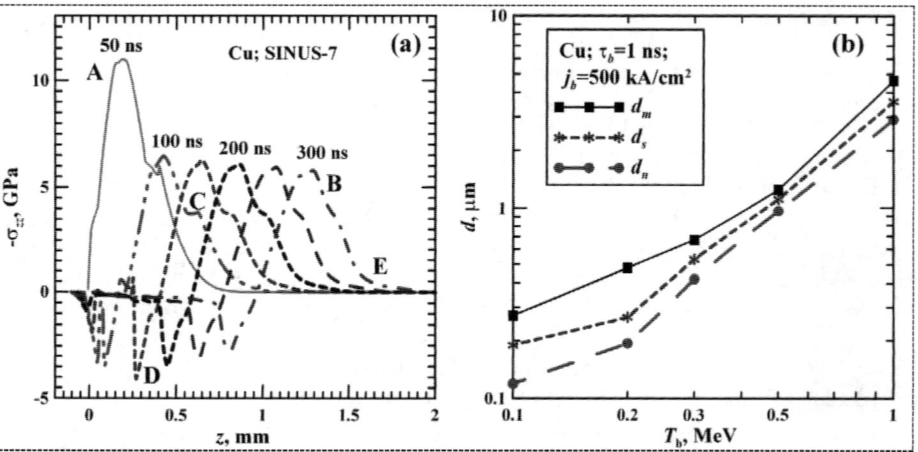

Figure 1. Irradiation of copper target by high-current electron beam: (a) stresses in target at irradiation with beam parameters corresponding to the SINUS-7 accelerator [2,9]; (b) diameters of liquid drops forming at ablation versus the initial energy of electrons (different curves correspond to different types of averaging).

4. Calculation results

Figure 1 shows the calculation results for irradiation of copper target by high-current electron beam. Figure 1(a) demonstrates the evolution of compression wave generated by irradiation (by the fast energy release): "A" is the high-pressure area in the energy release zone; "B" is the propagating compression wave evolving to the steady shock wave; "E" is the elastic precursor; "D" is the tensile wave, which amplitude is restricted by cavitation (non-equilibrium evaporation); "C" marks the melting boundary between the energy release zone and the main part of the target. Accounting of the non-equilibrium evaporation and the metastable state of expanded liquid in the energy release zone makes possible existence of negative pressures in the tensile wave, while in the frames of equilibrium approximation for the liquid-vapour transition [1] the pressure is limited below by the zero value.

Figure 1(b) presents the size of liquid drops forming at ablation of the surface layer versus the initial energy T_b of fast electrons; the average drops size grows from about 100 nm up to several micrometers with the increase of the fast electrons range in substance together with T_b.

Figure 2 shows the non-equilibrium nature of plastic deformations at the strain rates typical for the intensive irradiation problems. The amplitude of elastic precursor of the shock wave in iron is plotted versus the target thickness: experimental data [24-26] in comparison with our simulations. One can see that the elastic precursor can be very high–up to the amplitude of compression pulse itself (about 10 GPa in the considered case)–in thin targets, therefore, the elastoplastic properties is of high importance in this case. The dislocation plasticity model incorporated in the CRS can describe the experimentally observed evolution of the elastic precursor. Reduction of the target thickness and the thickness of the energy release zone increases the strain rates realized in the target material. High values of shear stresses near the loaded or irradiated surface are explained by restricted density of defects (dislocations)–the concentration of defects is not enough to provide an effective plastic relaxation of shear stresses at high strain rates. It can provoke activation of competitive plasticity mechanisms, such as twinning or the grain boundary sliding.

Figure 3 demonstrates the calculation results for volume fraction of twins and scalar density of dislocations in copper after the high-current electron irradiation (electrons energy is 1.5 MeV, current density is 100 kA/cm^2, and the pulse duration is 50 ns). The stress wave generated by irradiation initiates the structural changes in the target material–increasing of dislocations density and formation of twins. The twinning becomes an important only at very high intensities of irradiation for materials with low staking fault energy. Comprehensive numerical investigation of the dislocation density change in metals in the result of intensive electron and ion irradiation are presented in [4,27], while investigation of twining is a new result.

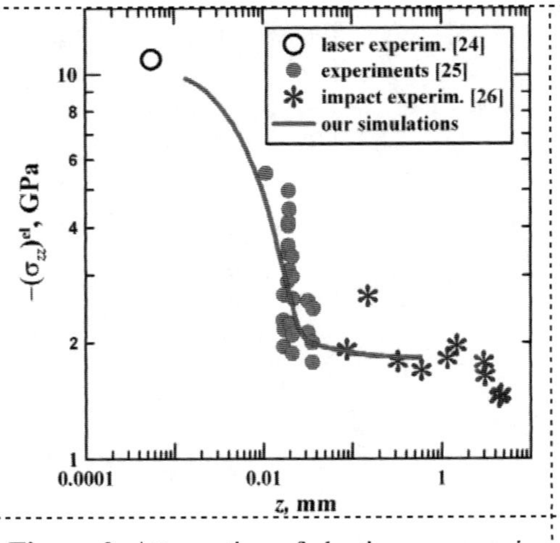

Figure 2. Attenuation of elastic precursor in iron: experimental data [24-26] versus our simulations with us of CRS program.

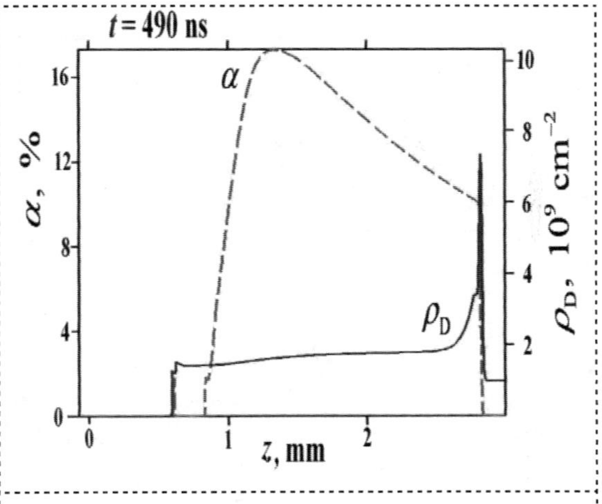

Figure 3. Spatial distributions of the volume fraction of twins α and scalar density of dislocations ρ_D in copper target after irradiation.

5. Conclusions

The mathematical model underlying the CRS computer code is briefly described. It allows one to account non-equilibrium nature of the plastic deformation, evaporation in the energy release zone and spall fracture near the back surface. In addition it determines characteristics of the material microstructure during and after the beam action, including the size of ablated drops, density of dislocations etc. It makes the CRS code a useful tool for analyzes of the metal irradiation by intensive electron and ion beams. Some calculation results are presented and discussed. For example, restricted rate of plastic deformation provides high values of shear stresses near the irradiated surface and action of several competitive plasticity mechanisms in general case.

Acknowledgments
This work is supported by the President of Russian Federation (grant MD-286.2014.1), the Russian Foundation for Basic Research (Grant No. 14-01-31454), and the Ministry of Education and Science of Russian Federation (scientific research of Chelyabinsk State University in the framework of competitive part of state assignment 3.1334.2014/K).

References

[1] Yalovets A P and Mayer A E 2002 *Proc. 6th Int. Conf. on Modification of Materials with Particle Beams and Plasma Flows (Tomsk, Russia)* ed G A Mesyats S D Korovin et al (Tomsk: Kursiv) pp 297–299
[2] Markov A B, Kitsanov S A, Rotshtein V P, Polenin S D, Proskurovskii D I and Dudarev E F 2006 *Russ. Phys. J.* **49** 758–65
[3] Volkov N B, Fen'ko E L and Yalovets A P 2010 *Techn. Phys.* **55** 1389–99
[4] Krasnikov V S, Mayer A E and Yalovets A P 2011 *Int. J. Plast.* **27** 1294–308
[5] Borodin E N and Mayer A E 2012 *Mater. Sci. Eng. A* **532** 245–8
[6] Borodin E N, Atroshenko S A and Mayer A E 2014 *Tech. Phys.* **59** 1163–1170
[7] Mayer A E, Khishchenko K V, Levashov P R and Mayer P N 2013 *J. Appl. Phys.* **113** 193508
[8] Mayer P N, Dudorov A E and Mayer A E 2013 Metallic melt fracture and fragmentation under the high-current electron irradiation *Proc. 13th Int. Conf. on Fracture 2013, ICF 2013 (Beijing, China, 16-21 June 2013)* Paper #S24-003
[9] Dudarev E F, Markov A B, Mayer A E, Bakach G P, Tabachenko A N, et al. 2013 *Russ. Phys. J.* **55** 1451–7
[10] Fortov V E, Khishchenko K V, Levashov P R and Lomonosov I V 1998 *Nucl. Instrum. Meth. Phys. Res. A.* **415** 604–8
[11] Kolgatin S N and Khachatur'yants A V 1982 *Teplofiz. Vys. Temp.* **20** 90–4 [in Russian]
[12] Nigmatulin R I 1987 *Dynamics of Multi-phase Mediums* (Moscow: Nauka) [in Russian]
[13] Van Wachem B G M and Almstedt A E 2003 *Chem. Eng. J.* **96** 81–98
[14] Evdokimov O B and Yalovets A P 1974 *Nucl. Sci. Eng.* **55** 67–75
[15] Skripov V P 1974 *Metastable Liquids* (New York: Wiley)
[16] Kuksin A Yu, Norman G E, Pisarev V V, Stegailov V V and Yanilkin A V 2010 *Phys. Rev. B* **82** 174101
[17] Landau L D and Lifshitz E M 1986 *Theory of Elasticity* (New York: Pergamon)
[18] Kosevich A M 1965 *Sov. Phys. Uspekhi* **7** 837–54
[19] Kittel C 2003 *Introduction to Solid State Physics* (New York: Wiley)
[20] Christian J W and Mahajan S 1995 *Prog. Mat. Sci.* **39** 1–157
[21] Allain S, Chateau J - P and Bouaziz O 2004 *Mater. Sci. Eng. A* **387–389** 143–7
[22] Yalovets A P 1997 *J. Appl. Mech. Tech. Phys.* **38** 137–50
[23] Khishchenko K V and Mayer A E 2014 Non-isentropic layers in matter behind shock and ramp compression waves / arXiv:1407.8381
[24] Ashitkov S I, Komarov P S, Agranat M B, Kanel G I and Fortov V E 2013 *JETP Lett.* **98** 384-8
[25] Smith R F, Eggert J H, Rudd R E, Swift D C, Bolme C A and Collins G W 2011 *J. Appl. Phys.* **110** 123515
[26] Kanel G I, Razorenov S V, Garkushin G V, Ashitkov S I, Komarov P S and Agranat M B 2014 *Phys. Solid State* **56** 1569-73
[27] Krasnikov V S and Mayer A E 2012 *Surf. Coat. Technol.* **212** 79–87

Journal of Physics: Conference Series **552** (2014) 012003 doi:10.1088/1742-6596/552/1/012003

Transport charge of gallium arsenide films synthesized on polycrystalline silicon by ion ablation

A V Kabyshev, F V Konusov*, G E Remnev and S K Pavlov
National Research Tomsk Polytechnic University, 30, Lenin Avenue, 634050, Tomsk, Russia

E-mail: konusov@hvd.tpu.ru

Abstract. Electrophysical and photoelectric properties of thin GaAs films deposited on polysilicon by pulse ion ablation using high-power ion beams have been investigated. The predominant charge carriers transfer mechanism in films and the type of dark and photoconductivity have been established. A vacuum annealing effect (10^{-2} Pa, 300–1000 K) on energetic and kinetic characteristics of dark and photoconductivity, the transfer mechanism and the type of charge carriers have been determined. The most probable causes of changes in the film electric and photoelectric characteristics have been discussed.

1. Introduction

Gallium-arsenide and silicon are prospective materials for electronics. Their production has been developed; they obtain a perfect combination of electric and photoelectric parameters, radiation hardness and thermal stability of properties [1–4]. Uniting these materials on one substrate is an upcoming trend in optimization of the optoelectronic device production technology [1–3, 5]. While growing GaAs on Si, there are some problems that are connected with the growth of a polar semiconductor on a non-polar substrate, with the difference in lattice parameters (up to 4 %) and the thermal expansion coefficient. Antiphase domains are formed, stress and high dislocation density near the boundary arise [1, 2]. Defects worsen the parameters of devices based on GaAs/Si [1–3]. The problems of the GaAs epitaxial deposition on Si are partially overcome by applying cycle annealing [3] and by creating thin joining transition layers [1].

Electric characteristics and charge carriers transfer mechanisms in crystalline epitaxial grown *c*–GaAs vary over a wide range [3–6]. The unique properties of amorphocrystalline *a*–GaAs films deposited from RF–plasma by impulse methods [6, 7], thermal deposition, co–evaporation of the elements [8] and laser deposition [9] enable to use them in the production of layers that match lattice parameters during the deposition of GaAs/Si and heterostructures on their basis. The characteristics of such films differ from c–GaAs due to their complex structure and strong defect disordering. Film deposition from plasma produced by a powerful ion beam (PIB) enables to implement high-speed deposition and to form films on significant areas and on various substrates [10–13]. The peculiarities of electromigration in the films synthesized by the pulse ion ablation (PIA) have been insufficiently studied.

Electric and photoelectric characteristics, the transfer mechanism, the type of charge carriers in thin GaAs films synthesized on polysilicon by PIA have been investigated in this article. The energy spectrum of defect levels, featuring material properties, has been specified.

Content from this work may be used under the terms of the Creative Commons Attribution 3.0 licence. Any further distribution of this work must maintain attribution to the author(s) and the title of the work, journal citation and DOI.
Published under licence by IOP Publishing Ltd

2. Experimental

The films have been obtained by means of the TEMP accelerator at sequential deposition on polysilicon and polycor. The method of film deposition has been described in [10]. The annealing was carried out in vacuum (pressure $P \leq 10^{-2}$ Pa, T_{ann}=300–1000 K). Surface σ_s and bulk σ_v dark and photoconductivity $\Delta\sigma_{ph}=\sigma_{ph}-\sigma$ (σ_{ph}–conductivity under lighting), photosensitivity $K(h\nu)=\Delta\sigma_{ph}/\sigma$ were measured at constant electrode voltage U=0.01–300 V, at temperature T=300–700 K and at photon energy $h\nu$=1.5–4.0 eV [12, 13]. The electrodes were deposited onto the film surface by silver thermal spraying. The temperature dependences σ, $\Delta\sigma_{ph}(T)$ were approximated by the equation for activation mechanism

$$\sigma_a(T) = \sigma_0 \times e^{(-\varepsilon_\sigma / k \cdot T)}, \quad (1)$$

where σ_0– a pre-exponential factor, ε_σ–an activation energy, k– Boltzmann constant and by the equation for transfer hopping mechanism on localized states (LS) near the Fermi level E_F within the Mott model

$$\sigma_p(T) = \sigma'_0 \times e^{(-(T_0/T)^{0.25})}, \quad (2)$$

where σ'_0 is a pre-exponential factor, T_0 is an activation energy [6, 12]. State density $N(E_F)$ near E_F was calculated according to T_0 in compliance with [6, 12]. The sign of dominant charge carriers was determined according to the method [12].

3. Result and discussion

Temperature dependences σ, $\Delta\sigma_{ph}(T)$ of films is due to LS with ε_σ=0.05–0.16 eV within the interval T=300–500 K at the charge carriers exchange between LS and the allowed bands. N–type of σ and $\Delta\sigma_{ph}$ predominate in GaAs/Si, whereas, the concentration of LS donors is higher in comparison with the one of acceptors. The effect of donors increases in the films deposited on the periphery [11, 12]. In the centre of deposition, defect contribution increases with their possessing shallow acceptor LS with energy $\varepsilon=E_v+(0.05–0.16)$ eV, where E_v is a valence band (VB) top. Values σ, $\Delta\sigma_{ph}$, K for σ_v are higher as compared to σ_s due to the influence of shallow acceptors (figure1). The comparison of film characteristics on silicon and polycor under the same deposition and measurement conditions showed the advantage of GaAs/Si in values σ, $\Delta\sigma_{ph}$ and their field excitation and heating stability (figure 1). This is due to the heterogeneous distribution of trapping centers and recombination near GaAs/Al_2O_3 boundary [14, 15]. In the case of Al_2O_3, the effect on the charge transfer barrier between GaAs VB and the conduction band (CB) bottom is more significant (a barrier Φ>3.2 eV) [15].

The relations between values ε_σ and σ_0 are conditioned by the high content of defects in films as compared to c–GaAs and by the influence of a hopping component σ_h involving LS defects similar to the approach [5–8, 12]. The approximation of curves σ, $\Delta\sigma_{ph}(T)$ indicates the predominance σ_h over σ_a within the interval T=300–500 K in PIA films. The most probable hopping length is R=4.5–7.6 nm with the calculated values of $N(E_F)=(0.74–3.2)\cdot 10^{18}$ eV^{-1} cm^{-3} (figure 2, curve 1). Values $N(E_F)$ of deposited GaAs/Si films are closer to the density of LS films obtained by co-evaporation of elements than to their values in films on polycor (figure 2) [8, 12]. The hopping conductivity is the main migration mechanism in LT–GaAs [4, 5]. The comparison with LT–GaAs enables to suggest that E_F is fastened near the middle of the band gap (BG) at $\varepsilon=E_v+(0.6–0.8)$ eV in GaAs films deposited by PIA [4–7]. This is due to high concentration of $N>10^{18}$ cm^{-3} of defects like EL2–centers ($As_{Ga}^{0,+}$, pairs $As_{Ga}V_{Ga}$, $As_{Ga}As_i$) creating LS deep within BG [4, 5, 12, 13]. The hole exchange between the shallow acceptor LS and VB dominates at T=300–500 K and the electron exchange between donor LS and CB – under light excitation, as K(U, T) show. Values ε_σ for $\Delta\sigma_{ph}(T)$ are lower than $\sigma(T)$ possesses on ~0.05 eV, whereas hopping length decreases from R=4.5–7.6 to 3.6–5.5 nm. LS occupation by charge carriers decreases, while the density $N(E_F)$ increases 1.8–4 times, which is typical of most materials [6, 12].

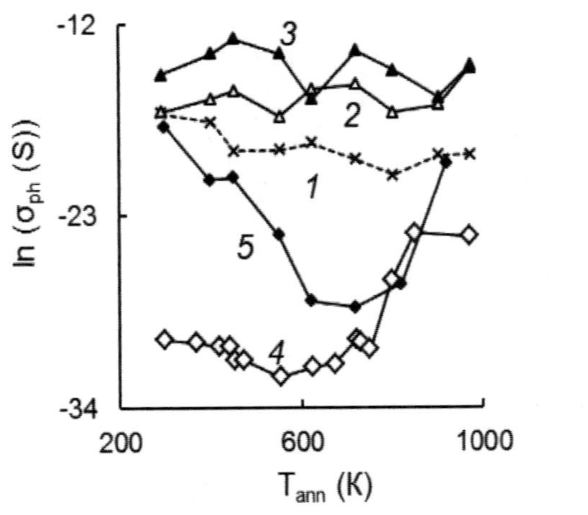

Figure 1. The annealing effect on photoconductivity σ_{ph} of silicon (curve 1), GaAs films synthesized by PIA on silicon (curves 2, 3) and on polycor (curves 4, 5) using the low-resistance (curves 2–4) and high-resistance (curve 5) target for surface (curves 1, 2, 4, 5) and bulk (curve 3) conductivity.

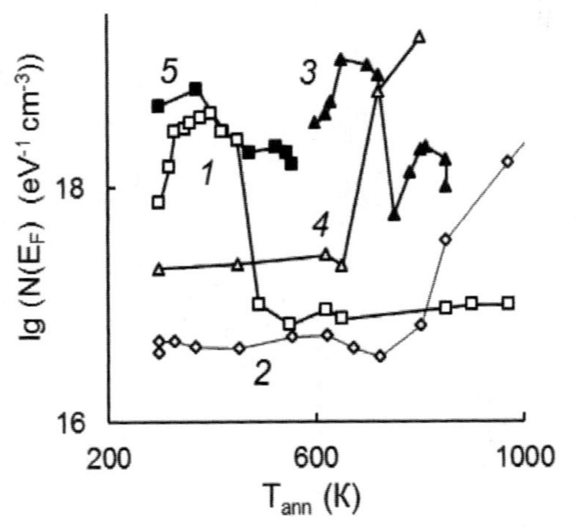

Figure 2. The annealing effect on the LS density of GaAs films on Si (curve 1) and polycor (curves 2–4) from the low-resistance (curves 1–3) and high-resistance (curve 4) target [12]. Data for a–GaAs dispensed by co-evaporation of elements (curve 5) [8].

Vacuum annealing modifies the electronic state of film materials in consequence of its partial crystallization, the change in the defect charge state and the formation of defect complexes [5, 7, 8, 11–13, 15]. LS density near the film surface and on the boundary with substrate changes, arsenic clusters form [4, 8, 12]. According to characteristic change, two annealing stages T_{ann}=300–600, 600–1000 K (figures 1–3) are distinguished. Correlation dependences $N(E_F)(\sigma_0)$ and $N(E_F)(\varepsilon_\sigma)$ divided into two branches for bulk and surface conductivity are implemented between parameters σ_h and σ_a, (figure 4, curves 1 and 2). The correlation between contributions to the electric conduction of the components σ_a and σ_h changes (figures 1–4).

After T_{ann}=300–600 K the change in GaAs/Si film characteristics differs from the films on polycor (figures 1–3). The decrease in σ of $GaAs/Al_2O_3$ films correlates with the increase in LS density (figures 1, 2) [12, 15]. At maxima σ, $\Delta\sigma_{ph}$, K at T_{ann}=350–450, K and ε_σ approaches maximum ε_σ=0.05–0.08 eV, while LS maximum density is $N(E_F)=(3–4)\cdot10^{18}$ eV^{-1}cm^{-3}, which indicates the increase in contribution of σ_h (figures 1–3). The relaxation of nonequilibrium charge carriers and their recombination between shallow LS and deep levels with $\varepsilon>1.4$ eV defects take place, which is confirmed by spectra K(hv, T_{ann}) [11–13]. At T_{ann}=450–600 K the correlation between the decrease in σ, $\Delta\sigma_{ph}$, the increase in K and ε_σ, the decrease in $N(E_F)=4\cdot10^{18}$ to $6\cdot10^{16}$ eV^{-1}cm^{-3} is observed, which indicates the increase in contribution to the activation mechanism electromigration (figures 1–3). As K(U) show, the effect of light-generated holes in VB on $\Delta\sigma_{ph}$(T, U) increases. The properties are affected by the defect annealing and recharging, as well as the dissociation of unstable complexes [11, 13]. The dependences $N(E_F)(\sigma_0, \varepsilon_\sigma)$ indicate less disordering of GaAs/Si films by defects in comparison with $GaAs/Al_2O_3$ and a–GaAs films (figure 4) [6–8, 12].

After T_{ann}=600–1000 K the change in parameters σ and $\Delta\sigma_{ph}$ indicates the increase in the influence on the properties of charge carrier processes in the substrate Si (figures 1–3). State density stabilizes at $N(E_F)=(5–10)\cdot10^{16}$ eV^{-1} cm^{-3}, whereas the hopping length increases from R=3.6–5.5 to 10–15 nm unlike the films on polycor, where $N(E_F)$ and contribution of σ_p increase and R decreases (figure 2)

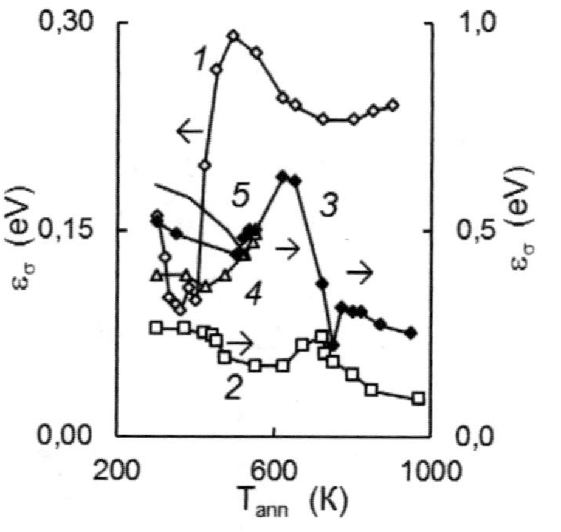

Figure 3. The annealing effect on activation energy ε_σ in GaAs films synthesized by PIA on silicon (curve 1) and on polycor (curves 2, 3) using the low-resistance target [12]. Data for a–GaAs films deposited from RF-discharge plasma (curve 4) [7] and co–evaporation (curve 5) [8] are presented.

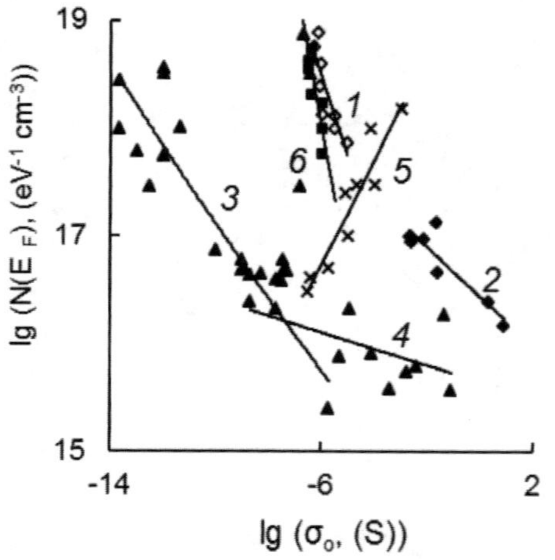

Figure 4. The correlation of $N(E_F)(\sigma_0)$ in GaAs films synthesized by PIA on silicon (curves 1, 2) and on polycor (curves 3, 4, 6) [12] from the low-resistance target for surface (curves 1, 3, 4, 6) and bulk (curves 2, 5) conductivity. Data for a–GaAs films deposited from RF-discharge plasma (curve 5) [6] are presented.

[12]. The recession of ε_σ in GaAs/Al_2O_3 films takes place at 130–230 K higher than in GaAs/Si films (figure 3). Ion implantation in polycor also has its influence [11, 12]. The relation typical of the strong defective GaAs is implemented between LS parameters, which allows to conclude that structural disorder in the films on polycor is higher than in the films on silicon, while its efficiency is determined by the deposition conditions (figures 2–4) [6–9, 12]. Fractional crystallization of film material, growth defect annihilation and recharge, clustering and enlargement of nanoparticles are probable reasons for changes in parameters σ, $\Delta\sigma_{ph}$ [11–13].

4. Conclusion

Electric conductivity and photoconductivity of GaAs films deposited by PIA using PIB on the silicon substrate has a mixed type. It is implemented by means of the activation and hopping mechanism involving conditions caused by defects localized in BG near the Fermi level. Two annealing stages at T_{ann} K are distinguished: 300–600 and 600–1000; they are characterized by different relations between parameters LS and transfer mechanisms. After 300–600 K the changes in conductivity characteristics are caused by charge carrier redistribution between the shallow and deep levels of donor and acceptor defects. At 600–1000 K the parameters of surface and bulk dark and photoconductivity, transfer mechanisms are caused by recharge, the annihilation of defects and their clustering.

Acknowledgments
The paper has been completed within megaproject of National Research Tomsk Polytechnic University «Materials for Extreme Conditions».

References

[1] Bolhovityanov Yu B and Pchelyakov O P 2008 *Usp. Sciences (In Russian)* **178** 459
[2] Bietti S, Scaccabarozzi A, Cesare F, Bollani M, Bonera E, Falub C V, von Känel H, Miglio L, and Sanguinetti S 2013 *Appl. Phys. Lett.* **103** 262106

[3] Shimizu Y and Okada Y 2004 J. Cryst. Growth. **265** 99
[4] Vilisova M D, Kunitsyn A E, Lavrentyeva L G, Preobrazhensky V V, Putyatov M A, Semyagin B R, Toropov S E and Chaldyshev V V 2002 *Semiconductors (In Russian)* **36** 1025
[5] Gregory S, Baker C, Tribe W R, Evans M J, Beere H E, Linfield E H, Davies A G and Missous M 2003 *Appl. Phys. Lett.* **83** 4199
[6] Reuter H, Schmitt H and Boffgen M 1995 *Thin Solid Films* **254** 96
[7] El Hadadi B, Carchano H, Seguin J.-L and Tijani H 2005 *Vacuum* **80** 272
[8] Abdel–Wahab F A and Kotkata M F 2005 *Physica* **368** 209
[9] Ullrich B and Erlacher A 2005 *Appl. Phys. Lett.* **87** 151115
[10] Saltymakov M S, Lie Zeng Feng and Remnev G E 2006 *Proc. of the universities. Physics (In Russian)* 8 371
[11] Kabyshev A V, Konusov F V and Remnev G E 2011 *J. Surf. Inves. X-ray, Synchr. and Neutr. Techn.* **5** 228
[12] Kabyshev A V, Konusov F V, Remnev G E and Pavlov S K 2013 *Proc. of the universities. Physics (In Russian)* **56** 30
[13] Kabyshev A V, Konusov F V and Remnev G E 2014 *J. Surf. Inves. X-ray, Synchr. and Neutr. Techn.* **8** 158
[14] Brammertz G, Lin H C, Martens K, Mercier D, Merckling C, Penaud J, Adelmann C, Sioncke S, Wang W E, Caymax M, Meuris M and Heyns M 2008 *J. Electrochem. Soc.* **155** H945
[15] Chou H Y, O'Connor E, Hurley P K, Afanas'ev V V, Houssa M, Stesmans A, Ye P D and Newcomb S B 2012 *Appl. Phys. Lett.* **100** 141602

Journal of Physics: Conference Series **552** (2014) 012004 doi:10.1088/1742-6596/552/1/012004

On the focused beam parameters of an electron gun with a plasma emitter

S Kornilov[1,*], N Rempe[1], A Beniyash[2] and N Murray[2]

[1] Tomsk State University of Control Systems and Radioelectronics, 40 Lenin Ave., 634050, Tomsk, Russia
[2] Institute of Materials Science, Leibniz University of Hannover, D-30823 Garbsen, Germany

E-mail: mail@elion-tomsk.ru

Abstract. The report presents the measurement results of the focused beam brightness in the electron gun with plasma emitter. The beam brightness was approximately 1010 A•m-2•sr-1 under the beam power up to 4 kW and an electron energy of 60 keV at the focal distance of 0.5 m. Qualitative assessment of the beam parameters was performed by welding test pieces. The results describing the possibility in principle of using the guns with a plasma emitter in non-vacuum technological devices are presented.

1. Introduction

Electron-beam guns with a plasma emitter based on the emission of electrons from a low-voltage discharge with a hollow cathode [1] have long been used in the beam technologies [2, 3]. In contrast to the widespread triode guns, guns with a plasma emitter are designed by a diode scheme. Current control in such guns is performed without grid electrode. The advantage of such a beam current control [4] is, in contrast to hot cathode triode gun, the electron-optical properties of the focused beam remain practically unchanged at the beam current variation. Within the broad range of the electron-beam experts is widely believed that the guns of this type provide a low beam current density because of the high electrons temperature by the emitting plasma. According to our estimates, based on the known formula of Langmuir [5], for typical parameters of the plasma emitter, the minimum size of a focused electron beam must not exceed a few tens of microns [6].

However, in experiments such a small diameter has not been reached for a long time for beam emitters using discharge processes. The reason was the underestimation of the influence of the magnetic field of the discharge chamber on the properties of the electron beam in the accelerating gap and the drift space of the gun. After the optimization of the magnetic field, electron beams with a power density of up to 10^7 W/cm² at 60 keV electron energy can be generated. In addition to the power density in the focal spot, the brightness of the electron beam is equally important to evaluate the electron-optical parameters of the gun. High brightness makes it possible to use guns with a plasma emitter for high-quality welding, applications involving pressure stage systems (non-vacuum, low-vacuum) or processes with high demands on beam quality such as EB drilling or rapid manufacturing or rapid prototyping. The results of our assessment of the beam generated by an electron gun with a plasma emitter are set out below.

Content from this work may be used under the terms of the Creative Commons Attribution 3.0 licence. Any further distribution of this work must maintain attribution to the author(s) and the title of the work, journal citation and DOI.
Published under licence by IOP Publishing Ltd

2. Test setup and measurement techniques

The measurement of the beam radius r_f at the focal spot and the convergence angle of the focused electron beam on guns with a plasma emitter was carried out at the installation shown in figure 1. The main parts of the installation are the electron-beam gun developed by Elion Ltd. (Russia) for Perndorfer Maschinenbau KG (Austria), the vacuum chamber and probing system (see figure 1*a*).

Electrons emitted from plasma caused by an electric field are required for the operation of a gun with a plasma emitter. The emitting plasma is produced by a low-voltage reflective discharge with a hollow cathode [1].

The basic component of a gun is the discharge chamber, a ceramic-metal unit. Its primary function is to produce charged particles in a Penning-type gas discharge. A schematic diagram of the electrode assembly of an electron gun with a plasma emitter is given in figure 1*b*.

The discharge chamber is formed by a hollow cathode (1), a cylindrical anode (2), and an emitter cathode (4). The voltage (DPS) is necessary for the discharge initiation and operation. It is applied between the cathodes and the anode. A permanent magnet (3) creates a magnetic field of $(0.8 \div 1) \cdot 10^{-1}$ T in the space between the hollow cathode and the emitter cathode.

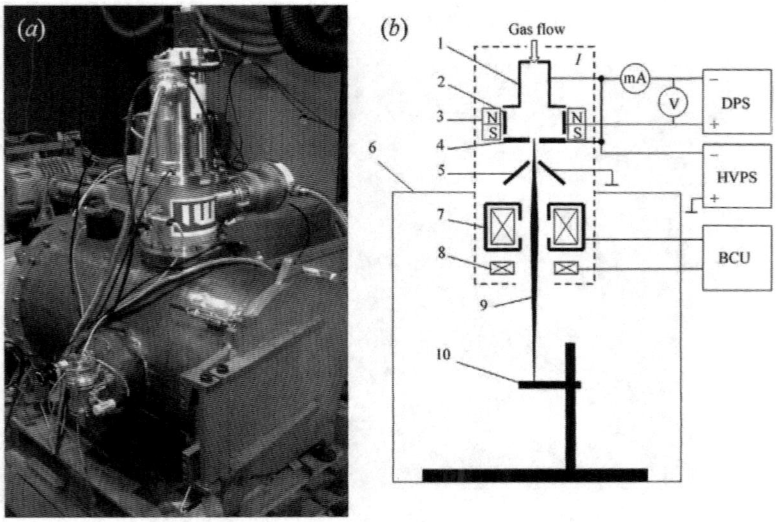

Figure 1. Appearance (a) and configuration (b) of experimental setup: 1 – hollow cathode; 2 – anode; 3 – permanent magnet; 4 – emitter cathode; 5 – accelerating electrode; 6 – vacuum chamber; 7 – focusing lens; 8 – deflection coils; 9 – electron beam; 10 – probing system; DPS – discharge power supply; HVPS – high-voltage power supply; BCU – beam control unit.

The working gas (usually air) is fed into the discharge chamber through the hollow cathode channel and is pumped out through the channel in the emitter cathode. The accelerating voltage (HVPS) up to 60 kV is applied to the emitter cathode (4) relative to the grounded accelerating electrode (5). The plasma electrons that leave the chamber through the emitter cathode channel accelerated in the high-voltage electric field and form a beam, which is focused by the magnetic field of the focusing lens (7). The deflection coils (8) are used for beam deflection. It is common to use a low-voltage form of glow discharge, i.e. with a voltage of 350÷450 V. The pressure in the discharge chamber is $(0.5 \div 1) \cdot 10^{-1}$ mbar. The value of the gas flow is small and usually does not exceed $(4.2 \div 5.6) \cdot 10^{-3}$ mbar·l/s. The maximum beam current is 150 mA.

3. Measurement of electron beam diameter

One of the most important parameters of an electron beam is its diameter. The current density of an electron beam is usually distributed in the form of a Gaussian curve. Depending on the application, a diameter is defined as the width of an area covered with a fixed fraction of the maximum value or the integral of the distribution. The half value width (FWHM, Full Width at Half Maximum), well-established in industry and science for the characterization of electron beams, is used in this work. This is the width of the range in which the power density exceeds at least half the maximum value. In this paper, measuring the diameter of the electron beam was carried out by stationary sensor with the deflection system (scanning slit).

For the first method (scanning slit), a system designed by TWI Ltd. was used (see figure 2).

The TWI EB probe system comprises the following:

- The two finger probe with integrated Faraday cup;
- A beam sink, or dump, capable of withstanding the maximum beam power for extended periods;
- A data acquisition system based around an industrial PC.

The probe heads measure a sample of the beam current as it passes through a narrow slit, as the beam is swept across the probe in a direction orthogonal to the alignment of the slit (see figure 2).

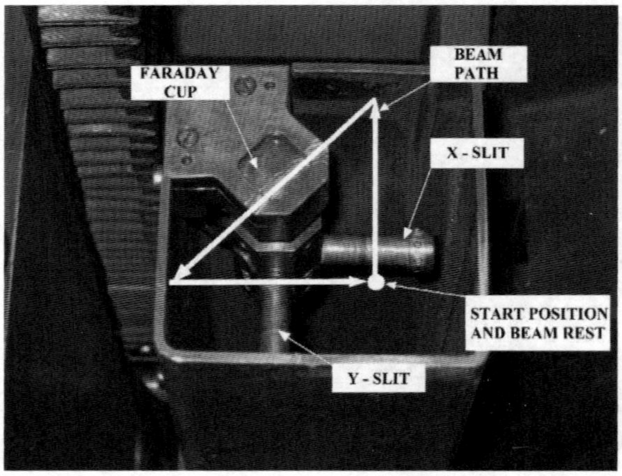

Figure 2. General view of the TWI EB probe device.

The main information gathered by the probe is the width of the beam in the direction of the sweep (see figure 3).

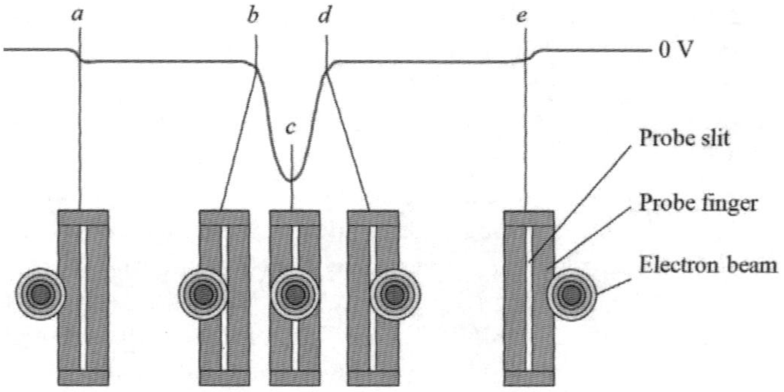

Figure 3. Scheme showing a passage of the beam over probe finger and resulting data trace.

Typically, this width is defined to be the full width of the beam at half of the maximum measured current (FWHM). The second probe is orientated perpendicular to the first and is used to measure the width in the orthogonal direction. In addition to the width of the beam, the beam speed can be measured by collecting current from the probe body. This allows an accurate measure of the velocity of the beam sweep, and enables the width measurements to be converted from a duration to a distance. The two-finger probe relies on the beam moving over the slit at a constant speed to acquire good quality data. The speed should be consistent across both the X and Y finger to ensure that the probe data is comparable. A Faraday cup is designed as an optional feature, to allow the total beam current to be measured. The sensor is a cup with an aperture size sufficient to capture the entire beam as it is swept over – thus enabling the total beam current to be measured.

Points *a* and *e* on the figure 3 correspond to the moments of the first and last contact of the beam with the probe finger. At the moment of time corresponding to points *b* and *d*, the beam first and last contacts the slit. At point *c*, most intense part of beam is over slit (Peak beam shape signal is generated).

The beam path was chosen as a triangular pattern moving out over the X finger, tracking across the Faraday cup and falling back to the free fall position across the Y finger as shown by the line in figure 2.

For a beam with Gaussian power distribution brightness can be written as [7]

$$B = \frac{i_{beam}}{(\pi r_f \theta)^2}, \tag{1}$$

where i_{beam} is the beam current, r_f is the beam radius at the focal plane and θ is the convergence angle.

The convergence angle θ of the electron beam can be calculated by

$$tg(\theta) = \frac{r_l - r_f}{b}, \tag{2}$$

where b is the distance of the sharp focus plane to the focusing lens and r_l is the beam radius in the focusing lens.

4. Conclusion

The brightness of the electron beam generated by the gun with a plasma emitter, made at the beam current range of 10÷60 mA was measured. The measurement was carried out at the accelerating voltage 60 kV and the focal plane of the electron beam was set to 520 mm. Table 1 shows the results of our measurement giving the diameter of the focused electron beam, as well as the calculated brightness and the convergence angle, respectively.

Table 1. The results of measuring the electron beam brightness.

Beam current (mA)	Beam diameter at focus (μm)	Convergence angle (rad)	Brightness ($A \cdot m^{-2} \cdot sr^{-1}$)
10	550	$1.6 \cdot 10^{-3}$	$0.5 \cdot 10^{10}$
20	560	$1.5 \cdot 10^{-3}$	$1.2 \cdot 10^{10}$
40	580	$2.5 \cdot 10^{-3}$	$0.8 \cdot 10^{10}$
60	460	$2.7 \cdot 10^{-3}$	$1.6 \cdot 10^{10}$
Tungsten hairpin filament cathode [8]			$5 \cdot 10^{8}$
Tungsten hairpin tip cathode, LaB_6 cathode [8]			$5 \cdot 10^{10}$
LaB_6 sharp tip cathode [8]			$5 \cdot 10^{11}$

The test pieces were made for qualitative evaluation of the measured parameters of the electron beam. Macrosection of the sample is shown in figure 4.

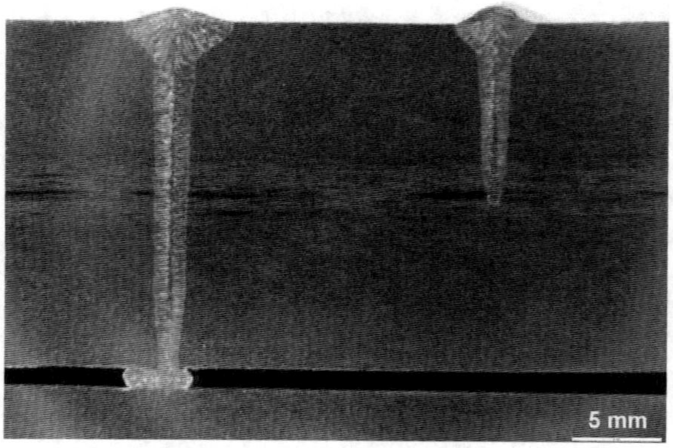

Figure 4. Crossection of the welds.

As evident from Table 1, the brightness of an electron beam from gun with the plasma emitter is not inferior compared to beams from thermionic cathodes of LaB_6 and of superior brightness compared to tungsten filament cathodes. Furthermore, a narrow and deep penetration profile with parallel walls (see figure 4) characterizes the beam with a small convergence angle, which correlates with the results of measurements the beam parameters.

High brightness and small convergence angle of the electron beam generated by the gun with a plasma emitter, have allowed to extract the focused electron beam in the atmosphere with an energy of 60 keV and a current 100 mA (see figure 5).

Figure 5. Electron beam from gun with a plasma emitter was extracted in the atmosphere.

The three-stage differential pumping system of Non Vacuum Electron Beam Welding machine (NV EBW 175-25 TU, PTR GmbH, Germany) was used as a prototype of the differential pumping system for gun with a plasma emitter. The pressure drop from the atmosphere to 10^{-4} mbar in the beam-forming region was provided by differential pumping system.

Acknowledgements

The research was funded by the Ministry of Education and Science of Russian Federation under Agreement number 14.577.21.0018 (RFMEFI57714X0018). We express gratitude to C. Ribton, the employee of TWI Ltd., for the measurements executed in this work.

References

[1] Galansky V L, Gruzdev V A, Osipov I V and Rempe N G 1994 *J. Appl. Phys.* **27** 953.
[2] Rempe N, Kornilov S, Beniyash A and Hassel T 2012 *Welding and Cutting* **11**(2) 122.
[3] Osipov I V and Rempe N G 2000 *Rev. Sci. Instrum.* **71**(4) 1638.
[4] Molokovsky S I and Sushkov A D 2005 *Intense electron and ion beams* (Berlin: Springer).
[5] Hawkes P W 1972 *Electron optics and electron microscopy* (Taylor and Francis).
[6] Kornilov S Yu, Osipov I V and Rempe N G 2009 *Instrum. Exper. Tech.* **52**(3) 406.
[7] Kornilov S Yu, Rempe N G, Beniyash A, Murray N, Hassel T and Ribton C 2013 *Tech. Phys. Lett.* **39**(10) 843.
[8] Reimer L 1989 *Transmission electron microscopy* (Berlin: Springer).

Journal of Physics: Conference Series **552** (2014) 012005 doi:10.1088/1742-6596/552/1/012005

Discharge in the saline solutions in a vicinity of the threshold voltages

Y D Korolev[1,2,3], I A Shemyakin[1,2*], R V Ivashov[1,2], V S Kasyanov[1,2], N V Landl[1,2], Y H Sun[4], T Shao[4], Y Gao[4]
[1]Institute of High Current Electronics RAS, 634055 Tomsk, Russian Federation
[2]Tomsk State University, 634050 Tomsk, Russian Federation
[3]Tomsk Polytechnic University, 634050 Tomsk, Russian Federation
[4]Institute of Electrical Engineering Chinese Academy of Science, 100190 Beijing, P.R. of China

E-mail: shemyakin@lnp.hcei.tsc.ru

Abstract. This paper deals with the discharge in saline solution (0.9 % NaCl) in coaxial electrode system at voltages up to several hundred volts. The discharge was powered by bipolar pulsed generator with a pulse repetition rate of 100 kHz and by a generator of negative voltage pulses. At voltages less than a threshold value (less than 300 V), different types of gas microcavities are formed at the surface of active electrode: the microbubble, the conglomerates of microbubbles (thin gas layers) and a macrobubble with diameter of about 1 mm and less. In a vicinity of threshold voltage the breakdown in the gas cavities occur. For negative polarity at the metal electrode the discharge burns in a form of microsparks of short duration. At positive voltage pulses, the glow-type discharge is sustained in the microcavities.

1. Introduction

In recent years, plasmas generated in and with liquids have attracted considerable interest because of their great potential for applications in various fields [1–4]. One of the progressing fields is the usage of discharges in the so-called plasma scalpel for plasma surgery [5–8]. In this device, the discharge is sustained in multielectrode or two-electrode system immersed in medical saline solution (0.9 % NaCl) [5, 6]. The electrode system is powered by a bipolar voltage with a pulse repletion rate of about 100 kHz and with voltage amplitude at the exit of pulsed generator up to 300 V. This voltage is insufficient to provide a complete breakdown of the gap. As shown in [5, 7, 8], the gas discharge plasma is generated in the vapor microbubbles or in the thin gas layers adjacent to an active electrode. In the rest part of the gap, the current flows through the saline solution that serves as a kind of ballast resistor for the discharge current. When an area of the discharge plasma is in contact with the soft tissues, it becomes possible to produce ablation, healing or the other kinds of the tissue treatment.

The saline solution represents an electrolyte in which, as distinct from the distilled water [1], the gas vapor cavities are generated at a low voltage applied to the electrodes (of about 100 V and even less). As soon as the voltage reaches some threshold value, the breakdown in the gas microcavities (in microbubbles of thin vapor layers) occurs; the discharge plasma appears in the cavities, and the discharge current starts flowing in the electrode system.

When speaking about the current flowing during a pulse of several microseconds, it is often implied that the discharge operation is accompanied by a certain average current value. However,

Content from this work may be used under the terms of the Creative Commons Attribution 3.0 licence. Any further distribution of this work must maintain attribution to the author(s) and the title of the work, journal citation and DOI.
Published under licence by IOP Publishing Ltd

within this time interval the current fluctuations on a scale less that 1 μs are superimposed on the above average current [7, 8]. Then the discharge in the microcavities seems to be essentially nonsteady-state phenomenon. The reasons for appearing of the current peaks and the current interruptions can be associated with the transitions from one forms of the discharge burning to the others, for example, with the glow-to-spark transitions [9–13]. Investigation of the nonsteady discharge behavior and elucidation of the nature of current and voltage fluctuations are the main goals of the present paper.

2. Experimental arrangement

Schematic of experimental arrangement is shown in figure 1. The discharge is sustained in the coaxial electrode system between active electrode *1* (from titanium or stainless steel) and the return grounded electrode *3*. The electrodes are placed in case *4* with saline solution (0.9 % NaCl). The case is equipped with the quartz windows that offer a possibility to take the discharge photographs and to obtain the time resolved waveforms of the discharge luminosity by means of photomultiplier *PMT*.

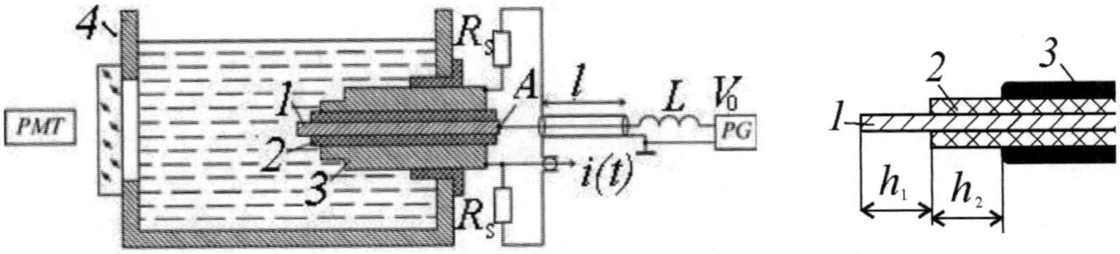

Figure 1. Schematic of arrangement for discharge investigation and sketch of electrode system. *1*–active electrode of diameter D_1 = 0.5 mm; *2*–ceramic insulating tube D_2 = 2 mm; *3*–grounded return electrode; *4*–case with the saline solution; *PG*–pulsed generator; V_0–output voltage of the pulsed generator; *L*–inductance of electric circuit; *l*–length of the coaxial connecting cable; R_s = 1.41 Ω – current shunt resistor; *PMT*–photomultiplier tube; h_1 = (0.7–1.5) mm, h_2 = 1.5 mm.

The discharge was powered by two types of the pulsed generators *PG*. One of them is the generator of bipolar voltage pulses with maximum amplitude at the open circuit of 300 V and with pulse repetition rate of 100 kHz. In this case a capacitance of connecting cable was C = 400 pF and the inductance L at the exit of pulsed generator was L = 13 μH. Due to the inductance, the regime of resonant charging of the cable capacitance was realized in some conditions, which allowed us to obtain at the gap a voltage pulse higher than 300 V.

Another electric circuit was based on the generator of rectangular negative voltage pulses with maximum voltage at the open circuit up to 400 V. The pulse duration was varied from 10 to 50 μs, a pause between the pulses was 10 μs and larger, so that a pulse repartition rate could be varied from 0.15 to 50 kHz. In most experiments, for this electric circuit, we used a coaxial cable of length l = 1.5 m (cable capacitance C = 150 pF), and a ballast resistor R_b = 490 Ω was inserted at the exit of pulsed generator instead of the inductance L.

Current in the discharge gap was recorded with a use of the low-inductance shunt R_s and the oscilloscope Tektronix TDS 3034. Voltage at the gap was recorded between the point *A* and the grounded electrode *3* with a use of probe P6139A.

3. Formation of the gas cavities at the surface of active electrode

When a voltage is applied to the gap, a current starts flowing via electrolyte and the gas microbubbles appear at the surface of active electrode *1*. The breakdown in the gas microbubbles and in the other types of vapour cavities occurs if only the voltage at the gap exceeds some threshold value [1, 7, 14]. Below we consider the characteristic features of behaviour of the gas cavities at low voltages. It should be noted that the cavities impede the current passage via electrolyte. Then the gap resistance turns out to be minimal when the microbubbles and the gas thin layers are absent at the electrode surface. Increasing a number of the gas cavities leads to increasing the gap resistance. Depending on the conditions in the gap, at each particular instant of time, different types of the cavities can exist at the electrode surface and in the bulk of electrolyte. Typical situations are illustrated in figure 2.

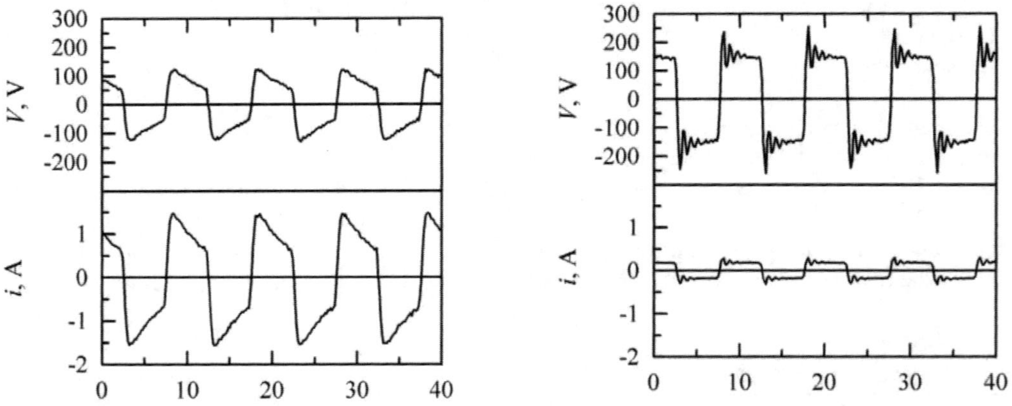

(*a*) (*b*)

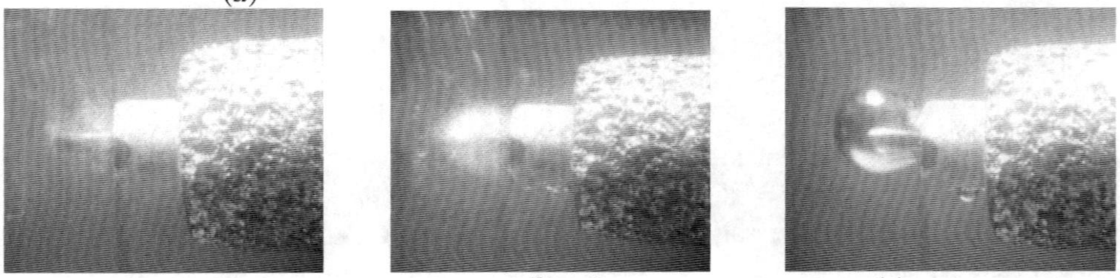

(*a*) (*b*) (*c*)

Figure 2. Voltage and current waveforms jointly with typical photographs of electrode system in the conditions when there is no discharge plasma in gas cavities. Active electrode: D_1 = 0.5 mm, h_1 = 1.5 mm. Output voltage of bipolar pulsed generator: *a*) V_0 = 100 V, *b*) V_0 = 150 V, *c*) V_0 = 175 V.

At a low voltage (of about 100 V and less), only small part of the electrode surface is covered by the cavities in a form of microbubbles (figure 2*a*). Some microbubbles detach from the surface and float upward. The current via the gap for this particular case is about 1 A and the gap resistance of about 80 Ω that is approximately equal to the resistance of electrolyte in absence of the microbubbles.

When voltage is increased, the microbubbles are merged with each other thus forming the conglomerate of the bubbles and thin gas layers (100 μm and less) at the electrode surface. Beside that, a lot of microbubbles can be clustered in a bulk of electrolyte near active electrode. Example of this situation is shown in figure 2*b*. The gap resistance is enhanced for such conditions up to typical values at a level of 1 kΩ. At a high gap resistance, we can see the oscillations at the beginning of voltage pulse that appears due to effect of resonant charging of cable capacitance.

One more pattern of the gas cavities is a single macrobubble with a size of about 1 mm which can be formed at the active electrode (figure 2*c*). In this case the most part of electrode is covered by the gas cavity so that the gap resistance becomes up to several kiloohms.

4. Regimes of the discharge burning

Further increase in the voltage of pulsed generator results in the conditions when a breakdown occurs and gas discharge plasma appears inside the microcavities. A threshold breakdown value is in a vicinity of 300 V and corresponds to minimum of Paschen's curve. Then the interelectrode gap can be classified as a system in which the discharge current flows between metallic and liquid electrodes. The

Journal of Physics: Conference Series **552** (2014) 012005 doi:10.1088/1742-6596/552/1/012005

photographs of the gap with discharge plasma for different types of the gas cavities are shown in figure 3. It is seen that the discharge plasma can be sustained both in the thin vapour layer (figure 3*a*) and in the macrobubble (figures 3*b* and 3*c*).

(*a*) (*b*) (*c*)

Figure 3. Typical photographs of the discharge in different regimes of its operation. Powering from generator of bipolar voltage. *a*) Discharge plasma is generated in a thin gas layer V_0 = 200 V; *b*) Discharge in a macrobubble V_0 = 175 V; *c*) Discharge in a macrobubble V_0 = 275 V.

Note that the photographs demonstrate an integral discharge images taken with a large exposition time. During this time the discharge burns in essentially nonsteady regime. The discharge properties are determined in a great extent whether the cathode is liquid of metallic. Let us consider the case when the pulsed generator of negative polarity is used, i.e. we deal with the metal cathode. An example of the current and voltage waveforms for this case is shown in figure 4.

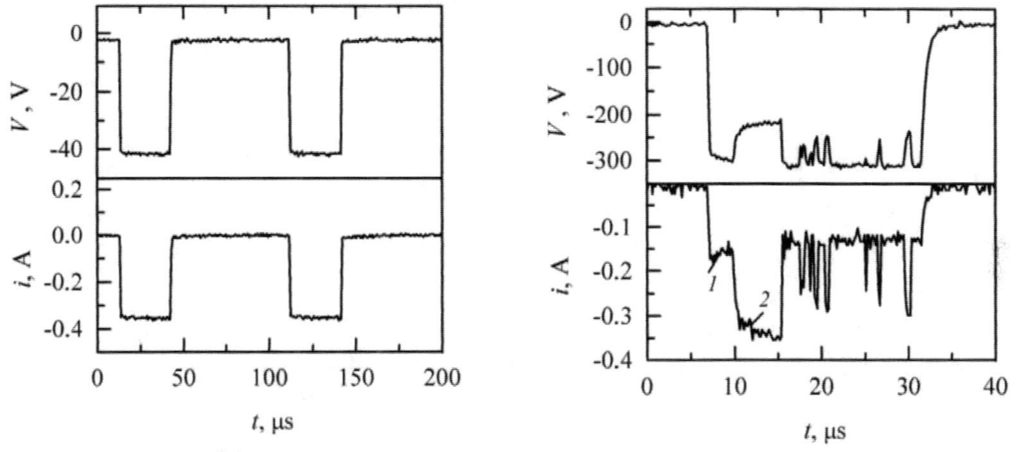

(*a*) (*b*)

Figure 4. The waveforms for the conditions when a voltage of negative polarity is used. Instead of the inductance *L*, a ballast resistor R_b = 490 Ω is inserted in electric circuit. Active electrode: D_l = 0.5 mm, h_l = 0.7 mm. Output voltage of the pulsed generator: *a*) V_0 = 210 V, *b*) V_0 = 400 V.

At low output voltage of the pulsed generator we have the situation in the gap shown in figure 2*a*. The gap resistance is *R*=115 Ω. The voltage waveform is the same as the output voltage of generator. When V_0 is increased to 400 V, gas discharge plasma is generated in the gas cavities, which is reflected at the voltage and current waveforms (figure 4*b*). It is seen that the discharge burns in a nonsteady regime.

At the temporal stage *1*, there is no discharge plasma in the cavities so that the gap resistance R_1 = 2000 Ω is determined by the area of active electrode covered with the gas microcavities. Once the breakdown occurs, the current via the gap sharply increases up to about 0.35 A, and the gap voltage decreases to 225 V (stage *2*). It seems that at the very beginning of stage *2*, a glow-type discharge appears in the cavity and after that the glow-to-spark transition occurs [9, 15–17]. As shown in [10, 15, 18], an energy for formation of the spark channel is delivered to plasma channel from the cable capacitance *C*.

Since the discharge current is limited by the ballast resistor and by the resistance of saline solution, the spark discharge is not able to be transformed into a steady-state arc. At the end of stage *2*, the spark current is interrupted and the gap resistance is increased again. After that new breakdown and new current peak is observed. Then the form of discharge sustaining in the case of metal cathode can be interpreted as a sequence of the partial discharges of a short duration.

When the bipolar voltage pulse is used, the forms of discharge burning with negative polarity at the active electrode and with positive polarity differ from each other. This case is illustrated in figure 5, where the voltage and current waveforms are supplemented with the waveforms of discharge luminosity obtained from photomultiplier tube V_{PMT}.

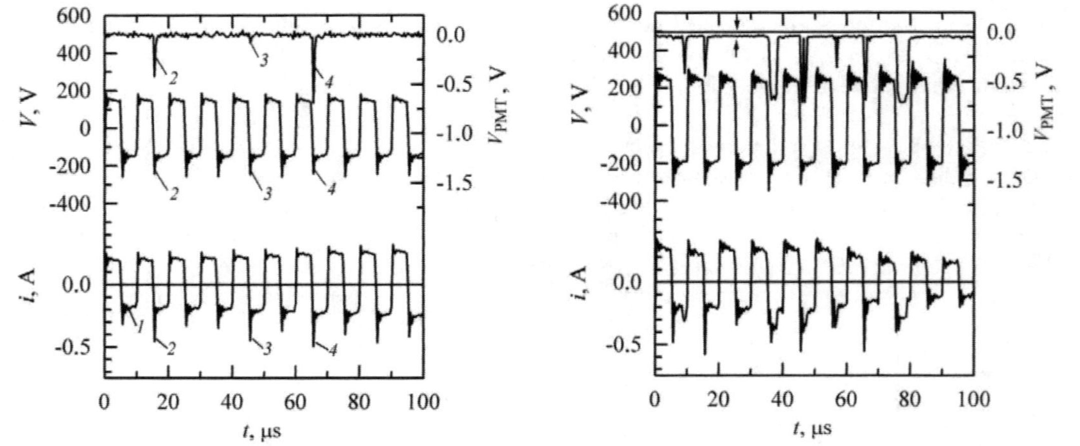

(*a*) (*b*)

Figure 5. Voltage and current waveforms jointly with the signal from photomultiplier for the conditions when the partial discharges appear in the gas cavities. Active electrode: D_1 = 0.5 mm, h_1 = 1.5 mm. *a*) V_0 = 150 V, *b*) V_0 = 225 V.

Figure 5*a* corresponds to the threshold voltage. Due to resonant charging of the capacitance *C*, a voltage value of 300 V is achievable at the gap at the beginning of voltage pulse. Then for negative polarity of the voltage, for some pulses the conditions are realized when the breakdowns in the microcavities occur. These breakdowns are accompanied by the current pulses of short duration. For example, in the particular case under discussion, during the pulse *1* there is no breakdown. The partial discharges are visible in the pulses *2*, *3* and *4*. These discharges are accompanied by the spikes in the voltage from PMT. For the most of negative pulses and for the positive pulses the signal from PMT is zeroes. This means that at the threshold voltage, the partial discharges occasionally appear only at negative polarity of voltage pulse. Even though a glow-type discharge arises in the microbubble, this discharge transforms into spark due to formation of the cathode spot on metal electrode [9, 17–20].

When voltage of pulsed generator is increased, the discharge is sustained not only in the negative pulses but also in the positive pulses (figure 5*b*). We can see a weak steady-state signal from photomultiplier which corresponds to glow-type discharge. With metal cathode the phenomenon of glow-to-spark transition manifests itself while with liquid cathode (at the positive pulses) the discharge burns in glow mode without formation of cathode spot and successive spark channel.

Acknowledgments
The work was supported by the Russian Foundation for Basic Research under the grant # 14-08-91153.

References

[1] Bruggerman P and Leys S 2009 Non-thermal plasmas in and in contact with liquids *J. Phys. D: Appl. Phys.* **42** 053001

[2] Locke B R and Shih K Y 2011 Review of the methods to form hydrogen peroxide in electrical discharge plasma with liquid water *Plasma Sources Sci. Technol.*, **20** 034006

[3] Sun Y H, Zhou Y X, Jin M J, Liu Q and Yan P 2005 New prototype of underwater sound source based on the pulsed corona discharge *Journal of Electrostatics* **63** 969–75

[4] Korolev Y D, Mesyats G A and Yarosh A M 1987 Film etching by particles produced by a pulsed bulk discharge in CF_4 *High Energy Chem.* **21** 389–92

[5] Schaper L, Stalder K R and Graham W G 2011 Plasma production in electrically conducting liquids *Plasma Sources Sci. Technol.*, **20** 034004

[6] Stalder K R, McMillen D F and Woloszko J 2005 Electrosurgical plasmas *J. Phys. D: Appl. Phys.* **38** 1728–38

[7] Stalder K R, Woloszko J, Brown I G and Smith C D 2001 Repetitive plasma discharges in saline solutions *Appl. Phys. Lett.* **79** 4503–05

[8] Woloszko J, Stalder K R and Brown I G 2002 Plasma characteristics of repetitively-pulsed electrical discharges in saline solutions used for surgical procedures *IEEE Trans. Plasma Sci.* **30** 1376–83

[9] Korolev Y D, Frants O B, Landl N V, Geyman V G and Matveev I B 2007 Glow-to-spark transitions in a plasma system for ignition and combustion control *IEEE Trans. Plasma Sci.* **35** 1651–57

[10] Korolev Y D, Frants O B, Landl N V, Geyman V G and Matveev I B 2009 Nonsteady-state gas-discharge processes in plasmatron for combustion sustaining and hydrocarbon decomposition *IEEE Trans. Plasma Sci.* **37** 586–92

[11] Korolev Y D, Frants O B, Landl N V, Geyman V G, Shemyakin I A, Enenko A A and Matveev I B 2009 Plasma-assisted combustion system based on nonsteady-state gas-discharge plasma torch *IEEE Trans. Plasma Sci.* **37** 2314–20

[12] Korolev Y D, Frants O B, Landl N V, Kasyanov V S, Galanov S I, Sidorova O I, Kim Y, Rosocha L A and Matveev I B 2012 Propane oxidation in a plasma torch of a low-current nonsteady-state plasmatron *IEEE Trans. Plasma Sci.* **40** 535–42

[13] Korolev Y D, Kuzmin V A, and Mesyats G A 1980 Nanosecond gas discharge in nonuniform field with explosive processes on electrodes *Zhurnal Tekhn. Fiz.* **50** 699–704

[14] Mukasa S, Nomura S, Toyota H, Maehara T and Yamashita H 2011 Internal conditions of a bubble containing radio-frequency plasma in water *Plasma Sources Sci. Technol.* **20** 034020

[15] Korolev Y D, Frants O B, Geyman V G, Kasyanov V S and Landl N V 2012 Transient processes during formation of a steady-state glow discharge in air *IEEE Trans. Plasma Sci.*, **40** 2951–60

[16] Korolev Y D, Frants O B, Landl N V and Suslov A I 2012 Low-current plasmatron as a source of nitrogen oxide molecules *IEEE Trans. Plasma Sci.* **40** 2837–42

[17] Korolev Y D, Frants O B, Geyman V G, Landl N V and Kasyanov V S 2011 Low-current “gliding arc” in air flow *IEEE Trans. Plasma Sci.* **39** 3319–25

[18] Kozyrev A V, Korolev Y D, Mesyats G A, Novoselov Y N and Shemyakin I A 1981 Contraction of a volume discharge, initiated by ultraviolet-radiation in Ar-SF_6 mixtures *Zhurnal Tekhnicheskoi Fiziki* **51** 1817–1822

[19] Genkin S A, Karlov N V, Klimenko K A, Korolev Y D, Kuzmin G P, Mesyats G A, Novoselov Y N and Prokhorov A M 1984 Application of mild X-ray-radiation for initiation of self-sustained spaced discharge in larger interelectrode intervals *Pisma Zhurnal Tekhn. Fiz.* **10** 641–45

[20] Kozhevnikov V Y, Kozyrev A V and Korolev Y D 2006 Drift model of the cathode region of a glow discharge *Plasma Phys. Rep.* **32** 949–59

Study of Hydrogen Accumulation in Palladium and Silver-Palladium Alloy

N.N. Nikitenkov*, Yu.I. Tyurin, V.S. Sypchenko , I.V. Dushkin , A.N. Nikitenkov, O.V. Vilkhivskaya
Tomsk Polytechnic University, 30, Lenin ave., Tomsk, 634050, Russia

E-mail: nikitenkov@tpu.ru

Abstract. This work presents the results of investigation of the hydrogen accumulation characteristics in palladium, silver and palladium-silver alloy ($Pd_{60}Ag_{40}$) at their saturation with hydrogen from different aggregate states of the ambient medium (plasma, liquid, hydrogen atmosphere).The presence of silver in the alloy of palladium-silver leads to the following features: the capture of hydrogen by Pd and $Pd_{60}Ag_{40}$ samples depends on the method of saturation. In the case, while plasma and electrolytic hydrogen saturation in both samples is being trapped in the same type of traps (not significantly different in binding energy). While the saturation by Sieverts method takes place, traps formed are characteristic for each type of the samples. In case of electrolytic and plasma saturation in palladium and palladium-silver alloy, the main type of trap is the binding of hydrogen in palladium hydride. Alloying palladium with silver does not lead to decreasing the sorption capacity of the alloy $Pd_{60}Ag_{40}$ compared to Pd.

1. Introduction

It is known that palladium is widely used for solving a wide range of problems in hydrogen energetics. However there is a deficiency of palladium which seriously impedes its use, particularly, for the membrane manufacture (e.g., in membrane reactors for the production of ultra-pure hydrogen [1-5]). The deficiency is that the alternate heating and cooling of palladium in a hydrogen atmosphere makes it gradually brittle and deformed. At the same time, some of palladium-silver alloys are able to work in an atmosphere of hydrogen without losing their mechanical strength for a long time [4].

The results presented below are development of the research [6].

The aim of this article is to clarify the influence of silver on the capture of hydrogen by palladium-silver alloy. First of all, it is necessary to examine the efficiency of hydrogen capture into palladium, silver and palladium-silver alloy (in this case a $Pd_{60}Ag_{40}$ alloy) by the different saturation methods:
1) From the plasma of high-frequency discharge (HFD),
2) From electrolyte (in an electrolytic cell at using of studied samples as cathodes),
3) From the gas phase at elevated temperature and pressure (Sieverts method).
The samples were investigated by thermally out gassing (TOG) (which is the same thermally stimulated desorption (TSD)) and X-ray phase analysis (XPA) methods.

2. Materials and experimental methods

Samples of palladium and palladium-silver alloy ($Pd_{60}Ag_{40}$) were made of sheet material in the form of rectangular plates 25×5×0.1 mm^3, and silver samples in the form of rectangular plates 25×5×2 mm^3

Content from this work may be used under the terms of the Creative Commons Attribution 3.0 licence. Any further distribution of this work must maintain attribution to the author(s) and the title of the work, journal citation and DOI.
Published under licence by IOP Publishing Ltd

(jeweler argentums). Each sample was saturated repeatedly both in the delivery condition and after "saturation-TOG" cycles. The obtained TOG temperature spectra were normalized to the sample volume.

To study the hydrogen accumulation during the electrolytic saturation of palladium samples, silver, and $Pd_{60}Ag_{40}$ alloy samples, 1M H_2SO_4 electrolyte on the basis of distilled water was used. Saturation was carried out for t=6 min (for Pd and $Pd_{60}Ag_{40}$ samples) and from 6 to 80 min (for Ag samples) at a current density J=0.24 A/cm^2.

Parameters of saturation regime by Sieverts method are: sample temperature T=100°C, pressure P=2 atm., saturation time t=2 hours.

Parameters of plasma saturation regime are: sample temperature T=150°C, the pressure in the saturation chamber $P=3\cdot10^{-1}$ Torr, the saturation time t=80 min.

The specified parameters of saturation are chosen so that intensity of the hydrogen yield at TOG was observed in one order of magnitude for all samples.

The installation for studying thermal- and radiation- stimulated outgassing of General Physics department of Tomsk Polytechnic University (TPU) and the international scientifically-educational laboratory "Technologies of hydrogen power engineering" TPU was used. It was also used to study the evolution of hydrogen thermally stimulated from the samples. During the study we carried out the analysis of dependences of the hydrogen yield intensities on temperature (the temperature spectra) of thermally outgassing. Linear heating of samples has carried out in the temperature range from room temperature to 1000°C.

Investigations of phase composition of the samples were carried out at the station "Anomalous Scattering" (channel number 2), Siberian Center for Synchrotron and Terahertz radiation, Novosibirsk. The parameters used in the system are: X-ray energy range: 5–20 keV; the size of the input beam: 5×(0.1–1.5) mm^2; the time to set diffractograms – from 0.2; resolution of the diffractometer: $\Delta d/d\sim10^{-4}$ in a high-resolution mode and $\Delta d/d\sim10^{-3}$– in the mode of one-coordinate detector.

The binding energies of hydrogen atoms in traps (E_b) shown in figures were estimated according to standard curves given in [6].

3. Results and discussion

Fig. 1 presents the TOG H_2 spectra from Pd and $Pd_{60}Ag_{40}$alloy samples saturated by electrolytic method. The figure 1 shows that the spectra are of about the same maximum intensity and the peaks are located at the same temperature ~ 360°C (which corresponds to the binding energy of hydrogen in traps E_b ~ 1.9 eV). In addition, the palladium-silver alloy is observed to have a feature at ~ 410°C (E_b~2 eV). The intensity of the peaks of TOG H_2 spectra of palladium and palladium-silver alloy is almost identical, and the values of the integrals under the curves are (shown in Fig. 4–5): $I_{Pd60Ag40}/I_{Pd}\approx6848/4994\approx1.37$.

Fig. 2 shows the TOG H_2 spectra from the Pd and alloy $Pd_{60}Ag_{40}$ samples saturated from the gas phase (Sieverts method). As it can be seen in the figure 2, after the saturation from the gas phase in the palladium and palladium-silver samples the binding energy of hydrogen in the traps is, as well as in the previous case, 2eV and 1.8 eV respectively. The spectrum of the hydrogen outgassing from the palladium-silver sample is shifted to the area of lower temperature by 71°C. In addition, the hydrogen spectrum of the outgassing from the palladium sample shows an unresolved peak (shelf) at 455°C, this temperature corresponds to the binding energy of 2.1 eV. It is also worth to note that the intensity at the peak of the TOG H_2 spectrum of palladium is lower than one of the palladium-silver alloy, and$I_{Pd60Ag40}/I_{Pd}\approx2.4$.

Fig. 3 shows the H_2 TOG spectra from Pd and $Pd_{60}Ag_{40}$ alloy samples, saturated from the plasma.

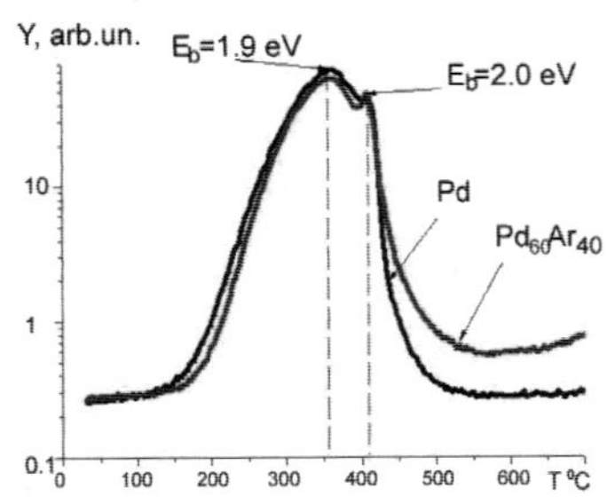

Figure 1. TOG H_2 temperature spectra from samples saturated from the electrolyte

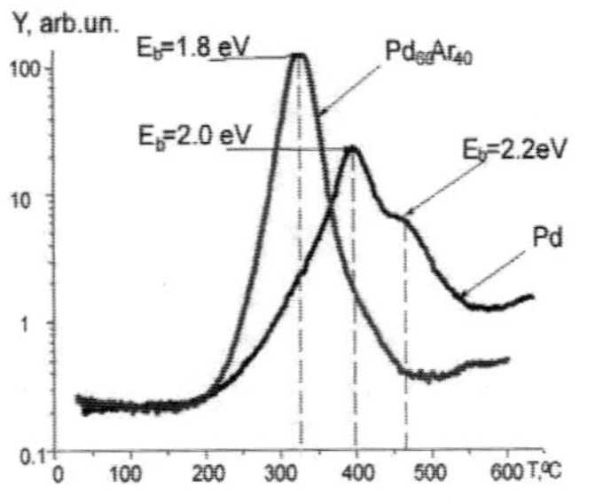

Figure 2. TOG H_2 temperature spectra of samples saturated by Sieverts method

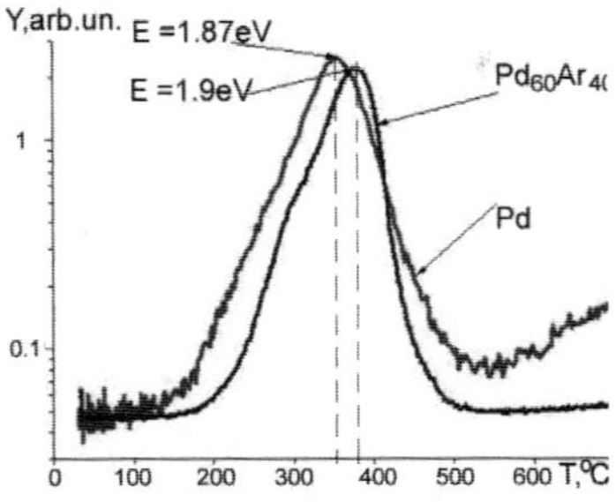

Figure 3. TOG H_2 temperature spectra from samples saturated from the plasma.

As it can be seen from the figure 3, while palladium and palladium-silver samples interact with plasma, hydrogen traps are formed with a binding energy of hydrogen 2eV and 1.8eV respectively. The peak intensity of the spectra from both types of samples is almost the same, and $I_{Pd60Ag40}/I_{Pd} \approx 2{,}9$. Concerning the temperature (T_m) at which there is a maximum of TOG H_2 spectra, we note the following: T_m of the electrolytic saturation is the same for both materials: $T_m(Pd)=T_m(Pd_{60}Ag_{40})$; while for saturation by Sieverts method $T_m(Pd)>T_m(Pd_{60}Ag_{40})$; with saturation of the plasma $T_m(Pd)<T_m(Pd_{60}Ag_{40})$.

Fig. 4–5 show TOG H_2 spectra of the Pd and $Pd_{60}Ag_{40}$ samples saturated by three methods: electrolytic, plasma and Sieverts.

It is seen that for Pd samples Tm in the temperature range of ways of saturation "electrolytic (el)–

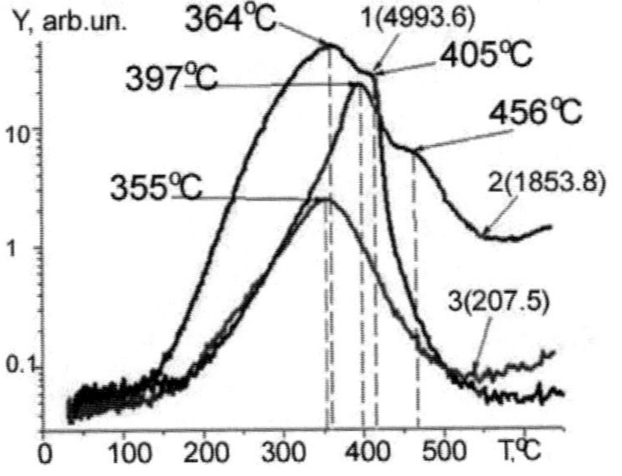

Figure 4. TOG H_2 temperature spectra from Pd samples, saturated by methods: 1 – electrolytic; 2 – Sieverts; 3 – plasma. Numbers in brackets are quantities of integrals (square under curves in standard units).

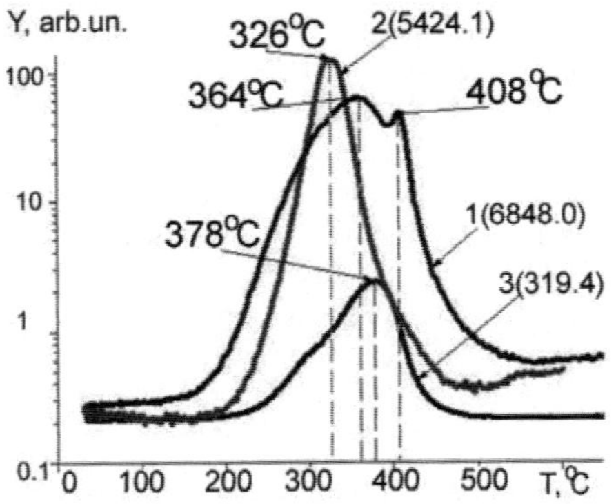

Figure 5. TOG H_2 temperature spectra from $Pd_{60}Ag_{40}$ samples, saturated by methods: 1 – electrolytic; 2 – Sieverts; 3 – plasma. Inscriptions at the curves is the same as in Fig. 4

Plasma (pl) – Sieverts (Siv)" are in ratio as T_m (el)$\approx T_m$ (pl)$>T_m$ (Siv), and for $Pd_{60}Ag_{40}$ samples as $T_m(el)\approx T_m(pl)<T_m(Siv)$. Thus, it can be inferred that the type of defects which capture hydrogen during plasma and electrolytic saturation prove to be the same in $Pd_{60}Ag_{40}$ and Pd. At the same time, the shift of T_m (Si) between the spectra of Pd and $Pd_{60}Ag_{40}$ reaches 71°C, which indicates that different types of traps are present in these samples at saturation by Sieverts method. The minimum amount of hydrogen is trapped in the plasma saturation, the maximum - in the process of electrolytic saturation. Fig. 6 presents the results of X-ray-phase analysis of Pd and $Pd_{60}Ag_{40}$ samples before and after hydrogen saturation.

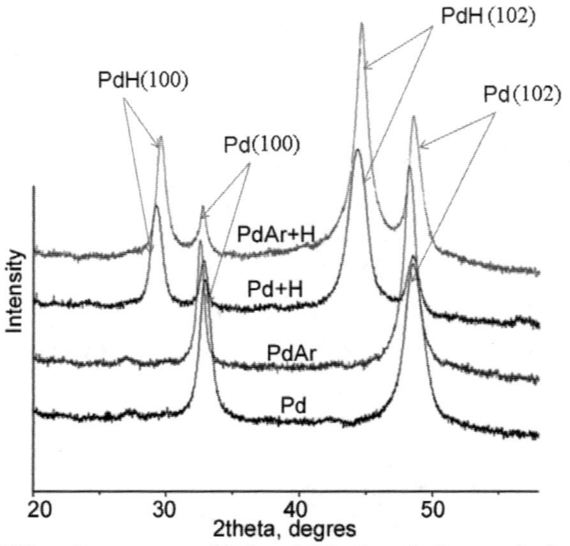

Figure 6. The X-ray diffraction spectra of Pd and $Pd_{60}Ag_{40}$ before and after hydrogen saturation by electrolytic method. The straight areas in spectra, respectively: Pd, PdAg – before saturation, Pd-H, PdAg-H – after saturation. Arrows specify peaks in spectra, corresponding to separate phases Pd and PdH. Energy of the probing X-ray: 10 keV.

Pd and PdAg curves show the phase compositions of samples before the saturation; Pd-H, PdAg-H – the same samples after hydrogen saturation. Two peaks in each spectrum correspond to the metallic phase of palladium. The presence of silver apparently corresponds to a small shift in the position of the Pd peaks relatively to their position in $Pd_{60}Ag_{40}$spectrum (silver displaces palladium atoms in the lattice and not significantly alters the interatomic distance). In the spectra of the saturated samples there is an additional peak (to the left of the peak corresponding to the metallic phase), which corresponds to a Pd-H phase. Note, that there are no peaks observed corresponding to Pd-Ag and Ag-H phases in the X-ray diffraction patterns (Fig. 6), thus, there are no the chemical bonds of silver with palladium or hydrogen in these samples. In other words, no silver hydrides are formed (for their formation high temperatures or (and) pressure are required). This result, first, suggests that the $Pd_{60}Ag_{40}$ alloy is a solid solution, and the introduced hydrogen (during the electrolytic saturation) bonds only with palladium.

4. Conclusions

1. The capture of hydrogen by Pd and $Pd_{60}Ag_{40}$ samples depends on the method of saturation. In that case when hydrogen saturation from electrolytic and plasma in both samples is being trapped in the same type of traps (not significantly different in binding energy). While the saturation by Sieverts method takes place, traps formed are characteristic for each type of the samples.
2. In electrolytic and plasma saturation in palladium and palladium-silver alloy the main type of traps is the binding of hydrogen in palladium hydride.
3. Alloying palladium with silver does not lead to a deterioration of sorption capacity of the alloy $Pd_{60}Ag_{40}$ compared to Pd.
4. The silver in the $Pd_{60}Ag_{40}$ alloy is in a solid-solution condition.

References

[1] Buranov G S, Roshan N R, Kol'chugina N B 2006 *Alternative Energy and Ecology* **7– 39** 36.
[2] Tereshchenko G F, Orekhova N V, Yermilova M M 2007 *Critical Technol. Membranes* **1–33** 4.
[3] Annual Review of Activities in Basic Areas Editor V. Parmov 2010 Novosibirsk 301.
[4] McLeary E E, Jansen J C and KapteijnF 2006 *Microp. & Mesop. Mat.* **90** 198.

[5] Sypchenko V S, Nikitenkov N N, Sigfusson T I, Tyurin Yu I 2012 *Bull. Russ. Acad. Scien. Physics.* **76–6** 712.
[6] Woodraff D P, Delchar T A, 1986 *Modern Techniques of Surface Science.* University Press.Cambridge 377.

Journal of Physics: Conference Series **552** (2014) 012007 doi:10.1088/1742-6596/552/1/012007

Unbalance magnetron plasma source for ion mass-separator

V L Paperny [1,*], V I Krasov[1], N V Astrakchantsev[2] and N V Lebedev [2]

[1] Irkutsk State University, Irkutsk, 664003, Russia
[2] Irkutsk State Technical University, Irkutsk, 664070, Russia
[3] E-mail: paperny.math.isu.runnet.ru

Abstract. The report presents the results of the preliminary studies characteristics of an unbalanced magnetron plasma source supplied with the transport system based on a curved magnetic field. The aim of these studies was to recognize if the system is suitable, in principle, for mass-separation of a multi-component plasma flow. The magnetron source has 50 mm diameter cathode manufactured of an alloy composed of Cu (64%), Pb (22.5%) and admixtures, about of 14% (Al, Zn, C). By means of an immersion time-of-flight spectrometer, a spatial distribution of ions of the cathode material was measured through the system output cross-section. Distribution of atom of these elements was measured here by the X-ray fluorescence spectrometry as well. Both methods showed that the ions of the lighter element (Cu) were concentrated in the inner part of the plasma flow deflected by the magnetic field while the distribution of the heavy element (Pb) was shifted toward the outer area of the flow. The similar effect was observed for each couple of the elements. Such a system is promising for use in plasma technology of reprocessing spent nuclear fuel, namely for separation heavy radioactive fission product from nuclear waste.

1. Introduction

Plasma techniques have long been used for various applications in nuclear researches. In particular, a few types of rotating magnetized plasmas are a convenient way to separate particles by mass. Namely, a concept of centrifugal confinement produced a set of the circular electrodes in a magnetized rotating plasma was proposed in paper [1]. Concept of a filter to separate particles of different mass using rotating plasma with both centrifugal and magnetic confinement was examined in paper [2].

The mass-separators mentioned above are based on magnetic system with straight field lines. Nevertheless, mass-separation effects from centrifugal force may be anticipated when a plasma flows along curved magnetic field lines. Various electro-magnetic guiding structure designs based on curved magnetic fields, for instance, are used frequently as filters to remove macroparticles from cathodic vacuum arc plasma jet. Recently, the present authors have suggested using a vacuum arc supplied with a transport system based on the curved magnetic field for mass separation a multi-component cathodic plasma flow [3,4].

It is interesting to perform the similar measurements with a plasma flow produced with a magnetron source because the source produces an intensive and stable plasma flow. The objective of the study is to find out whether it is possible to use such a system as an element of a technological system for plasma reprocessing of atomic industry waste [5].

Content from this work may be used under the terms of the Creative Commons Attribution 3.0 licence. Any further distribution of this work must maintain attribution to the author(s) and the title of the work, journal citation and DOI.
Published under licence by IOP Publishing Ltd

2. Experimental details

2.1. Experimental setup

Experiments were carried out using a modified device similar to that described in [3,4], where the vacuum arc plasma source was substituted with a magnetron source. The schematic of the device is shown in figure 1. The discharge burned at the surface of 50 mm diameter and 4 mm thickness target as a cathode, the rear end of which was cooled with water. The target was made of a composite material consisting of light and heavy components, specifically, alloy of Cu (m_{Cu} = 64 amu) and Pb (m_{Pb} = 204 amu) with mass abundances of 64% and 22%, respectively, and the admixture (C –7%, Al, Zn – each about of 2%).The low-current (100 mA, 500 V) duty magnetron discharge was sustained with a dc power supply. To ensure the high ionisation degree in the plasma flow, a high power impulse magnetron sputtering (HiPIMS) was used [6]. The main pulsed discharge was initiated by means of application to the cathode a high voltage (- 1500V) pulse from a capacitor of 100 μF so that the discharge current length was of 3.5 ms and the average discharge current was about of 20 A. The grounded vacuum vessel was as an anode. The vessel was evacuated down to a residual pressure not

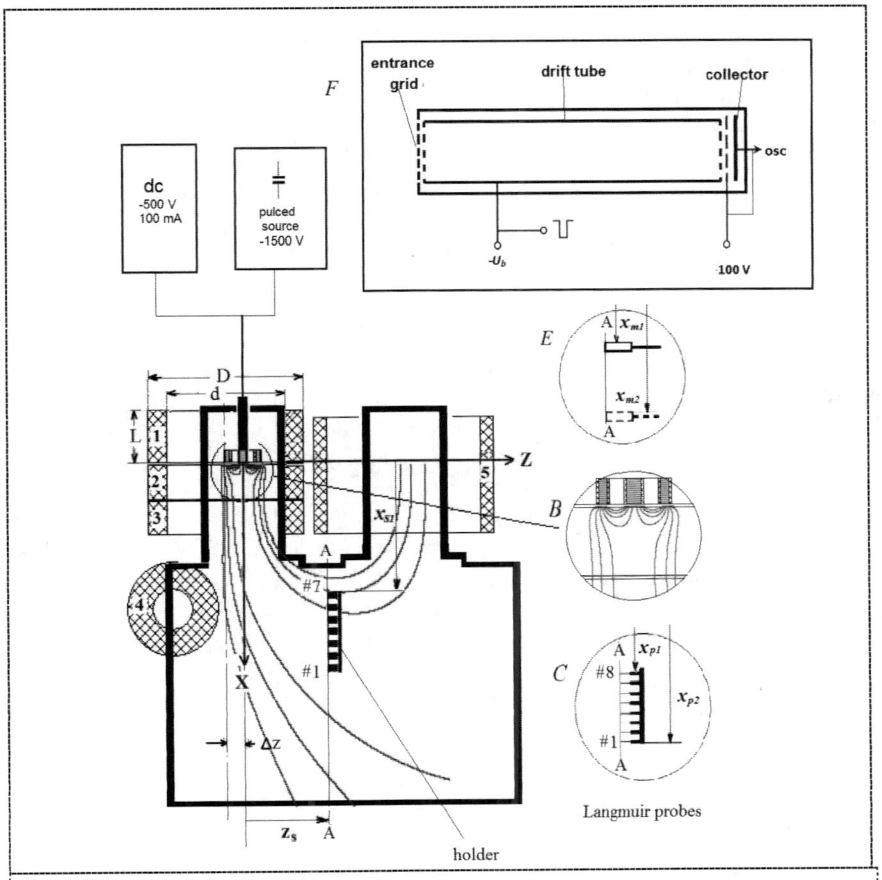

Figure 1. Schematic diagram of the experimental setup. The setup is in the horizontal plane and is viewed from above. 1 – 5 are solenoids of the magnetic transport system; z_s = 11 cm, x_{s1} = 15 cm, x_{m1}= 18 cm, x_{m2} = 24 cm, x_{p1}= 17 cm and x_{p2}= 28 cm; inset *B* depicts the enhanced map of magnetic field line near the cathode, inset *C* depicts the probes holder and inset *E* depicts the ITOFS positions in the A – A plane

exceeding of 2×10^{-3} Pa and the after that the working gas (Ar) was introduced into the vessel up to about 2×10^{-2} Pa pressure, which was held constant. The pulse repetition rate was 10 pps; the current was measured with the Rogovski coil. A flow of metallic plasma was produced from the cathode surface and was guided through the working vessel by a transport system based on a curved magnetic field.

The magnetic field was formed with a set of magnetic solenoids, so that three of them (1, 2, 3 in figure 1) were co-axial to each other, and the co-axis of the other two (4 in the figure) was perpendicular to the axis of the former. In this series of experiments, the standard magnetic system was modified by introducing an additional solenoid 5. The characteristics of the solenoids, namely their inner (*d)* and outer (*D*) diameters and length (*L*), which are depicted in figure 1a, as well as the number of coils (*N*) and the solenoid currents (*I*) are shown in Table 1. Solenoids1and 2 were intended for magnetic stabilization of the discharge on the working surface of the cathode. Solenoids 3 and 4 were placed so that their magnetic field affects the structure of the total field in the travelling area of the cathodic plasma flow; hence it helped us to govern the flow magnetically. Solenoids 4 were supplied with currents of equal values but opposite polarity so that their magnetic fields were also oppositely directed. Solenoid 5 was designed to reduce the curvature of lines of the guiding magnetic field, hence to increase the centrifugal force providing the mass-separation effect.

The magnetic field configuration presented in this paper was calculated by a procedure similar to that applied in [3,4] and it was described previously in detail there, hence it is presented here briefly. The configuration was calculated by applying Biot–Savart's law to an array of current carriers representing the magnetic solenoid coils used in the transport magnetic system. A code to set up the current carrier array and then to trace the magnetic field line is written using MATLAB™. A comparison of the calculated and measured magnetic fields in the reference points shows that they differ by less than 15% from each other. Figures 1 depicts a calculated map of the magnetic field lines at a set of the solenoid currents presented in the Table 1.

Due to the specific design of the device, the cathode axis was shifted relative to the common axis of solenoids 1 – 3 by $\Delta z = 30$ mm (see figure 1). As a result, the magnetic field lines passing through the cathode surface were deflected in the +*Z* direction. In the A–A plane, where the parameters of the cathode flow were measured, the field lines were directed almost perpendicular to their initial direction. The plasma flow propagating along the field lines was deflected in a similar manner. In this magnetic configuration, the field strength varies substantially as the plasma propagates from the cathode to the measurement area. The maximum field strength in the vicinity of the cathode was about of 70 mT, and it drops with distance from the cathode so that in the A – A plane it was 2 – 3 mT.

To establish the effect of the magnetic field on the plasma components, take into account that the electron temperature in the magnetron plasma was of about 2 eV and the plasma density estimated from the Langmuir probes measurements was of the order of 10^{12} cm^{-3}. From these data one may estimate the electron Hall parameter in the plasma flow: $\omega_{Be}\tau_{ei} >> 1$, where $\omega_{Be} = eB/m = 10^{10} - 3\times10^{8}$ c^{-1} is the electron cyclotron frequency and $\tau_{ei} \approx 10^{-8}$ c is the electron-ion collision time. The similar estimation for ions shows that $\omega_{Bi}\tau_{ei} << 1$. Hence, a conclusion may be drawn that the plasma electrons

Table 1. Characteristic of solenoids of the magnetic transport system

	D (mm)	*d* (mm)	*L* (mm)	*n*	*I* (A)
1	240	180	85	1570	1.05
2	240	180	55	1100	1.05
3	240	180	55	1100	2.0
4	145	65	110	1270	1.4
5	240	180	220	1050	0.72

in the transport system are magnetized and the plasma ions are non-magnetized. This implies that the travel of a plasma flow through the system is governed by varying the magnetic field by the solenoids.

2.2. Langmuir probe measurements

We used a set of Langmuir probes as a diagnostic technique. The set consisted of eight identical pieces biased by −50V that provided operation of the probes in the ion current saturation mode. The probes were intended for measurement of the ion saturation current profile throughout the horizontal diameter of cross-section A–A of the plasma flow. For this purpose the probes were inserted at a holder, where they were spaced 15mm from each other and the holder was located as shown in figure 1. The probe had a tungsten wire of 10 mm length and 1mm diameter. The probe data were acquired by a computer. Shape of the probe signals were near the same as that of the discharge current (see figure 2), and the figure 3 presents the profile of probe ion currents across the horizontal diameter of the plasma flow cross-section on the A–A plane. One can see that there a maximum at the probes 4 and 5 is observed that corresponds to 21 – 24 cm distance from the *Z*-axis and the current decreases sharply at the plasma flow area approaching the rear vacuum chamber wall.

2.3. Mass spectrometer measurements

We used an immersion time-of-flight spectrometer (ITOFS) for measurement of the plasma flow mass composition at exit of the magnetic transport system. The detailed description of the spectrometer is presented in [7] and we describe it briefly. The plasma flow components are separated in the gap between the entrance grid and grid on the drift tube (see inset *E* in figure 1). These fine grids have a 100 μm mesh size that is close to the plasma Debye length; hence we suppose that the ion-electron separation occurs in the inter-grid gap. A high-voltage (-800 V) pulse of rectangular shape and 200 ns duration was applied to the drift tube that formed an ion bunch at the tube entrance. This bunch travelled through the tube and when passing the tube, the ions were separated by the time of flight τ in accordance with parameter *m/Z*. Thus, for ions of m_1 and m_2 masses and Z_1 and Z_2 charge states, the relation is valid as follows,

$$\frac{\tau_1}{\tau_2} = \left[\frac{m_1}{m_2}\frac{Z_2}{Z_1}\right]^{1/2}.$$

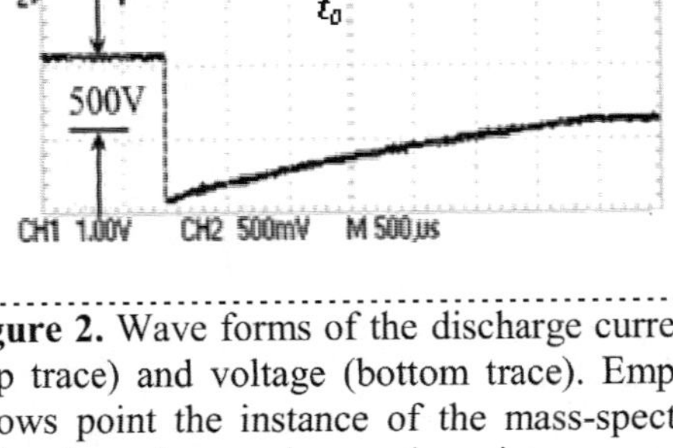

Figure 2. Wave forms of the discharge current (top trace) and voltage (bottom trace). Empty arrows point the instance of the mass-spectra and the Langmuir probe ion current measurements

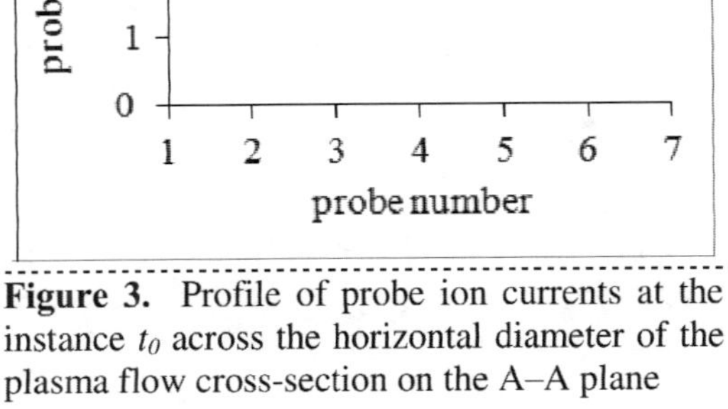

Figure 3. Profile of probe ion currents at the instance t_0 across the horizontal diameter of the plasma flow cross-section on the A–A plane

After travelling the drift tube, the ions passed through the exit grid and were captured with the collector.

Figure 4 presents the ITOFS collector signals obtained at positions x_{m1} and x_{m2} on the A – A plane, depicted in figure 1, at the instant t_0 pointed in figure 2. One can see from figure 4 that the plasma flow ion compositions differ significantly in these positions. Namely, there are observed a few peaks corresponding to the main plasma component, Cu^+ and Cu^{++} ions, as well as to the light ions Ar^+ (m_{Ar} = 40 amu) in position x_{m1} . As the ITOFS is displaced to position x_{m2} , then the Cu^{++} and Ar^+ ion peaks disappear and a peak corresponding to the Pb^+ ions appears.

2.4. Measurements of ion component distributions in the plasma flow with X-ray fluorescence spectrometry

The separating characteristics of the magnetic transport system were determined from the distribution of the cathode material components in the A–A plane. To do this, a series of seven identical collectors fixed in a holder was placed in the A–A plane so that the holder was parallel to the *X*-axis (see figure 1) and placed in the plane *XZ*. The collectors were spaced by distances of 2 cm from one another along the *X* axis. Collector no. 1 was placed at a distance of 27 cm from the *Z*- axis, while collector no. 7 was placed at a distance of 15 cm from this axis. The collectors were 12 mm diameter polished discs from Ti foil faced the plasma jet. The collector surface was cleaned in a glow discharge and then irradiated by the plasma jet for 30 min. As a result, a metal film consisting of the elements of the cathode material formed on the surface of the collectors.

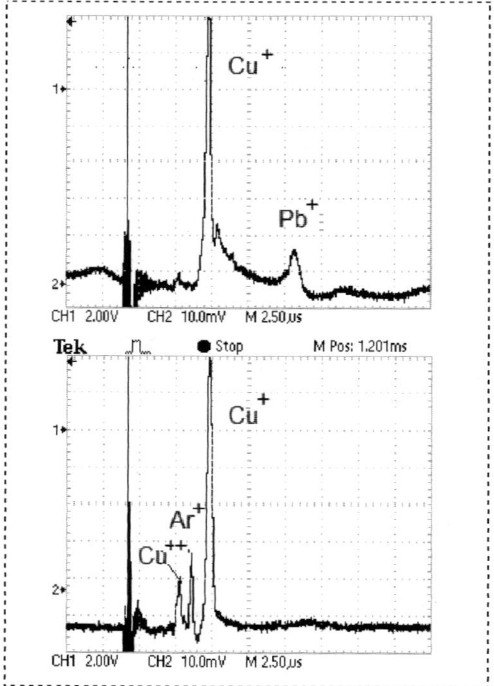

Figure 4. Wave forms of the ITOFS signals at position x_{m1} (bottom panel) and x_{m2} (top panel)

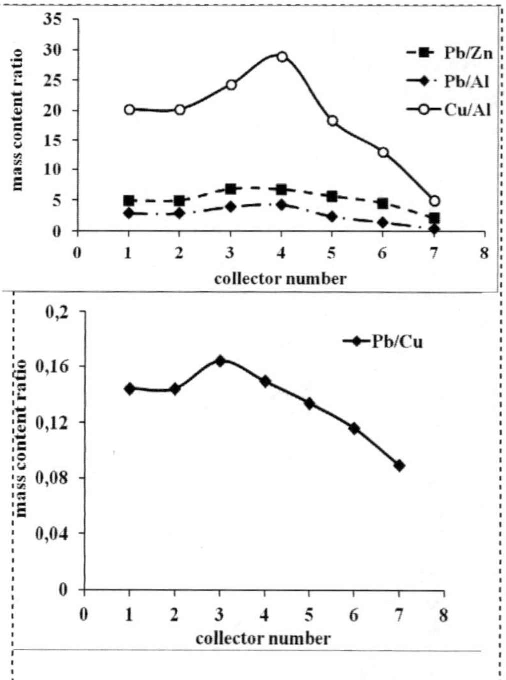

Figure 5. Ratio of mass content of the main cathode components (bottom panel) and that of the admixtures (top panel) in the films deposited on the surfaces of the collector array on the A – A plane

The elemental composition of the film was studied by means of electron probe X-ray spectrum analysis by using a JXA 8200 microanalyzer and an EX-84055MU energy dispersive spectrometer (JEOL Ltd., Japan). X-ray emission was generated by an electron beam with a 1×1 μm cross section and contains radiation of the elements of the cathode material (Cu, Pb and admixtures), as well as the substrate material (Ti) and gas admixtures (O_2, C, N_2). The data were used to determine the relative mass content of elements in the film as functions of the collector number, i.e., the collector position in the A–A cross section of the cathode jet.

The obtained distributions of the mass content ratios are shown in figure 5. It can be seen that the ratios of the heavy to light elements depend on the collector position. The ratios decreases with the collector number increases, i.e. as position of the collector is displaced towards the internal area of the curved plasma flow. This trend was observed both for the main components, Pb and Cu, and for admixtures as well (see figure 5). Thus, after the cathode jet has passed through the magnetic transport system, its components turn out to be spatially separated along the curvature radius of the jet.

3. Discussion and conclusions

First, note that as figure 4 shows, the working gas (Ar) is found in a minor amount only in the inner area of the curved plasma flow. Obviously, that is due to the low pressure of the working gas, and a conclusion may be drawn, as well, that the magnetron, in fact, operates in a HiPIMS mode. Furthermore, it is seen from the figure that a presence of a small amount of the Cu^{++} ions is found only in the inner area of the flow. This finding implies that an electric field arises in the plasma flow that is due to the separation of magnetized electrons "attached" to the magnetic field lines and ions, which mainly follow the direction of the curvilinear magnetic field because of the relatively strong charge separation electric field. This field directed to the centre of curvature of the bent flow and it strongly affects the Cu^{++} ions. The authors observed recently the similar effect in the curved plasma jet produced with a vacuum arc plasma source [8].

Finally, the results presented in figure 5 show that a significant enrichment of the outer area of the plasma flow with the massive ions is observed for any couple of the elements consisting the cathode material This effect, obviously, is due to influence of the centrifugal force on the ions.

Thus, a conclusion may be drawn that the presented device, in principle, is promised as an ion mass-separator for any applications. For instance, it can be used in the plasma technology of reprocessing spent nuclear fuel, namely for separating heavy radioactive fission product from nuclear waste.

Acknowledgment

The results of this study were obtained in part using the equipment of the Baikal Center of Nanotechnolgies. This work was supported by the Russian Foundation for Basic Researches, project no. 14-02-00943-a

References

[1] Grossman M W and Shepp T A 1991 *IEEE Trans. Plasma Sci.* **19** 1114
[2] Fetterman A J and Fisch N J 2011 *Physics of Plasmas* **18** 094503
[3] Paperny V L, Krasov V I and Lebedev N V 2011 *Plasma Sources Sci. Technol.* **20** 35005
[4] Paperny V L and Krasov V I 2011 *Technical Physics Letters* **37** No.1 75
[5] Zhil'tsov V A, Kudryavtsev E G, Kulygin V M et al *At. Energ.* 2006 **101** 102
[6] Oks E and Anders A *J. Appl. Phys.* 2009 **105** 093304
[7] Ryabchikov A I, Ryabchikov I A, Stepanov I B et al *Rev. Sci. Instrum.* 2006 **77** 303
[8] Korobkin Yu V, Lebedev N V and Paperny V L *Techn. Phys. Lett.* 2012 **38** No. 3 254

Journal of Physics: Conference Series **552** (2014) 012008 doi:10.1088/1742-6596/552/1/012008

Boron carbide nanopowder synthesized using electrical discharge plasma

I.A. Rakhmatullin[1], A.A. Sivkov, A.F. Makarova
[1]Assistant, Tomsk Polytechnic University, Tomsk, Usova st. 7, Russia

E-mail: riam@tpu.ru

Abstract. In this paper we present the possibility to realize the direct plasmadynamic synthesis of dispersed crystalline boron carbide in the supersonic jet of boron-carbon electrical discharge plasma flowing into the chamber filled with argon at ambient pressure and temperature. The phase composition was determined using X-Ray diffraction method. The predominance of boron carbide crystalline phase with average particle size of 50 nm was determined. Particles morphology was shown using transmission electron microscopy. The obtained nanosized powder was used to produce the bulk ceramic material using Spark Plasma Sintering method. The hardness and crack resistance values were measured and equal to 35 GPa and 6.5 MPa*$m^{1/2}$, respectively.

1. Introduction

Boron carbide is one of the superhard materials and attracts an interest due to its high hardness, refractory properties, chemical and wear resistance and low density. Therefore it is used as reinforcing additives in ceramic materials, in cutting tools, nozzles of jet adhesive machining and also as neutron absorber material [1, 2]. In the temperature range above 1000 °C the hardness of this material is higher than the hardness of cubic boron nitride and even diamond [2]. Due to low density and low weight it can be used for low-weight armour ceramic [3].

Synthesis of boron carbide can be realised by different methods, such as carbothermic reduction [1, 2], mechanosynthesis [3], sol-gel method [4] or gas-phase laser synthesis [5]. These methods have a number of well-known disadvantages such as the duration of process, the necessity of implementing high temperature and as a result high energy consumptions. Plasmadynamic synthesis method is one of the alternative ways to solve these problems. This synthesis method is realized in the system based on a pulse high current coaxial magnetoplasma accelerator (CMPA).

This paper shows the experimental results on possibility of plasmadynamic synthesis of nanosized boron carbide in the supersonic electrical discharge plasma jet, generated by CMPA, which is powered from the pulse capacitive energy storage.

2. Experimental

The synthesis was realized with the use of a high current (~100 kA) pulse (~500 μs) CMPA, which generates the supersonic (~2 km/s) electrical discharge plasma jet, flowing into the chamber, filled with argon at ambient pressure and room temperature. The initial powder mix of amorphous boron and carbon black in ratio 4:1 with mass of 1.0 g was put into the plasma formation zone electrodes gap and

Content from this work may be used under the terms of the Creative Commons Attribution 3.0 licence. Any further distribution of this work must maintain attribution to the author(s) and the title of the work, journal citation and DOI.
Published under licence by IOP Publishing Ltd

slightly pressed. Capacitive energy storage with values of capacity 6.0 mF and charging voltage 3.0 kV was used to supply the system connected to electrodes. The voltage and current waveforms registered during the experiment and calculated curves of power and released energy are shown in figure 1. The opening and collecting of obtained powder product were done after one hour after the experiment. The weight of the collected powder was 0.6 g when 1.0 g of initial precursor mix was loaded.

The synthesized product was investigated using XRD (Shimadzu XRD 7000S, 0.15406 nm), TEM (Philips CM12) and sintered to bulk ceramic material with Dr. Sinter SPS system.

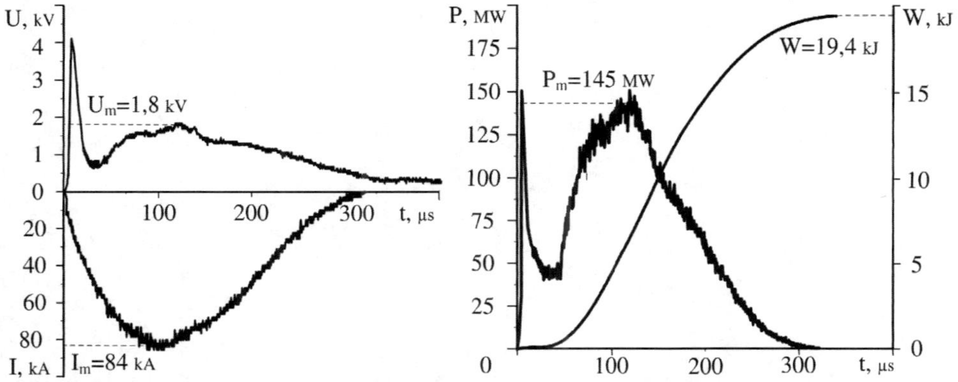

Figure 1. Voltage, current, power and released energy waveforms of the synthesis process

3. Results and discussions

The XRD pattern shown in the figure 2 presents that product is highly crystalline and heterophase. The phase analysis shows that product mainly consists of boron carbide B_4C and has some impurities such as graphite, silicon carbide and boron oxide. The calculated lattice parameters *a* and *c* for synthesized boron carbide are 0.5601 nm and 0.1209 nm, respectively. It is in a good accordance with the data of ICDD card [01-075-0424 PDF4]. The presence of small amount of impurities can be caused by several reasons. Graphite, eroded from the accelerating channel surface under plasma influence, reacts not only with boron but with silicon, the presence of which can be explained by the electro erosion of the electrodes gap insulator. Boron oxide is a residue from initial amorphous boron powder precursor. But it worth noting the level of impurities does not exceed 5 %. Average value of the boron carbide particle`s CSA is about 45 nm.

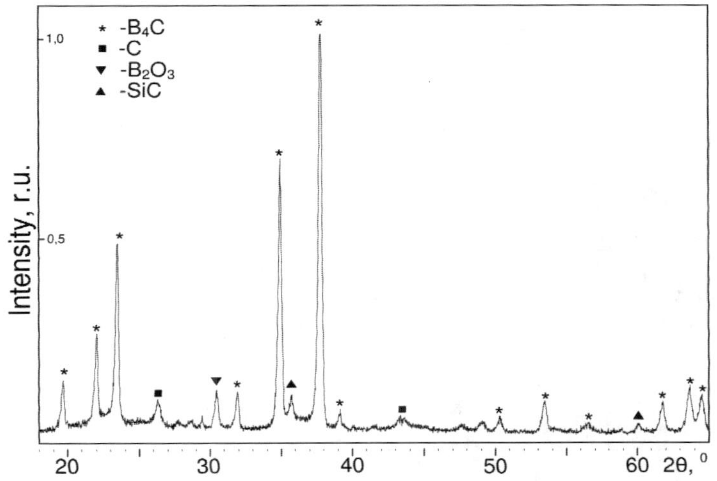

Figure 2. XRD pattern of the synthesized product

Particles have rounded plate shape and according to TEM data (figure 3) their sizes are in the range of 30 nm up to 200 nm.

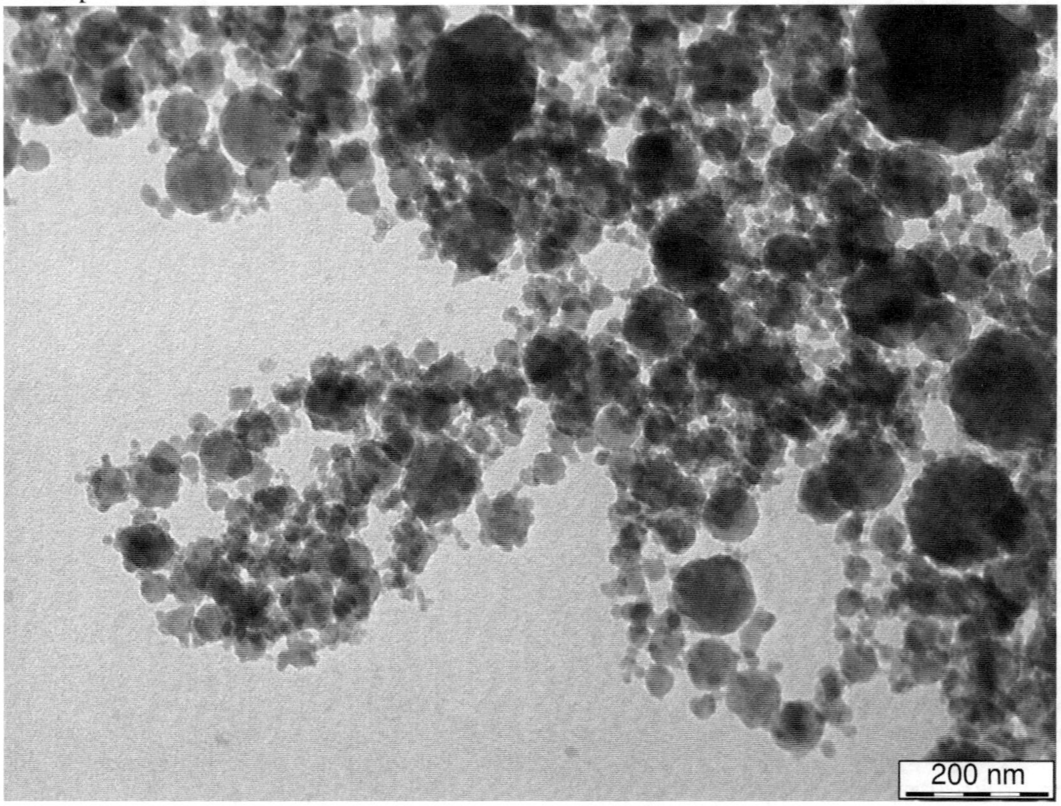

Figure 3. TEM picture of the synthesized product

Bulk ceramic material was sintered using SPS system with following conditions: temperature 1950 ^{0}C, pressure 60 MPa, graphite die diameter 15 mm, rate of temperature rise 100 ^{0}C/min, holding time at maximum temperature 5 min. After sintering the ceramic tablet (figure 4) was obtained and polished to measure the density and investigate its phase composition.

Figure 4. Bulk ceramic material of the synthesized boron carbide base

The relative density of sintered ceramic was measured by geometric and hydrostatic methods. Both methods show that density value is similar and close to 98 % from theoretical density of the boron carbide monocrystall. The XRD pattern of polished boron carbide ceramic sample is shown in the figure 5.

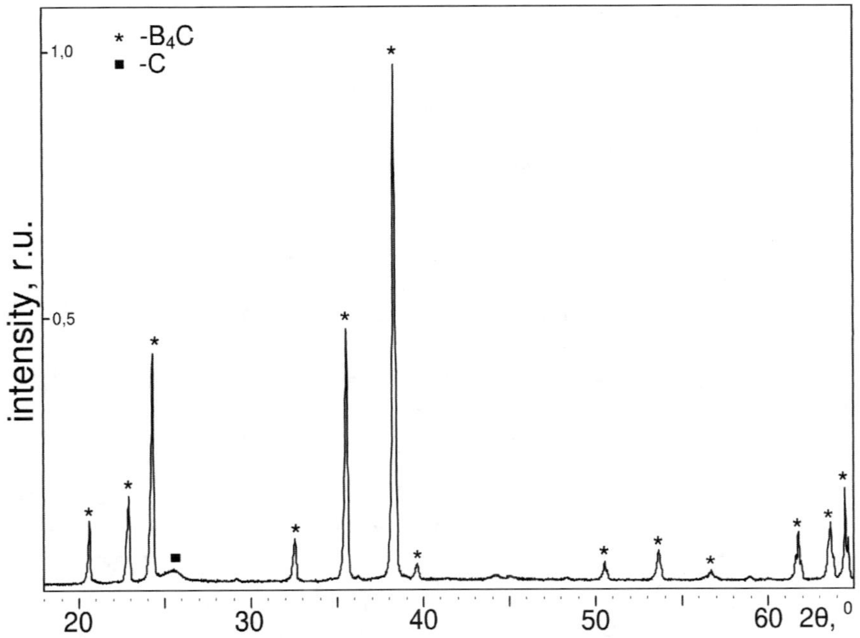

Figure 5. XRD pattern of the polished boron carbide ceramic sample

The XRD analysis shows that the lattice parameters of boron carbide ceramic *a* and *c* became 0.5606 nm and 0.12084 nm, respectively. Average value of CSA became about 85 nm that indicates on low level of recrystallization of grains during sintering using SPS method. The hardness and crack resistance were measured for the synthesized boron carbide ceramic using Vickers indentation with diamond pyramid. Average values of hardness and crack resistance were 35 GPa and 6.5 MPa*$m^{1/2}$, respectively, that evidence on the increasing of crack resistance in comparison with ceramic produced from micron particle sizes powder with silicon carbide additives intended for improving of the poor crack resistance of pure boron carbide ceramic [6].

5. Conclusions

The possibility of the synthesis of nanosized boron carbide with low amount of impurities using a high current pulse CMPA is shown. The improving of mechanical properties of the boron carbide ceramic produced by SPS method using nanosized powder confirms the suitability of method for producing boron carbide powder.

References

[1] Chen S, Wang D Z, Huang J Y, Ren ZF 2004 Applied Physics A **79** 1757
[2] Kakiage M, Tahara N, Yanase I, Kobayashi H 2011 Materials Letters **65** 1839
[3] Sharifi E M, Karimzadeh F, Enayati M H 2011 Advanced Powder Technology **22** 354
[4] Najafi A, Golestani-Fard F, Rezaie H R, Ehsani N 2012 Ceramics International **38** 3583
[5] Oyama T, Takeuchi K 1999 Carbon **37** 433
[6] Tkachenko Yu G *et all.* 2005 Powder Metallurgy and Metal Ceramics **44** 196

Ion trajectories calculation for negatively biased needle cathode in volume discharge plasma

A G Remnev[1,3,*], K Uemura[1], A V Kozyrev[2] and V V Lopatin[3]
[1]ITAC Ltd., Shinmaywa 1-1, Takarazuka 665-0052, Japan
[2]National Research Tomsk State University, Lenin Avenue 36, Tomsk, 634050, Russia
[3]Institute of High Technology Physics, Lenin Avenue 2a, Tomsk, 634050, Russia

E-mail: remnev.a@shinmaywa.co.jp,

Abstract. Ion trajectories were simulated for the case of multi-needle negatively biased electrode immerged into the volume type plasma. The model was simplified to the 2d case with planar plasma boundary. The electrical field distribution was calculated with the FEA method. Resulting piece wise function was then used to predict ion trajectories emitted from the plasma sheath boundary. Series of the ion trajectories were simulated for different plasma and accelerating gap parameters using single particle analysis. Distribution of the ion current density along the needle surface and angles of the ion incidence were obtained from the simulation. Experimental and theoretical etching profiles are consistent.

1. Introduction

Ion sputtering is widely used for solving numerous technological problems. One of the most important application is sputter deposition, where the incoming energetic ions are producing the flow of targets material atoms that are consequently deposited on the substrates surface [1]. It is commonly used in the magnetron, ion beam and other sputtering systems. Another very important application for ion sputtering of materials is sputter etching, where the target itself is being processed by the ion flow in order to modify surface properties of the target [1].

Previously we have reported a method in which the ion-plasma sputtering (IPS) was used for sharpening of medical needles by means of sputter etching [2]. The needles were orderly placed in two-dimensional array and underwent bias ion sputtering in the plasma of hot filament hollow cathode plasma generator. Argon was used as the plasma forming gas. Addition of active gas allowed for smoother surface and sharper cutting edges of the resulting needles [3].

Besides the sharpening, an unexpected distinctive non-linear needle shape was observed as the result of IPS treatment [2]. In order to understand the mechanism of such shape modification, the ion flow distribution on the needle surface can be simulated.

There are numerous studies devoted to ion flow simulation for various geometries and plasma properties. The mainstream method is particle-in-cell (PIC) method [4, 5]. It is often used for plasma simulations and can be applied for wide range of the problems. However utilization of existing modeling codes can be time and resource consuming. In the present study we investigate a simplified approach for particular case of multi-needle electrode in plasma, which can give some of the sputtering process characteristics without high computational load and within shorter code

Content from this work may be used under the terms of the Creative Commons Attribution 3.0 licence. Any further distribution of this work must maintain attribution to the author(s) and the title of the work, journal citation and DOI.
Published under licence by IOP Publishing Ltd

development period. The method includes FEM field calculation, Newton's equations solving and TRIM calculation for sputtering rate angular dependence.

2. Simulation

Mathematically the IPS process can be described with the motion equations for the ions starting from the plasma boundary with near zero velocity and arriving at the needle surface with certain flux density and angle. Ion flux density and angle of incidence at the impact site defines the local sputter rate, which can be interpolated for the whole surface of the target needle by cumulating the local values. To calculate the ions trajectories, one needs to know electric field distribution $\boldsymbol{E}(x,y,z)$ which in this case cannot be calculated directly, so the numerical FEA approach was selected to simulate the distribution. The model was simplified to 2d case and needles were substituted with thin rods in order to reduce the computational load.

The plasma boundary from which the ions are extracted is generally of complex shape, dictated by the plasma density and the voltage drop across the plasma sheath.

However we can consider two particular cases: [i] the plasma density is relatively high and the voltage is relatively low, so the plasma leaks into the gaps between the rods; [ii] the plasma density is relatively low and the voltage is relatively high, so the double layer is getting large and the ion emission surface become nearly planar. In the present study, only the second case was investigated. Space charge in the plasma sheath was neglected. The width for the planar plasma sheath can be estimated from the Child-Langmuir equation.

2.1. Electric field distribution

Electric field distribution across the acceleration gap was evaluated with FEA software Maxwell 2d. Input data for the analysis consists of following: needle electrode geometry, acceleration gap (i.e. distance from the plasma boundary to the needles tips) and applied voltage. The values were taken from the range of actual parameters used during the IPS (see Table 1).

Table 1. Input parameters

Parameters	Values
Bias voltage U, V	-400
Pitch between rods p, mm	5
Length L, mm	10
Rod diameter d, mm	0.2
Acceleration gap width D, mm	2, 4, 6, 8

The acceleration gap width D was varied from 2mm to 8mm; different values for D represent different plasma densities, according to Child-Langmuir equation. The working pressure during the actual process is P=0.4Pa, so the acceleration gap was considered collissionless. Domain of study dimensions were 70mm×42mm. Resulting electrical field distribution is shown on figure 1 for the case of D=4mm.

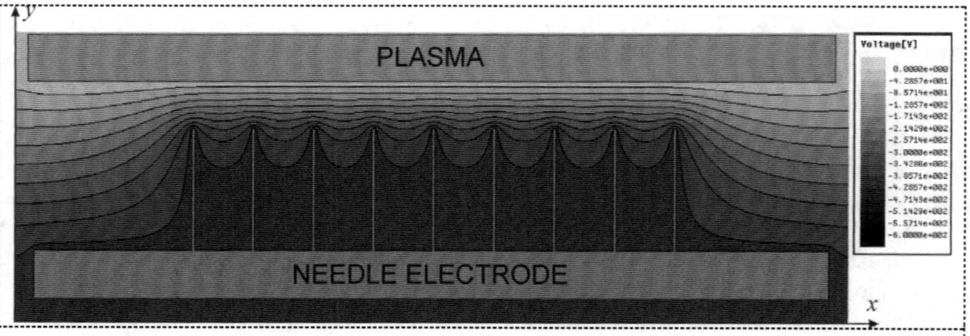

Figure 1. Electric field intensity calculated with U=-400V, p=5mm, L=10mm, d=0.2mm, D=4mm.

2.2. Ion trajectories

In order to calculate the probe ion trajectory the domain of study was divided into rectangular shape cells with the nodes *[X[i], Y[j]]*. Then the electric field distribution obtained with the FEA (see figure 1) was mapped on the rectangular mesh. The electric field within one cell is assumed to be an average of the nodular values.

Having two matrixes of coordinates *[X[i], Y[j]]* and electric field $[E_x[i,j], E_y[i,j]]$components one can analytically calculate whole trajectory of an ion injected from the plasma boundary into the acceleration gap. Once the ion leaves the boundary towards the needle cathode it will unavoidably enter a cell with known constant electric field. Then within the cell the motion equations can be analytically solved:

$$X[i \pm 1] = \frac{E_x[i, j] \cdot e}{2 \cdot m} \cdot t^2 + v_x \cdot t + x$$
$$Y[j \pm 1] = \frac{E_y[i, j] \cdot e}{2 \cdot m} \cdot t^2 + v_y \cdot t + y, \qquad (1)$$

where m – ion mass, e – electron charge, v_x, v_y – ion velocity components which it had on point of entering the current cell, $X[i\pm1], Y[j\pm1]$ coordinates of the cells boarders, where the probe ion leaves the current cell and enter the next cell. For the next cell similar equations (1) are solved with different input parameters. The procedure repeats until the ion reaches the needle electrode or leaves the domain of study. By selecting different injection points for the probe ions, a series of trajectories can be obtained and consequently ions individual incidence angle and local linear current density at the rods surface can be derived.

2.3. Ion current and incidence angle

As can be understood from the figure 1 the ion trajectories will have the same pattern for all the needles except the first and the last needle in the array. Thus, the ion current density distribution was simulated for the needle inside the array in order to rule out the edge effect. Small step size of 7.5um between the probe ion injection points was used in order to achieve sufficient accuracy. The distance from the plasma boundary to the needle tips has been varied (see figure 2). Resulting y-coordinate of the ions impact and angles of incidence θ were plotted (see figure 3). The distributions are calculated along the vertical coordinate on the selected rod – y_r, with y_r=24mm coordinate indicating the upper end of the rod and y_r=14mm – the bottom end.

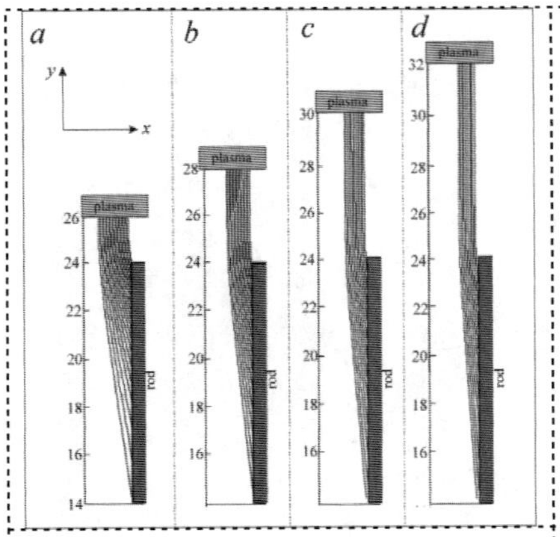

Figure 2. Ion trajectories series for *U*=-400V and *D*=2mm (a), *D*=4mm (b), *D*=6mm (c), *D*=8mm (d).

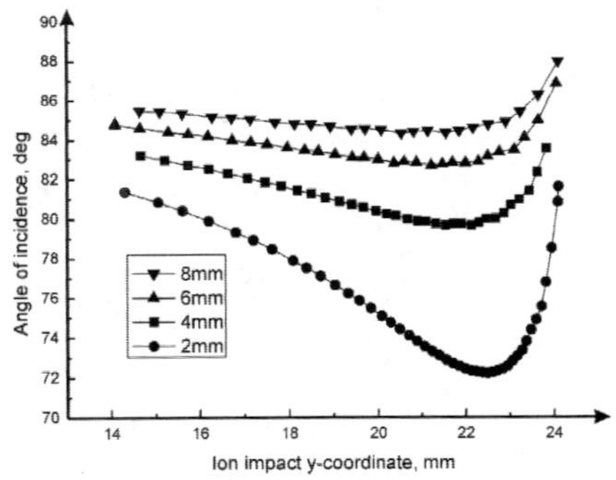

Figure 3. Ion angle of incidence distribution.

2.4. Sputter rate

Sputtering rate distribution along the rod can be evaluated using combination of ion current density and sputtering yield distributions. The latter is angle dependent value, therefore $Y(\theta)$ for 0.4KeV Ar^+ striking Fe target was calculated with SRIM software at several angles within 70 - 90 degree range (see figure 4) and then interpolated with 2 degree polynomial function as:

$$Y(\theta) = -4.14 + 0.18 \cdot \theta - 0.0014 \cdot \theta^2 \quad (2)$$

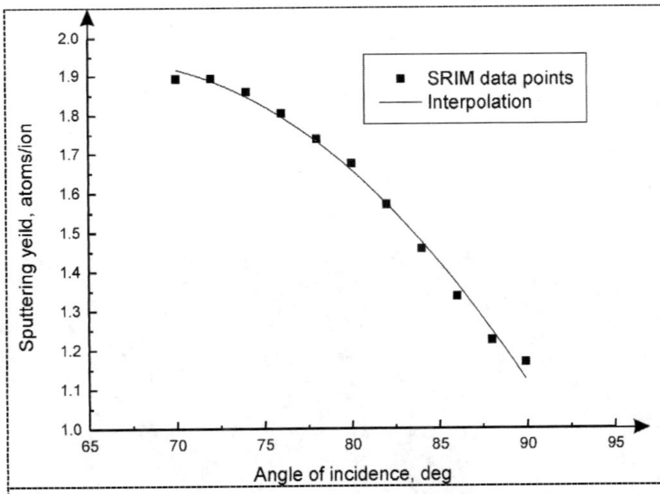

Figure 4. Sputtering yield as function of incidence angle

The yield distribution $Y(y_r)$ along the rod can be found as:

$$Y(y_r) = Y(\theta(y_r)), \quad (3)$$

while the angle of incidence can be found as:

$$\theta(y_r) = arctg(v_{xf}(y_r) / v_{yf}(y_r)), \quad (4)$$

where $v_{xf}(y_r)$ and $v_{yf}(y_r)$ are final velocity components at the ion impact point.

Relative linear current density $j(y_r)$ was interpolated from the density of the calculated trajectory lines on the rods surface; the value depends on the step size between the emitted probe ions. Resulting dimensionless sputtering rate:

$$S(y_r) = Y(y_r) \cdot j(y_r) \quad (5)$$

Simulated $S(y_r)$ is given on the figure 5.

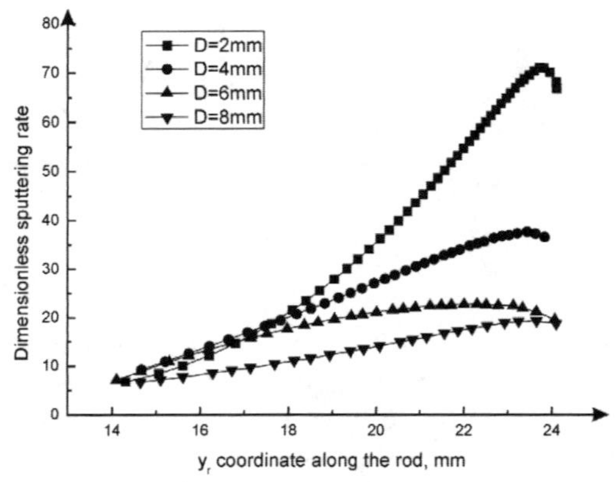

Figure 5. Dimensionless sputtering rate distribution.

3. Experiment

Model experiment with thin rods was carried out in order to validate the simulation approach. The rods were set in 2d array with pitch of 5mm. The plasma was generated with hollow cathode hot filament plasma source; negative 400V bias was applied to the array of the rods. The acceleration gap thickness D was visually evaluated to be about 4mm. The SEM images of the rod before and after the process are compared on the figure 6.

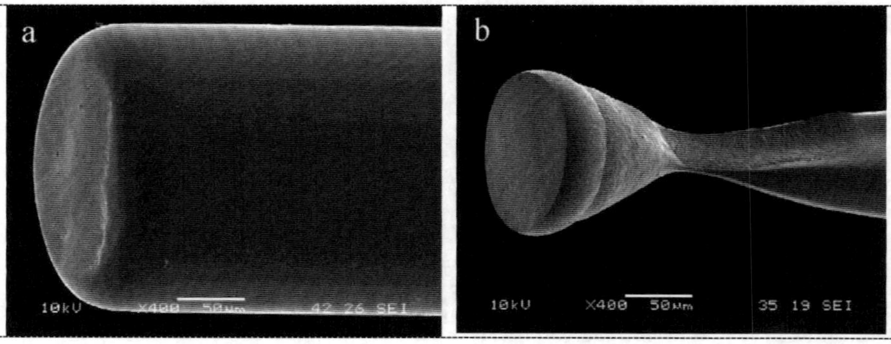

Figure 6. Scanning electron micrographs of stainless steel rods before (a) and after (b) sputtering, with U=-400V and D≈4mm, p=5mm, L=10mm and d=0.2mm.

4. Discussion

Calculated sputtering rate distribution (see figure 5) in case of 2mm, 4mm and 6mm *D* has prominent peak near the upper end of the rod in the similar manner to the experimentally etched profile. Having the results of the simulation one can say that the etched profile is formed due to the [i] angular sputtering yield dependence, which is monotonously decreasing with the incidence angles in the range from 70 to 90 degree (see figure 4), [ii] incidence angle distribution along the rods vertical axis which has a peak near the end of the rod (see figure 3) and [iii] ion current density distribution along the rods vertical axis. Thus, the simulation clarified the reasons of the non-linear shape formation obtained in the experimental IPS on the rods and needles. This computational approach can also give a good opportunity for understanding in which way the process needs to be adjusted when specific shape is required.

5. Conclusion

In the present study simplified 2d simulation method for the bias IPS process was suggested and carried out using combination of FEA and analytical motion equations solving. The results of the simulation are showing the same trend as the experimental results, thus this rather simple computational method can be used for estimation of the ion flow and sputtering distribution. Further model refining and experimental data accumulation can give a deeper understanding of the ion flow driven shape modification of sputtered needles as well as more numerical results. The case of non-planar plasma boundary is good candidate for following study.

References

[1] Behrisch R and Eckstein W 2007 *Sputtering by Particle Bombardment* (New York: Springer).
[2] Remnev A G, Shalnov K V, Uemura K and Nagato K 2012 *ISPlasma: Proceedings of the 4th International Symposium on Advanced Science and its Applications for Nitrides and Nanomaterials* (Chubu University, Aichi, Japan) 79.
[3] Alghuraira O, Remnev A G, Nagato K, Ikeshima S, Yasunoria O, Uemura K and Nakao M 2013 *Procedia CIRP* **5** 53-56.
[4] Ledvina S A, Ma Y J and Kallio E 2008 *Space Science Reviews* 139.
[5] Birdsall C K 1991 IEEE Transitions on Plasma Science 19 65-85.

Journal of Physics: Conference Series **552** (2014) 012010 doi:10.1088/1742-6596/552/1/012010

Bias-assisted magnetron sputtering of yttria-stabilised zirconia thin films

A A Solovyev[1,*], S V Rabotkin[2], I V Ionov[2], A V Shipilova[2], A N Kovalchuk[1] and A O Borduleva[1]

[1] Tomsk Polytechnic University, 30 Lenin Ave., Tomsk, 634050, Russia
[2] Institute of High Current Electronics SB RAS, 2/3 Akademichesky Ave., Tomsk, 634055, Russia

E-mail: andrewsol@mail.ru

Abstract. Cubic yttria-stabilized zirconia (YSZ) thin films were deposited by reactive magnetron sputtering on NiO–YSZ fuel cell anodes under different conditions. The influence of substrate bias voltage, temperature and porosity on texture and morphology of the deposited films were investigated. Comparing film morphology of YSZ grown on NiO–YSZ anodes, it was found that films deposited on a substrate with a large porosity were columnar and contained voids regardless of the deposition parameters. It was shown that only using of anode support with very high surface quality, substrate heating up to 500 °C and pulsed substrate bias (about -30 V MF) during film deposition is necessary for deposition of non-columnar thin films without voids and cracks that may be suitable for SOFC application.

1. Introduction

Solid oxide fuel cells (SOFCs) convert chemical energy directly into electrical energy with high efficiency and low emission of pollutants. However, before fuel cell technology can gain a significant share of the electrical power market, the operation temperature needs to be reduced in order to decrease costs, improve the durability of the cells and achieve greater flexibility in the choice of material for stack construction. Application of thin film electrolytes is a way of achieving this goal. Yttria-stabilized zirconia (YSZ) is the preferred material for electrolytes in solid oxide fuel cells as it is an acceptable ionic conductor, electrical insulator, chemically inert and relatively cheap to produce.

Typically, a tape-casting method is applied for the production of the electrolyte on top of the anode support which results in electrolyte thicknesses below 50 μm [1]. However, by applying a physical vapor deposition (PVD) technique it is possible to produce thin film electrolytes with thicknesses down to a few microns and thereby lower the operation temperature of the SOFC to an intermediate temperature range of 650–800 °C [2]. Previous investigations have demonstrated deposition of YSZ thin films by techniques such as pulsed laser deposition [3], electron beam evaporation [4], atomic layer deposition [5] and magnetron sputtering [6,7].

Implementation of magnetron sputtering in large-scale SOFC manufacturing was shown by Sønderby et al. [8]. However problem often encountered when depositing YSZ thin films by magnetron sputtering at relatively low synthesis temperatures is the formation of insufficiently dense and columnar microstructures. Such microstructures are unfavorable as they may result in internal leakage in the cell leading to decreased open cell potential and electrochemical performance. Nédélec

Content from this work may be used under the terms of the Creative Commons Attribution 3.0 licence. Any further distribution of this work must maintain attribution to the author(s) and the title of the work, journal citation and DOI.
Published under licence by IOP Publishing Ltd

et al. have deposited electrolytes without visible continuous columnar structure and has achieved significant improved electrochemical performance at low temperatures [9].

It is reported in the literature that bias-assisted sputtering is capable of producing films with better quality compared to normal sputtering [10]. Under optimum conditions, the positively charged Ar atoms will provide the necessary displacement energy to enhance the growth process towards layer growth. Thus, using bias the morphology of the deposited layer can be changed from porous to dense.

In the present study deposition of YSZ films by pulsed DC magnetron sputtering is carried out on SOFC NiO-YSZ anodes under the influence of substrate bias voltage. The morphology and texture dependencies of substrate bias voltage, substrate temperature and substrate type are investigated.

2. Methodology and Materials

YSZ thin films are deposited on commercial NiO-YSZ SOFC anodes (Ningbo Materials Institute of Technology and Engineering (NIMTE), China) 20 mm in diameter. To determine the influence of the structure and porosity of the anode substrate on the structure of deposited electrolyte reduced anodes composed of 60 vol.% Ni and 40 vol.% $Zr_{0.9}Y_{0.1}O_{1.95}$, made of a ceramic anode tapes 42420 and 42421 (ESL ElectroScience, USA) also used as substrates. Last substrates had a greater volume of open pores due to their reduced state. Experiments were performed using a 100 mm diameter, 8 mm thick $Zr_{0,86}Y_{0,14}$ target equipped in a circular magnetron source placed in a cylindrical stainless steel vacuum chamber. Before deposition substrates were cleaned by sputter etching for 10 min. In this case, the magnetron power was 200 W and negative bias voltage with amplitude of 600 V was applied to the substrate holder. After that samples were heated up to the deposition temperature approximately 500 °C. Sputtering was carried out in the oxygen-argon atmosphere at a working pressure of 0.2 Pa. Power supplies produced by Applied Electronics. Ltd. (Russia) were used for applying a voltage to the substrate and sputtering target. DC or a pulsed DC bias (MF) with a frequency of 100 kHz was applied to the substrates. Bipolar magnetron sputtering was carried out at 80 kHz with reverse pulses of 4 μs and the amplitude of 15% of the nominal negative sputtering voltage. The average discharge power was 2 kW. The distance between the magnetron target and the substrate was about 80 mm. The maximal deposition rate of stoichiometric YSZ films was 12 μm/h, which was about 70% of the film deposition rate in pure argon at the same power (17 μm/h).

X-ray diffraction (XRD) measurements in the θ–2θ geometry were performed with a XRD 7000S (Shimadzu, Japan) diffractometer using CuKα radiation. Scanning electron microscopy (SEM, Hitachi TM3000, Japan) was used to determine the film morphology as well as film thickness.

3. Results and Discussion

The microstructure of the YSZ electrolyte layers deposited on NiO-YSZ anodes at different substrate temperatures and at floating potential was investigated using a SEM. Figure 1 shows micrographs taken from cross-sections.

From the viewpoint of industry applications heating of the substrate is undesirable because it complicates the equipment, extends the process of electrolyte deposition due to the additional steps of heating and cooling of the substrate and result in additional expenditure of electrical energy. But Figure 1a shows that after YSZ deposition at room temperature electrolyte delamination, i.e. debonding between the film and substrate is observed. This is due to poor adhesion and strong compressive stresses generated in the film during its growth. Poor adhesion of the film to the substrate, despite the fact that the latter was subjected to plasma pre-cleaning, may be caused by desorption of residual gas from the substrate pores during film growth due to the heating of the substrate in a magnetron discharge. Plasma cleaning removes contaminants located on the surface only, but in the case of porous substrates this is not enough. Film growth with residual stresses only aggravates the problem of insufficient adhesion. Formation of internal stresses in the films deposited by magnetron sputtering is typical [11] and caused by bombardment of a growing film by high-energy particles. In the case of magnetron sputtering, such particles can be neutral atoms scattered from the cathode, ions of a working gas (Ar) that are generated in the discharge, and neutralized ions reflected from the cathode.

The micrographs for the deposition with substrate temperature 300 and 500°C (figures 1b and 1c) show already good layer adhesion without film delamination. This is due to the desorption of residual gas from the porous anode and adatom mobility increase with temperature increasing. The using a substrate temperature of about 500 °C from a technological point of view is preferential, since it is known that the higher the ratio between a substrate temperature during the deposition and a melting point of the material, the denser the deposited layer. Increasing the substrate temperature above 500 °C is not rational, since it leads to considerable complication of the heating system.

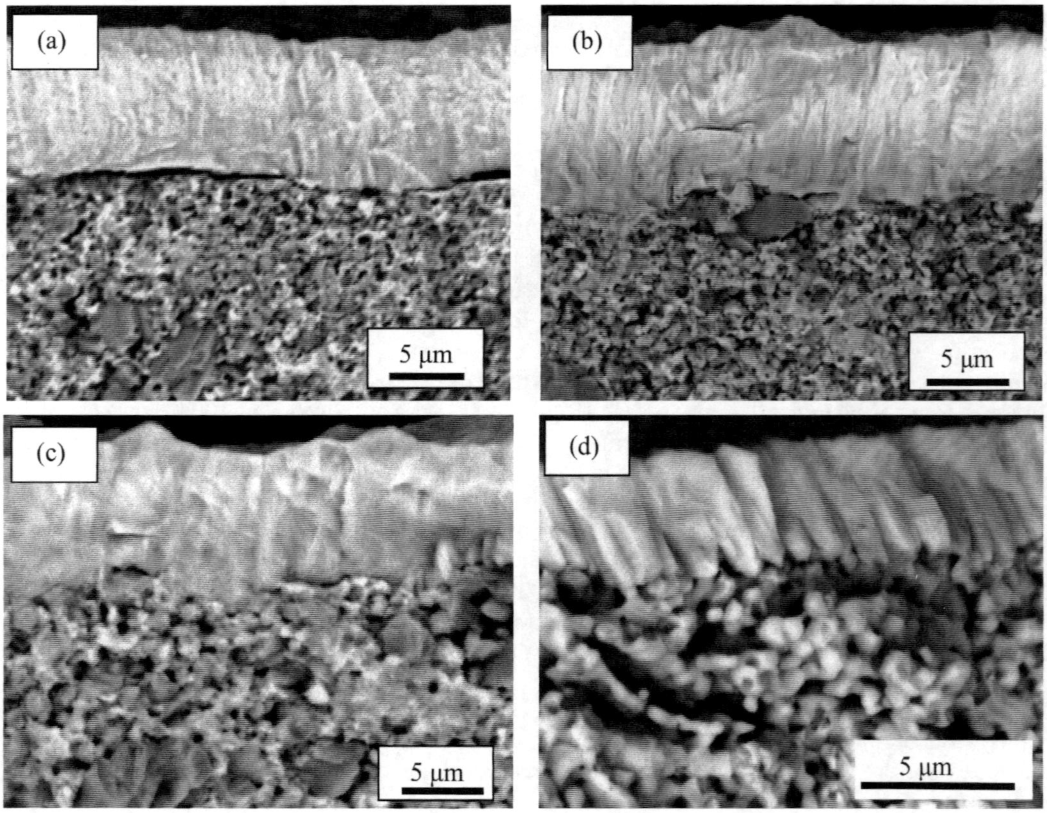

Figure 1. Cross section SEM micrographs of YSZ films deposited at room (a), 300°C (b), and 500°C (c) substrate temperature and at floating potential on NiO-YSZ anodes produced by NIMTE. Micrograph of YSZ film deposited at substrate temperature 500°C and at floating potential on Ni-YSZ anode produced from ESL ElectroScience anode tape (d).

It is interesting, the YSZ films deposited on unreduced NiO-YSZ anodes produced by NIMTE (figures 1a-1c) with sufficiently dense structure and small pore size, also possess a dense structure in which columnar crystals are practically indistinguishable. However, when depositing an YSZ film on the more porous anode, for example, produced from ESL ElectroScience anode tape (figures 1d) columnar structure of the electrolyte becomes inevitable. This is caused by the fact that initial islands of the YSZ do not attain the coalescence stage completely, thus, limiting the continuity of the deposited layer [10]. These discontinuities will have a clear negative effect on gas-tightness of deposited layers. Therefore very high surface quality of anode support is essential for the formation of thin layers of gas-tight electrolyte.

Another way of increasing adatom mobility besides heating the substrate is application of negative substrate bias voltage to draw positive ions to bombard the surface and provide energy to the film. Figures 2 and 3 show cross section and surface SEM micrographs of the YSZ films deposited at different bias voltages.

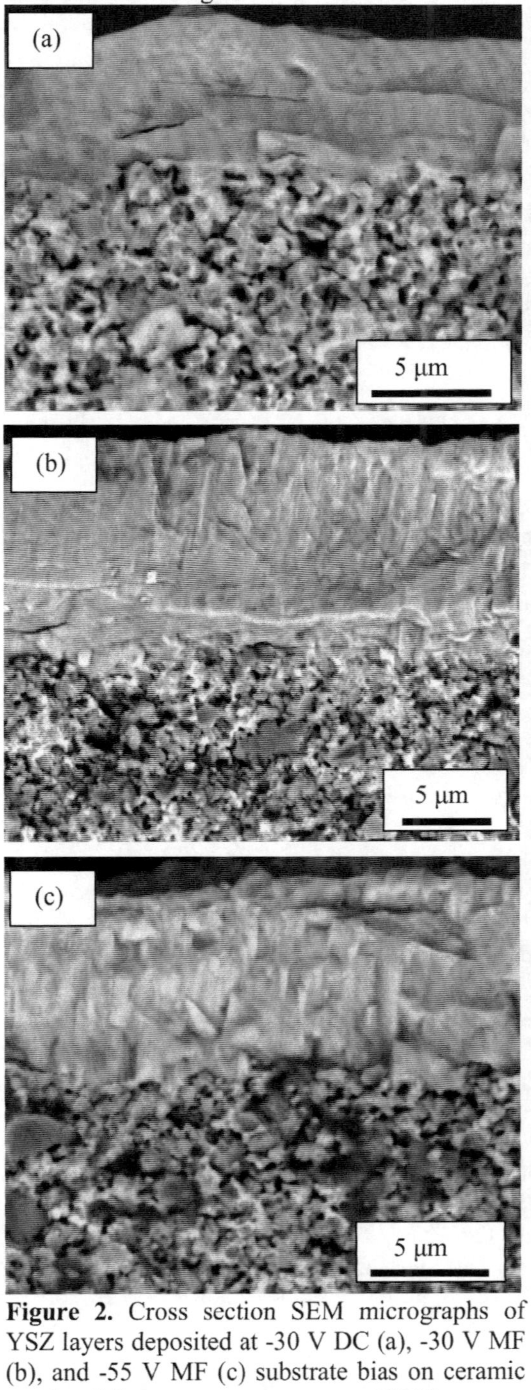

Figure 2. Cross section SEM micrographs of YSZ layers deposited at -30 V DC (a), -30 V MF (b), and -55 V MF (c) substrate bias on ceramic anodes. Substrate temperature was 500°C.

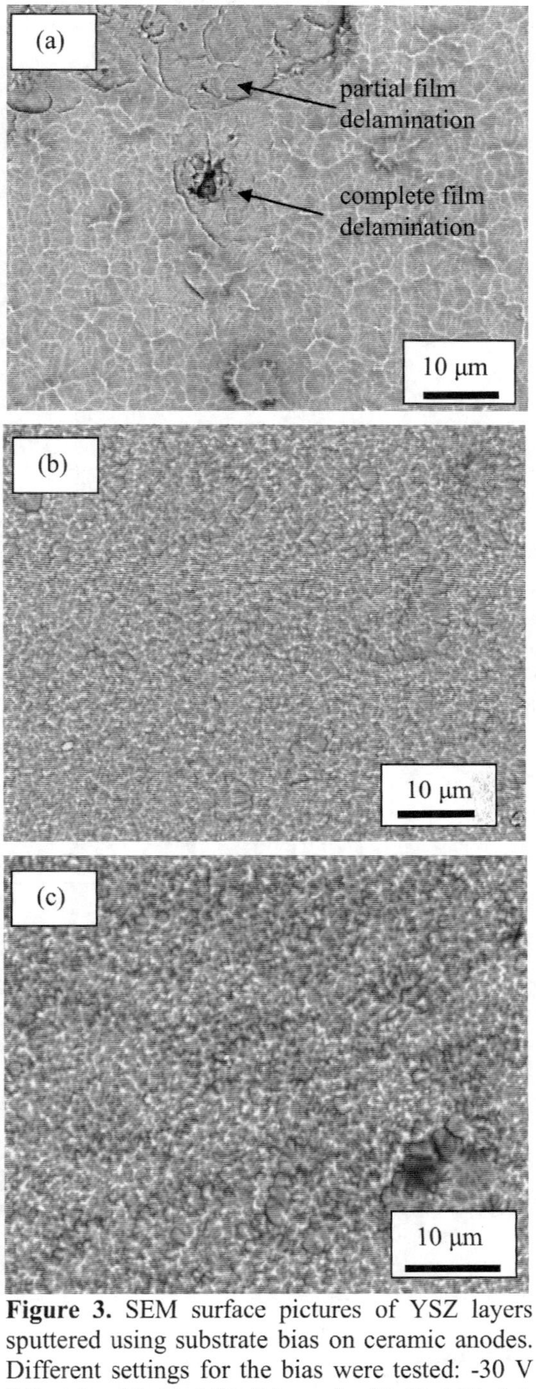

Figure 3. SEM surface pictures of YSZ layers sputtered using substrate bias on ceramic anodes. Different settings for the bias were tested: -30 V DC (a), -30 V MF (b), and -55 V MF (c). Substrate temperature was 500°C.

At DC bias voltage of −30 V applied to the substrate YSZ electrolyte (figure 2a) has a layered structure with longitudinal cracks. In addition, there is a partial or complete YSZ film delamination on some parts of the substrate surface (figure 3a). Apparently, the use of DC bias voltage during deposition of dielectric coating is not effective method due to charging of nonconductive film by positive ions. Therefore it is preferably to use RF or pulsed DC bias during the deposition of YSZ electrolyte.

At a small pulsed DC bias voltage (-30 V) YSZ film has non-columnar structure without voids and cracks (figure 2b). The film surface is clearly facetted (figure 3b), and delamination of the film from the substrate was not observed. By applying -55 V pulsed DC bias the film obtains a homogeneous and fine grained microstructure (figure 2c) but minor delaminates were observed due to the build-up of stress.

Fig. 4 shows the X-ray diffraction patterns of the YSZ films deposited without and at -30 V MF bias applied to the substrate. In all cases, the cubic YSZ film phase and metallic Ni were generally detected. No peaks of monoclinic, tetragonal or amorphous phases were detected. It should be noted that films deposited with substrate bias have broader peaks on the diffractograms indicating smaller grain size.

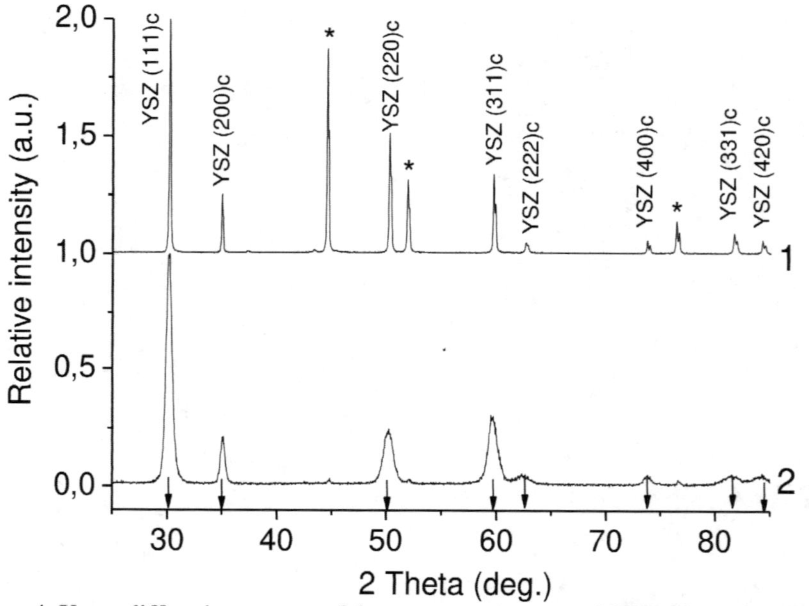

Figure 4. X-ray diffraction pattern of the magnetron sputtered YSZ films: 1 – without substrate bias, 2 – at -30 V MF substrate bias. Intensities are normalized to the (111) reflection. Reflections of the Ni are marked with in asterisk. The Miller indices and peak positions for the 8 mol.% Y_2O_3-doped ZrO_2 powder reference sample (ICDD powder diffraction database entry # 030–1468) with a cubic fluorite structure are included for comparison.

Ion bombardment has been seen to increase nucleation rates and film density, to decrease the average grain-size, to inhibit formation of columns, and to control the defect density and orientation of a wide range of sputtered coating [12]. The smaller grain size of a film deposited at substrate bias than at floating potential is presumably due to the formation (by ion bombardment) of more nucleation centers.

4. Conclusion

It was shown that it is possible to deposit non-columnar YSZ thin films without voids and cracks by bias-assisted magnetron sputtering that may be suitable for SOFC application. The key parameters for the formation of YSZ films with the desired characteristics are quality of substrate surface, the substrate temperature and low-energy ion bombardment of the growing film. Since the gastight YSZ film can be obtained only if its thickness significantly larger than the maximum pore size of the substrate, the substrate should not have such large defects as big voids, cracks, etc. The substrate surface to be coated should be as smooth as possible. Heating of porous substrates to a temperature above 300 °C is a prerequisite for achieving good film adhesion, while ion assisting necessary to obtain dense and non-columnar structure of films.

Acknowledgments
This research was carried out within the State assignment of the Institute of High Current Electronics SB RAS and under financial support of Russian Foundation for Basic Research (grant № 14-08-31679).

References

[1] Tucker M 2010 *J. Power Sources* **195** 4570
[2] Pederson L R, Singh P, Zhou X D 2006 *Vacuum* **80** 1066
[3] Kosacki I, Rouleau C M, Becher P F, Bentley J, Lowndes D H 2005 *Solid State Ionics* **176** 1319
[4] Karthikeyan A, Chang C, Ramanathan S 2006 *Appl. Phys. Lett.* **89** 183116
[5] Shim J H, Chao C C, Huang H, Prinz F B 2007 *Chem. Mater.* **19** 3850
[6] Smeacetto F, Salvo M, Ajitdoss L C, Perero S, Moskalewicz T, Boldrini S, Doubova L, Ferraris M 2010 *Mater. Lett.* **64** 2450
[7] Sillassen M, Eklund P, Pryds N, Johnson E, Helmersson U, Bøttiger J 2010 *Adv. Funct. Mater.* **20** 2071
[8] Sønderby S, Nielsen A J, Christensen B H, Almtoft K P, Lu J, Jensen J, Nielsen L P and Eklund P 2012 *Surf. Coat. Technol.* **206** 4126
[9] Nédélec R, Uhlenbruck S, Sebold D, Haanappel V A C, Buchkremer H P, Stöver D 2012 *J. Power Sources* **205** 157
[10] Chapman B N 1980 *Glow Discharge Processes: Sputtering and Plasma Etching* (New York: John Wiley and Sons)
[11] Solov'ev A A, Sochugov N S, Oskomov K V 2010 *Phys. Met. Metallogr.* **109** 111
[12] Sillassen M B 2010 Synthesis and ionic conductivity properties of stabilized ZrO_2 thin films. PhD Dissertation. University of Aarhus. Denmark.

Journal of Physics: Conference Series **552** (2014) 012011 doi:10.1088/1742-6596/552/1/012011

Experimental study of the neck formation in an X pinch

A P Artyomov*, S A Chaikovsky, A V Fedunin, N A Labetskaya, A G Rousskikh, A S Zhigalin, and V I Oreshkin
Institute of High Current Electronics SB RAS, 2/3 Akademichesky Ave., Tomsk, 634033, Russia

E-mail: aap545@gmail.com

Abstract. X-pinch experiments have been performed on a compact 250 kA, 180 ns pulsed power generator specially designed for this purpose at the Institute of High Current Electronics (Tomsk, Russia). The X pinches were composed of two molybdenum wires of diameter 25 μm making an angle of 36° with the *z*-axis. The X-pinch dynamics was recorded with a 3 ns exposure time using an HSFC Pro four-frame camera. Axial plasma jets propagating toward both the anode and the cathode were observed. The jets became noticeable within 10 ns after the onset of current flow, which approximately corresponded to the time at which the electrical explosion of the X-pinch wires occurred. The velocity of the anode-directed jet reached 10^7 cm/s, which was about 1.5 times the velocity of the cathode-directed jet. These high jet velocities are inconsistent with the plasma temperature resulting from the wire explosion. Hence, these jets seem to develop due to implosion of the light plasma layer stripped by magnetic forces from the wire surface, and the increase in their velocities is perhaps due to cumulative effects taking place at the X-pinch axis. The X-pinch neck formed as a rule above the initial wire cross point (closer to the anode). In this region, the plasma diameter gradually increased with time and then drastically decreased 10–15 ns prior to the x-ray pulse. Immediately before the x-ray pulse, in the (250–300 μm long) plasma neck, a lower scale constriction developed, forming a "hot spot". It has been confirmed that the anode-directed plasma jet could take some part of the X-pinch wire current because of the evident jet pinching in the anode region. This process seems to determine the neck length.

1. Introduction

Two or more metal wires 10–50 μm in diameter crossed like the letter "X" are used as the so-called X-pinch loads for pulsed power generators. When a current pulse is passed through an X pinch, a plasma hot spot several micrometers in size can be produced. The hot spot efficiently radiates soft x-rays in a 1–2 ns pulse. Hot spots of this type are currently one of the brightest soft x-ray laboratory sources. The parameters of the X-pinch-based soft x-ray sources make them attractive for projection radiography of various short-living physical objects. Although X-pinches have been studied for long, since 1982 [1], they remain an object of intense fundamental research because of a number of interesting physical features. The emission properties of hot plasmas at near-solid densities, the radiative collapse, and the formation of supersonic jets are of most interest.

The sequence of physical processes leading to the formation of a hot spot can be revealed using a nanosecond time resolved diagnostics. Relevant experiments were carried out on different pulse generators using optical and x-ray frame imaging, interferometry imaging, and soft x-ray backlighting [2–11]. Nevertheless, the results obtained could not be properly interpreted because of strong

Content from this work may be used under the terms of the Creative Commons Attribution 3.0 licence. Any further distribution of this work must maintain attribution to the author(s) and the title of the work, journal citation and DOI.
Published under licence by IOP Publishing Ltd

sensitivity of the X-pinch dynamics to the current pulse waveform. Since 2006, X-pinch experiments using a set of self-made compact pulsed power generators have been performed at the Institute of High Current Electronics (Tomsk, Russia). This paper presents the results of experiments aimed at studying the dynamics of X-pinch necking from the wire explosion stage to the formation of a hot spot.

2. Experimental setup

The experiments were carried out on a compact, 250-kA, 180-ns pulse generator specially designed for X-pinch studies [12–16]. The generator consists of four self-made 250-nF capacitors connected in parallel. Each capacitor is equipped with a built-in triggered gas switch. The X pinches were composed of two 25-μm molybdenum wires each making an angle of 36° with the z-axis.

The X-pinch dynamics was recorded with a 3-ns exposure time using an HSFC Pro four-frame camera. The frame interval was varied in the range 20–40 ns. Figure 1 shows a set of typical frames. Horizontal lines indicate the initial position of the wire cross point, which is indicated by X_0. The cathode position is at the bottom of each image. The fourth frame in Fig. 1 corresponds to the time of occurrence of an x-ray pulse.

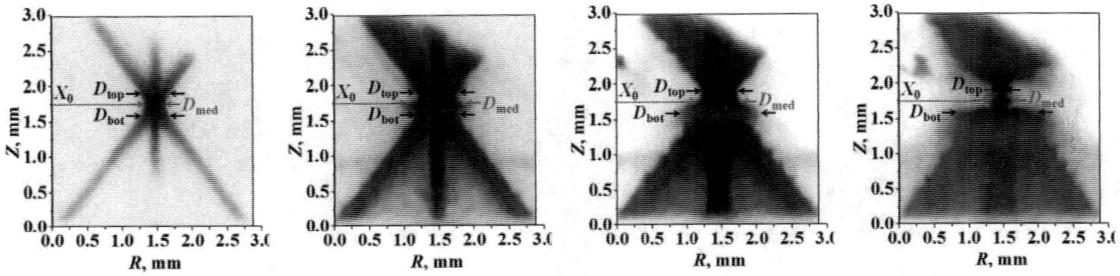

Figure 1. Typical optical pictures of an X pinch made of two 25-μm molybdenum wires taken in one shot: X_0 indicates the initial position of the wire cross point.

The X-pinch current was measured using a Rogowski coil. The x-ray emission was detected with a copper-cathode x-ray diode (XRD) filtered to view the spectral region $h\nu > 1$ keV. Figure 2 presents typical current and XRD signal traces. In all experimental shots, the time of x-ray pulse generation, $t_{x\text{-}ray}$, measured from the onset of current flow was 140 ± 5 ns.

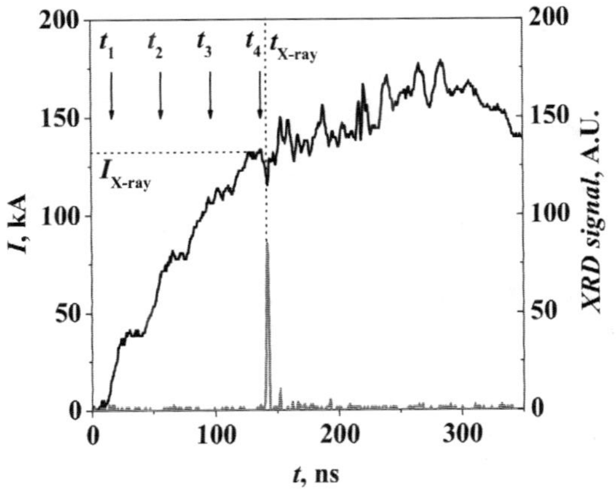

Figure 2. Typical current and XRD signal traces

The small spread in the time of x-ray pulse generation allowed us to compare X-pinch images taken in different shots. To do this, all pictures captured with different camera delays were combined to form an integrated sequence of frames which would cover all stages of the X-pinch dynamics from the wire explosion to the disruption of the plasma neck. This integrated picture was used to determine the evolution of the diameter of an individual wire, the velocities of the axial plasma jets, and the plasma expansion/implosion velocity near the wire cross point.

3. X-pinch wire expansion and axial jet formation

The integrated sequence of frames captured in several shots is presented in Fig. 3. Each image is marked by relative time $t/t_{\text{x-ray}}$ ($t_{\text{x-ray}}$ = 140 ns).

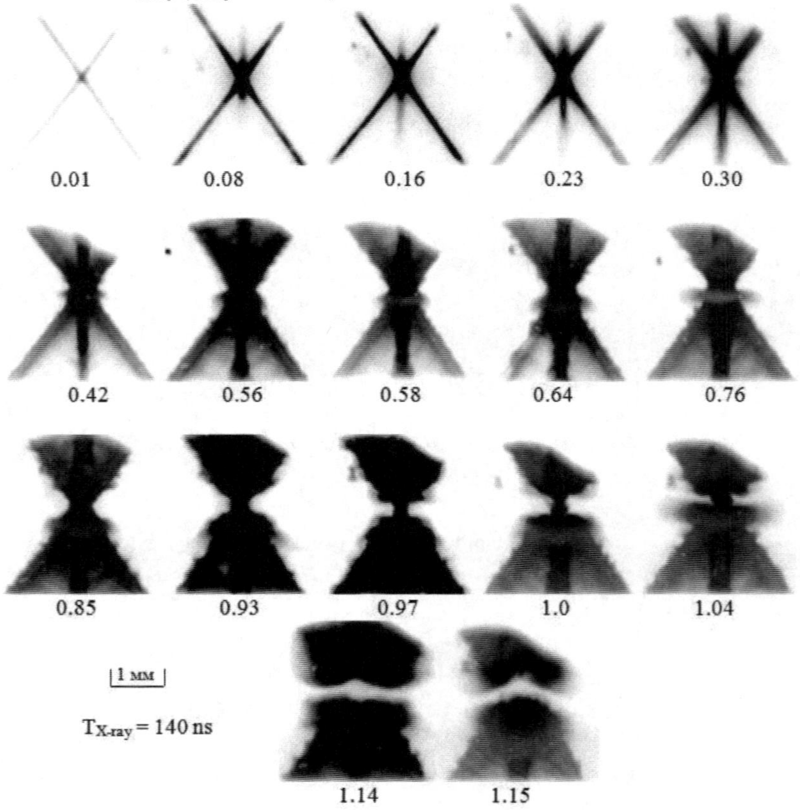

Figure 3. Integrated sequence of frames captured in several shots. The relative time $t/t_{\text{x-ray}}$ is given beneath each image.

At the initial stage of X-pinch dynamics, the wires explode and expand. Figure 4a shows the time variation of the distances R_{in} and R_{out} from the initial position of the axis of an individual wire to the expanding plasma edge, measured, respectively, toward and outward the other wire. If a wire would expand uniformly from its axis, the values of R_{in} and R_{out} should be equal and could be considered to be the wire radius. As can be seen in Fig. 4, the wire plasma expands symmetrically about the wire axis until the relative time $t/t_{\text{x-ray}} \approx 0.4$ ($t_{\text{x-ray}}$ = 140 ns). The expansion velocity, slightly varying with time, is $(2.6\text{–}3.0)\cdot 10^5$ cm/s. The subsequent behavior of the wire plasma features a slightly rising inward velocity and full cessation of the outward expansion.

The wire radius increases about fourfold within the first 10 ns of the process due to thermal expansion of the surface plasma (see Fig. 4a). Clearly seen axial plasma jets propagating toward both the cathode and the anode are observed as early as within 2 ns. The velocity of the anode-directed jet, V_{top}, reaches $\sim 10^7$ cm/s, which is about 1.5 times the velocity of the cathode-directed jet (see Fig. 4b). These high jet velocities are inconsistent with the plasma temperature resulting from the wire explosion. Hence, these jets seem to develop due to implosion of the light plasma layer stripped by magnetic forces from the wire surface, and the increase in their velocities is perhaps due to cumulative effects taking place at the X-pinch axis.

The X-pinch wires become nonuniform in structure at $t/t_{\text{x-ray}} \approx 0.3$. A highly luminous plasma corona is observed which surrounds a wire core whose brightness is considerably lower. The core diameter is approximately half that of the coronal plasma.

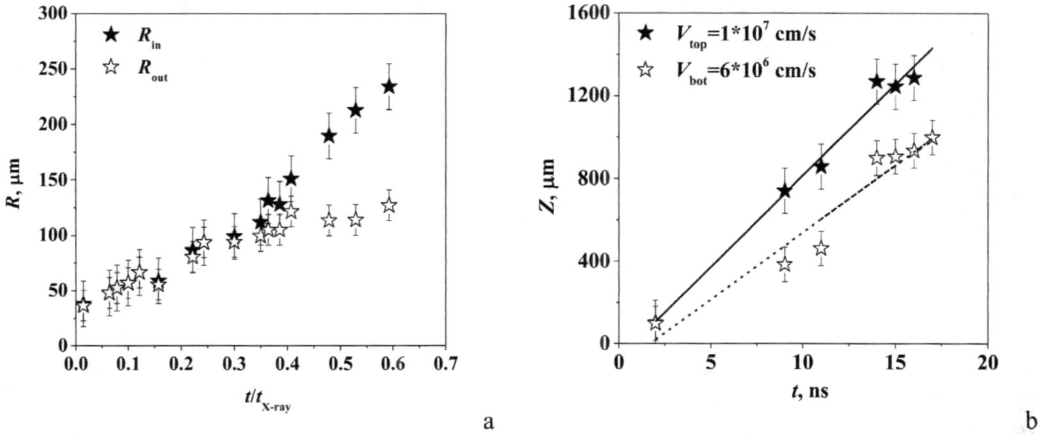

a b

Figure 4. Time variation of R_{in} and R_{out} (a) and the axial plasma jet lengths measured from the wire cross point toward the anode (bold asterisks) and toward the cathode (open asterisks) (b): $t_{\text{x-ray}} = 140$ ns.

4. X-pinch necking

The main features of the X-pinch dynamics can be inferred from Fig. 3. At the initial stage ($t/t_{\text{x-ray}} < 0.16$), the plasma surrounding the cross point slightly expands, remaining shaped like a cone. A few nanoseconds later, a visible sausage-like perturbation develops near the cross point, and this is just the onset of neck formation. Along with the occurrence of the neck, a clearly visible plasma plume (or a radially expanding jet) is observed a little bit below the point X_0. This plasma plume and the plasma located below the cross point show no evidence of implosion. Thus, the implosion occurs in the region that extends to about 500 µm above and including the cross point.

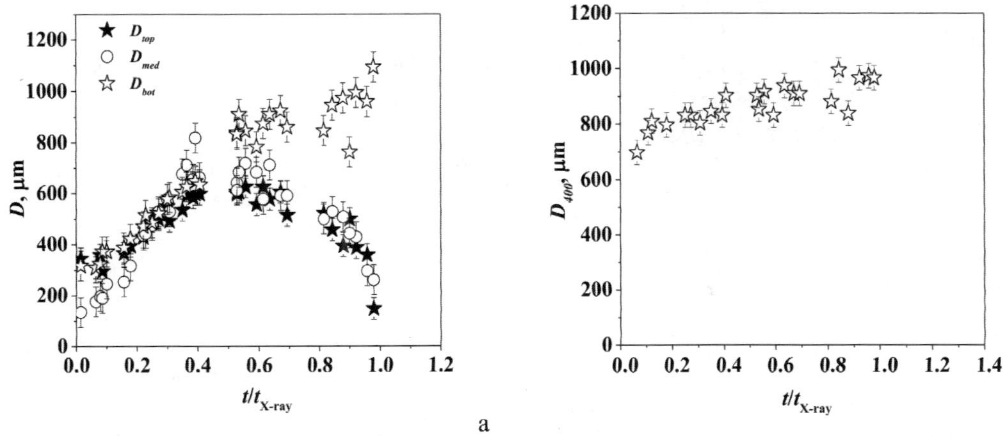

a b

Figure 5. Plasma size measured in the horizontal cross-sections at the point X_0 (D_{med}), 150 µm above (D_{top}) and 150 µm below the point X_0 (D_{bot}) (a), and 400 µm above the point X_0 (D_{400}) (b).

Immediately before the X-ray pulse is generated ($t/t_{\text{x-ray}} = 0.97$), a visible plasma pinch of length 200 300 µm is formed in which a constriction of significantly smaller size can be seen. This constriction forms the X-pinch micrometer-size "hot spot". The hot spot formation takes only a few nanoseconds or less.

The hot spot is located closer to the anode, 50–150 μm above the point X_0. In this region, the plasma diameter gradually increases with time, showing an abrupt decrease 10–15 ns before the X-ray pulse (see Fig. 5a). Figure 5a presents the plasma size measured in the horizontal cross-sections at the point X_0 (D_{med}), 150 μm above (D_{top}) and 150 μm below the point X_0 (D_{bot}) (see Fig. 1). The plasma expansion in the neck region goes with a velocity of $3 \cdot 10^5$ cm/s until $t/t_{x\text{-ray}} \approx 0.4$, proceeding in fact in step with the expansion of an individual wire. The plasma expansion velocity at the point X_0 (see the data for D_{med} in Fig. 5a) is about half that in the case of purely radial jet formation (see Fig. 3, $t/t_{x\text{-ray}}$ = 0.3).

A few nanoseconds later, the size D_{med} abruptly decreases. The reason is that the jet position is shifted downward, giving rise to the mentioned plasma plume on the cathode side. Subsequently, a pronounced implosion is observed in the region between the point X_0 (D_{med}) and the point located 150 μm above (D_{top}). The implosion velocity reaches (3–5)$\cdot 10^6$ cm/s at a radius of 100 μm.

It can be seen that the bottom part of the plasma (D_{bot} in size) does not implode. This is just the region of the plasma plume. The X-pinch plasma in the region 400 μm above the cross point also does not implode (see Fig. 5b). This seems to be due to the current switching to the axial plasma jet whose diameter is 400–600 μm. The switching can reduce the current flowing through the wires and slow down the implosion of the wires at a diameter greater than the jet diameter, resulting in a shorter neck.

The supposition that the current could partially flow through the axial jets is additionally supported by the observed implosion (pinching) of the top axial jet. This can clearly be seen in Fig. 6a. This image was taken for another X pinch made of four Mo wires of diameter 25 μm. Pinching of the near-anode (top) part of the axial plasma jet is evident.

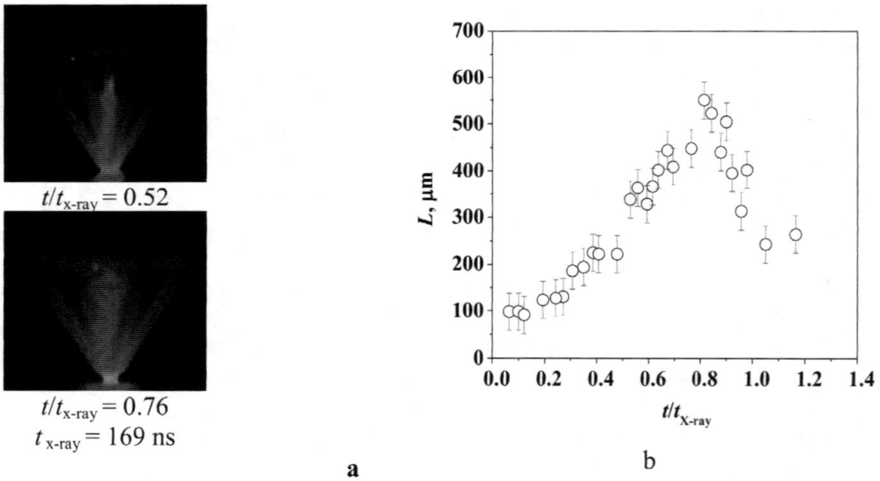

Figure 6. Typical optical images of an X pinch made of four molybdenum wires of diameter 25 μm, taken above the wire cross point (a) and the time variation of the neck length (L) in the cross-point region (b).

Figure 6b presents the measurements of the neck axial dimension, which could also be termed the neck length. It can be seen that neck length L strongly depends on time, increasing from 100 to 550 μm within 110 ns. Thereafter, the axial dimension of the perturbation decreases during 30 ns and becomes equal to 250–300 μm nearly at the time the x-ray pulse is generated. The increase of the neck length can be accounted for by the action of the axial component of the $\boldsymbol{J}\times\boldsymbol{B}$ force. The subsequent elongation of the neck can be accounted for by the progressive loss of linear mass in the middle of the neck due to the axial plasma outflow.

5. Conclusion

X-pinch experiments have been performed on a compact 250 kA, 180 ns pulse generator specially designed for this purpose at the Institute of High Current Electronics (Tomsk, Russia). The X pinches were composed of two molybdenum wires of diameter 25 μm making an angle of 36° with the *z*-axis. The X-pinch dynamics was recorded with a 3-ns exposure time using an HSFC Pro four-frame camera. Axial plasma jets propagating toward both the anode and the cathode were observed. The jets became noticeable within 10 ns from the onset of current flow, which approximately corresponded to the time at which the electrical explosion of the X-pinch wires occurred. The velocity of the anode-directed jet reached 10^7 cm/s, which was about 1.5 times the velocity of the cathode-directed jet. These high jet velocities are inconsistent with the plasma temperature resulting from the wire explosion. Hence, these jets seem to develop due to implosion of the light plasma layer stripped by magnetic forces from the wire surface, and the increase in their velocities is perhaps due to cumulative effects taking place at the X-pinch axis.

The X-pinch neck formed as a rule above the initial wire cross point (closer to the anode). In this region, the plasma diameter gradually increased with time and then drastically decreased 10–15 ns prior to the X-ray pulse. Immediately before the X-ray pulse, in the (250–300 μm long) plasma neck, a lower scale constriction developed, forming a "hot spot". It has been confirmed that the anode-directed plasma jet could take some part of the X-pinch wire current because of the jet pinching in the anode region. This process seems to determine the neck length. We believe that the results obtained can be useful in theoretically modeling the X-pinch dynamics.

Acknowledgments
This work was supported in part by Presidium of the Russian Academy of Sciences under a program of "Fundamental Problems of Pulsed Power" and by the Russian Foundation for Basic Research (grant No. 12-08-00868-a).

References

[1] Zakharov S M, Ivanenkov G V, Kolomenskii A A *et al.* 1982 *Sov. Tech. Phys. Lett.* **8** 456
[2] Kalantar D H, Hammer D A, Dangor A E, Bayley J M and Beg F N 1994 *Citation: AIP Conf. Proc.* **299** 604
[3] Choi P, Christou C and Aliaga R 1991 *SPIE* **1552** 270
[4] Mitchell I H, Aliaga-Rossel R, Saavedra R, Chuaqui H, Favre M and Wyndham E S 2000 *Phys. Plasmas* **7** 5140
[5] Pikuz S A, Shelkovenko T A, Sinars D B *et al.* 2001 *J. Quant. Spectr. Radiat. Trans.* **71** 581
[6] Shelkovenko T A, Sinars D B, Pikuz S A and Hammer D A 2001 *Phys. Plasmas* **8** 1305
[7] Green J S, Bland S N, Collett M, Dangor A E, Krushelnick K, Beg F N and Ross I 2006 *Appl. Phys. Lett.* **88** 261501
[8] Beg F N, Ciardi A, Ross I, Zhu Y, Dangor A E and Krushelnick K 2006 *IEEE Trans. Plasma Sci.* **34** 2325
[9] Chittenden J P, Ciardi A, Jennings C A, Lebedev S V, Hammer D A, Pikuz S A and Shelkovenko T A 2007 *Phys. Rev. Lett.* **98** 025003
[10] Madden R E, Bott S C, Haas D, Eshaq Y, Ueda U, Collins G and Beg F N 2008 *Phys. Plasmas* **15** 112701
[11] Collins G W, Valdivia M P, Zick T, Madden R E, Haines M G and Beg F N 2013 *Phys. Plasmas* **20** 042704
[12] Artyomov A P , Fedyunin A V, Chaikovsky S A, Zhigalin A S, Oreshkin V I, Ratakhin N A and Rousskikh A G 2013 *Instrum. Exp. Tech.* **56** 66
[13] Ratakhin N A, Fedushchak V F, Erfort A A *et al.* 2007 *Izv. Vyssh. Uchebn. Zaved., Fiz.* **50** 87
[14] Artemov A, Labetskaya N, Fedyunin A and Chaikovskii S 2010 *Bull. Leb. Phys. Inst.* **37** 31
[15] Rousskikh A G, Oreshkin V I, Chaikovsky S A *et al.* 2008 *Phys. Plasmas* **15** 102706
[16] Mesyats G A, Shelkovenko T A, Ivanenkov G V *et al.* 2010 *Russ. Phys. JETP* **111** 363

Journal of Physics: Conference Series **552** (2014) 012012 doi:10.1088/1742-6596/552/1/012012

Electrical discharge phenomena application for solid fossil fuels in-situ conversion

A A Bukharkin*, **V V Lopatin, S M Martemyanov and I A Koryashov**
High Voltage Department, Institute of High Technology Physics, Tomsk Polytechnic University, Tomsk 634050, Russia

E-mail: amplexor@ya.ru

Abstract. The application of high voltage to oil shale initiates partial discharges (PDs) with the following treeing like in insulating dielectrics. Critical PDs and treeing with a high propagation rate occur under the low electrical intensity ~10^2 V/cm due to oil shale's high porosity, heterogeneity and anisotropy. The completed discharge occurs as a result of these phenomena. Carbonization is initiated around a plasma channel at the treeing stage and extended during electromagnetic action time. Carbonized rock electrical resistance decreases by 8÷10 degrees to 10 ohm•cm, and shale and coal could be heated by Joule heat in carbonized volume and discharge plasma. A high-current supply is necessary for this heating stage. Also, a high-voltage supply with steep-grade characteristics can be used for PDs and treeing initiating and heating the carbonized rock with low resistance. Thus, these phenomena allow in-situ processing in order to produce a flammable gas and synthetic oil from inferior solid fossil fuels by pyrolytic conversion. Computations show that the ratio between energy derived from gas flaming and energy for shale conversion is more than fifty. Therefore, oil shale conversion with the help of electrical discharge phenomena application can be very efficient, as it needs little energy.

1. Introduction

Technologies of solid fuels underground conversion into liquid and gaseous energy feedstock have been attracting the attention of investigators and engineers for a long time. These technologies allow refusing mines and quarries building. This will decline solid fuels reserves development costs and give the possibility of effective inferior solid fossil fuels mining, which are unprofitable for developing by conventional methods. Many countries with poor conventional hydro carbons reserves, for example, China, are carrying out intensive investigations of underground oil shale processing after the successful experience of some USA companies in shale gas and oil extraction from oil shale through bed hydraulic fracturing [1]. Widespread in a significant amount oil shale deposits seem to be prospective energy feedstock. Hydrofrac application allows elaborating only 3÷5 % of a rock organic component. An extremely negative ecological impact calls into question this technology application, and the moratorium is imposed on fracturing in the majority of European countries. A highly efficient and ecological friendly technology of inferior solid fossil fuels underground pyrolytic conversion into liquid and gaseous energy feedstock can be made by using electrical discharge phenomena in a rock.

2. Partial discharge and treeing

Content from this work may be used under the terms of the Creative Commons Attribution 3.0 licence. Any further distribution of this work must maintain attribution to the author(s) and the title of the work, journal citation and DOI.
Published under licence by IOP Publishing Ltd

Partial discharges (PDs) are incepted under the high voltage impact in oil shale and brown coal and induce treeing with the following breakdown. PDs are local surface discharges, foreign inclusions breakdowns, which are mainly presented by macro- and microscopic pores and cracks in material. PDs arise in high local electrical stress zones, which are created due to the difference between inclusions and enclosing material dielectric constant [2, 3]. As the size of the inclusion is less than the interelectrode space to several degrees, the electrical stress on inclusion can be found by the formula [3, 4]:

$$\frac{E_{inc}}{E_{av}} = \frac{\varepsilon_d}{\varepsilon_{inc}} \tag{1}$$

E_{inc} – electrical stress on the inclusion; E_{av} - average electrical stress; ε_d – enclosing material dielectric permittivity; ε_{inc} – inclusion dielectric permittivity.

Long-term PD activity induces treeing in some materials. Treeing is the growth of stochastically branching discharges structures, which are called dendrites [5]. PDs and treeing inception electrical intensity in oil shale does not exceeds 40 V/cm, when the interelectrode space is about 20 cm, however, for bigger interelectrode space PD inception the voltage from one to several tens kilovolts is needed. Under the discharges plasma impact rock thermal destruction occurs in the oil shale, like in some insulation materials [6]. Hence, the carbon concentration increases on dendrite sidewalls, so carbonization occurs. The carbonized volume electrical resistivity is about 10 Ohm•cm, what is less than the initial shale resistivity in 8÷10 degrees. The voltage in several tens kilovolt is needed for PD initiation and treeing propagation in oil shale with big interelectrode spaces. The current, flowing through the rock, includes displacement current and PD current, so it has small values. Therefore, the power of the supply may not be high.

3. Breakdown

Electrothermal breakdown with the formation of a plasma channel surrounded by a low-resistance carbonized area occurs, when the dendrite straps the interelectrode space. This channel resistance is significantly lower than the original rock resistance. However, it remains high due to a small dendrite diameter (about 200 micrometers). The current, flowing through the channel, grows sharply under the high voltage impact, and it heats the rock due to joule heat. The volume subjected to carbonization increases and the low resistance zone widens radially from the dendrite. The power supply, which provides the voltage to several kilovolts and the current up to several amperes, is appropriate to use for reaching low resistance due to channel widening. The resistance decreases to 0.1÷1 ohm, when the carbonization zone reaches a specified cross-section. A low voltage and high current power supply is necessary for the purpose of oil shale further heating.

4. Conversion

The conductive channel temperature influences on the ratio between gaseous and liquid conversion products and depends on the power that is input in the bed. The channel resistance decreases as a result of carbonization volume widening due to heating. Hence, the control on the power that is input in the bed is needed for specified conversion products obtaining. The regulation has to be carried out in a wide range because of the significant resistance decrease.

5. Experiment

A step-up transformer with the maximum root-mean square voltage up to 10 kV and the power-to-size ratio 2 kW is used in experiments for PD and treeing initiation with the following electrothermal breakdown. The breakdown electrical stress is about 45 V/cm, when the interelectrode space is about 50 cm. The rod electrodes with a 6 mm diameter inserted in shale on the 50 mm depth are used for electromagnetic energy input in the rock. The transformer with the maximum root-mean square voltage up to 1 kV and power-to-size ratio 2 kW is used at the channel formation stage. Voltage regulation is carried out by a transformer with variable magnetic coupling, which simultaneously performs the function of a high-current supply for pyrolytic conversion. This transformer power-to-size ratio is 100 kW with the regulation possibility from 0 V to input voltage and the maximum current about 264 A. The experimental functional circuit is shown in Figure 1. A three-stage power supply scheme is realized by high-voltage high-current contactors. This voltage-current characteristic has a sectionally approximate hyperbolic form (Figure 2). It allows providing constant input power up to 2 kW in a wide load resistance range. Contactors are rated at the maximum voltage of 10 kV and current 400 A. A pressure tight chamber with a 60 L volume is used to imitate fluid pressure in the underground conditions. Liquid and gaseous conversion products are sampled from the chamber. Nitrogen is used as the inert environment.

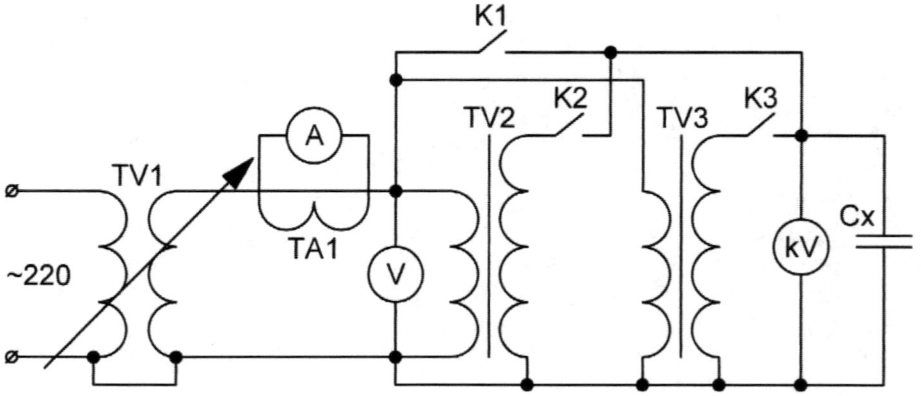

Figure 1. Experimental scheme. TV1 – the transformer with variable magnetic coupling; TV2 – the step-up transformer 1 kV; TV3 – the step-up transformer 10 kV; K1, K2, K3 – high voltage contactors; TA1 – the current transformer; C_x – a rock specimen.

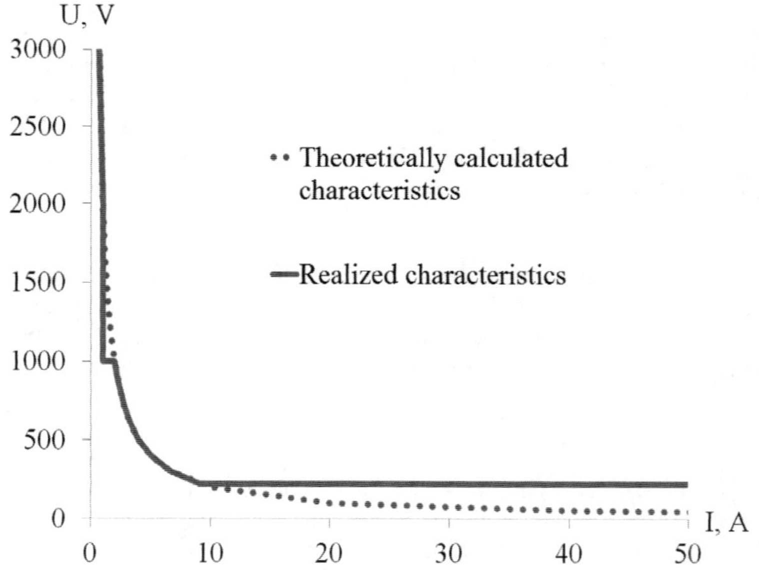

Figure 2. Power supply voltage-current characteristics for providing about 2 kW in load resistance wide range. A dotted line – theoretically calculated characteristics, a solid line – realized in experiment characteristics.

Pyrolytic conversion products composition depends on a temperature and heating rate. The resistance of the conductive channel, which is simultaneously a heating element, decreases because of the channel cross-section increase. Therefore, the power input in the rock should be regulated for specified conversion products obtaining. This task is necessary to solve for a very wide load resistance range. A three-stage power supply is used in the implemented experiments. The supply voltage-current parameter is a sectionally approximate hyperbolic function. Theoretically calculated for 2 kW and realized power supply characteristics are shown in Figure 2.

6. Conclusion

The highly efficient technology of oil shale organic component pyrolytic conversion into liquid and gaseous energy feedstock can be created due to electrical discharge phenomena in the rock. A power supply for the implementation of this conversion method has to provide high voltage and high current at different stages. Conventionally, the power source characteristics can be divided into three ranges: before the breakdown, immediately after the breakdown, and heating. Therefore, ranges discrete switching with fine control into each of the range is advisable for weight and size parameters of the power supply and its cost-efficiency.

References

[1] Hughes J. D. 2013 *Nature*. **494** 307
[2] Boggs S. and Densley J. 2000 *IEEE Electrical Insulation Magazine*. **16** 13
[3] Dissado L.A. and Fothergill J.C. 1992 *Electrical degradation and breakdown in polymers* (London: Peter Peregrinus Ltd.)
[4] Kuchinskiy G. S. 1979 *Partial discharges in high voltage constructions.* (L.: Energiya. Leningr. dep-t)
[5] Chen G. and Tham C. H. 2009 *IEEE Transactions on Dielectrics and Electrical Insulation.* **16** 179
[6] Dodd S.J., Chalashkanov N.M., Fothergill J.C. 2010 *IEEE International Conference on Solid Dielectrics.* **1** 1

Journal of Physics: Conference Series **552** (2014) 012013 doi:10.1088/1742-6596/552/1/012013

Dynamics of large-scale instabilities in conductors electrically exploded in strong magnetic fields

I M Datsko[1,*], S A Chaikovsky[1], N A Labetskaya[1], V I Oreshkin[1,2], and N A Ratakhin[1]

[1]Institute of High Current Electronics SB RAS, 2/3 Akademichesky Avenue, Tomsk, 634055, Russia

[2]National Research Tomsk Polytechnic University, 30 Lenin Avenue, Tomsk, 634050, Russia

E-mail: datsko@ovpe.hcei.tsc.ru

Abstract. The growth of large-scale instabilities during the propagation of a nonlinear magnetic diffusion wave through a conductor was studied experimentally. The experiment was carried out using the MIG terawatt pulsed power generator at a peak current up to 2.5 MA with 100 ns rise time. It was observed that instabilities with a wavelength of 150 μm developed on the surface of the conductor hollow part within 160 ns after the onset of current flow, whereas the surface of the solid rod remained almost unperturbed. A system of equations describing the propagation of a nonlinear diffusion wave through a conductor and the growth of thermal instabilities has been solved numerically. It has been revealed that the development of large-scale instabilities is obviously related to the propagation of a nonlinear magnetic diffusion wave.

1. Introduction

Interest in the electrical explosion of conductors (EEC) is related to various applications, such as generation of superstrong magnetic fields, implosion of heavy metal liners, energy transmission using vacuum transmission lines, etc. The current flow in an electrically exploded conductor can occur in three modes: a mode of uniform current distribution in the load, a skin effect mode, and a transient mode. The basic phenomena characteristic of EEC occurring in the skin-effect mode are a shock wave (SW) and a nonlinear magnetic diffusion wave (NMDW) propagating in the conductor, formation of low-temperature plasma at the conductor surface, and thermal instabilities [1, 2, 3]. The nonlinear diffusion of an electromagnetic field into a conductor features anomalously high speed compared to conventional diffusion. The higher diffusion speed is related to a decrease in conductivity of the metal due to its heating by electric current. Nonlinear diffusion takes place in rather high magnetic fields with the induction at the conductor surface reaching several tens of Teslas (for aluminum, the characteristic field $B_0 \approx 27$ T). The upper limit of a magnetic field for the propagation of an NMDW is its value at which plasma generation begins at the conductor surface and, hence, the temperature dependence of conductivity changes in nature. This limiting field is approximately equal to 10 B_0. The critical field for a conductor can be estimated by the formula:

$$B_0 = \left(\frac{6\mu_0 \rho_0 k}{\alpha m_a} \right)^{1/2} \tag{1}$$

Content from this work may be used under the terms of the Creative Commons Attribution 3.0 licence. Any further distribution of this work must maintain attribution to the author(s) and the title of the work, journal citation and DOI.
Published under licence by IOP Publishing Ltd

where α is the temperature coefficient of resistance, k is Boltzmann's constant, m_a is the atomic mass, ρ_0 is the density of the conductor, and μ_0 is the permeability of vacuum.

The development of large-scale instabilities that arise in a conductor during the propagation of a nonlinear magnetic diffusion wave was studied experimentally. The experiment was carried out with the use of the MIG terawatt pulsed power generator at a peak current up to 2.5 MA with 100 ns rise time.

2. Experiment

The growth of large-scale instabilities during the propagation of a nonlinear magnetic diffusion wave through a conductor was studied experimentally. The experiment was carried out with the MIG terawatt pulsed power generator at a peak current of up to 2.5 MA with 100 ns rise time. The diagnostic complex of the setup consisted of Rogowski coils, magnetic probes, voltage dividers, vacuum x-ray diodes (XRD) and an HSFC-Pro four-frame optical camera capable of shooting pictures with a minimum exposure time of 3 ns. The generator load was a cylindrical aluminum conductor 3 mm in diameter. With this diameter, the field strength at the conductor surface was not above 10 B_0; that is, plasma formation processes were ineffective. The load is sketched in figure 1. It consisted of two parts: a solid rod of external diameter D = 3 mm with a cavity of diameter 2.5 mm drilled from the rod end face to a certain depth, so that the solid rod of length 7.5 mm served as a load, and a hollow tube of the same diameter and length with 250 μm wall thickness.

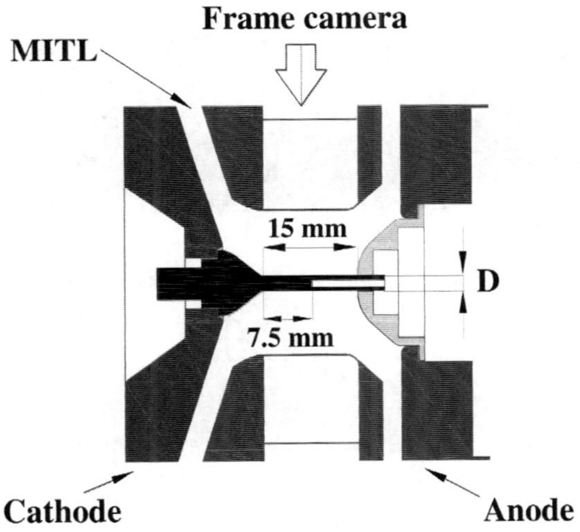

Figure 1. Sketch of the load of the MIG generator.

The choice of this design was dictated by the different times it takes for a nonlinear diffusion wave to propagate through a rod and through a hollow tube. We expected that this difference would allow us to elucidate whether an NMDW is related to the growth of large-scale instabilities.

Figure 2 presents pictures of an aluminum load taken in the optical range at different times from the onset of current flow. It can be seen that after the 160th nanosecond, instabilities of wavelength 150 μm developed on the surface of the hollow part of the conductor, whereas the surface of the solid rod remained in fact unperturbed.

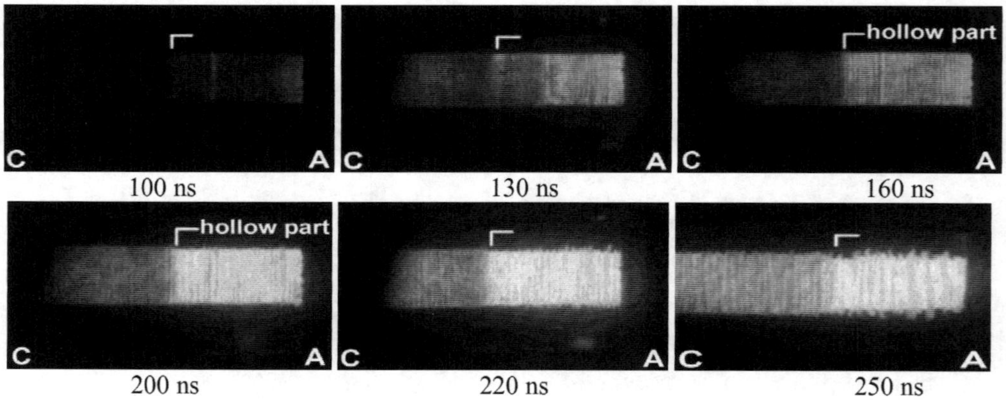

Figure 2. EOS images of an exploded aluminum conductor of external diameter 3 mm with 250 μm wall thickness of its hollow part.

It should be noted that for a 2 mm diameter aluminum conductor of the same design, which was exploded when the magnetic field induction at the conductor surface was above $10B_0$, surface instabilities occurred after the onset of plasma generation and developed over the hollow and the continuous part of the conductor simultaneously. This can clearly be seen in figure 3.

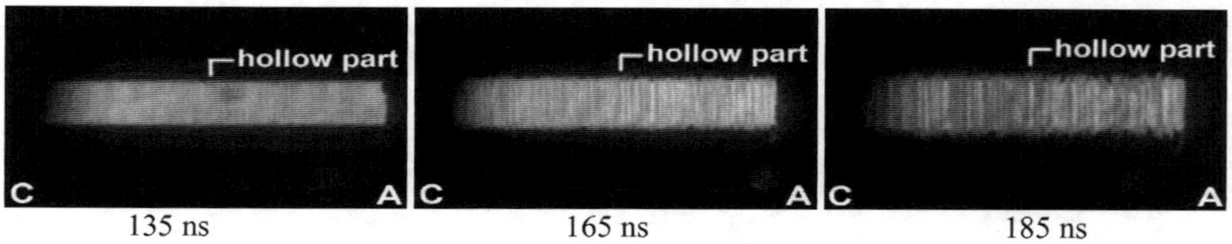

Figure 3. EOS images of an exploded aluminum conductor of external diameter 2 mm with 350 μm wall thickness of its hollow part.

3. Results and discussion

To interpret the experimental results, a numerical model was used which describes the propagation of a nonlinear magnetic diffusion wave [1, 5]. Nonlinear diffusion of a magnetic field is described by the Maxwell equations written in a quasi-stationary approximation (not taking into account displacement currents) and complemented with Ohm's law and the Joule–Lentz law [1, 4, 5]:

$$\nabla \times \mathbf{B} = \mu_0 \mathbf{j} \qquad \nabla \times \mathbf{E} = -\frac{\partial \mathbf{B}}{\partial t} \tag{2}$$

$$\mathbf{E} = \mathbf{j}\delta \tag{3}$$

$$\frac{\partial Q}{dt} = \mathbf{j}^2 \delta \tag{4}$$

where **E** is the electric field strength, **B** is the magnetic field induction, **j** is the current density, Q is the thermal energy density, δ is the resistivity, and μ_0 is the permeability of vacuum. The temperature dependence of resistivity is given by:

$$\delta(T) = \delta_0\left(1 + \frac{\partial \delta}{\partial T} T\right) \tag{5}$$

where δ_0 is the resistivity under normal conditions and $\frac{\partial \delta}{\partial T}$ is the derivative of resistivity with respect to temperature. As $Q = \rho c_v T$, where ρ is the density of the conductor and c_v is its heat capacity at constant volume, the resistivity can be expressed in terms of thermal energy density as:

$$\delta(Q) = \delta_0\,(1 + \beta Q) \tag{6}$$

where $\beta = \frac{1}{\rho c_v \delta_0}\frac{\partial \delta}{\partial T}$.

The system of equations (2)–(6) allows us to describe not only the propagation of an NMDW through a conductor, but also the growth of thermal instabilities whose structure is determined by the functional form of the temperature dependence of resistivity. In cases where resistivity increases with temperature, as this takes place for most of the metals in liquid and solid states, thermal instabilities are responsible for the formation of layered structures (strata) with the layers arranged normal to the direction of current flow. In a skin effect mode, the increments of thermal instabilities peak when the NMDW arrives at the boundary of the conductor.

The system of equations (2)–(6) was solved numerically, and the results are presented in figure 4. Curve 1 in this figure depicts the evolution of the magnetic field induction at the conductor surface observed in the experiment; curves 2 and 3 show the time-varied location of the NMDW front in the solid rod and in the hollow conductor, respectively. The location of the wave front was determined by the position of the current density maximum. In the hollow conductor, once the NMDW front had arrived at the internal boundary of the conductor, a reflected NMDW was generated which propagated outward.

As can be seen from figure 4, the reflected nonlinear diffusion wave arrived at the external boundary of the conductor at about the 160th nanosecond.

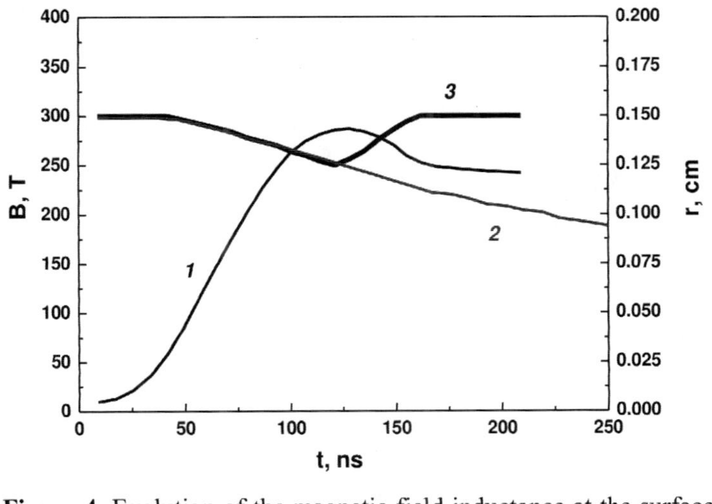

Figure 4. Evolution of the magnetic field inductance at the surface (curve 1) and of the location of the NMDW front in the solid rod (curve 2) and in the hollow conductor (curve 3).

The experiment has shown (see figure 2) that large-scale instabilities (strata) started developing on the conductor surface after the 160th nanosecond; that is, the relation between the growth of instabilities and the propagation of an NMDW is evident.

As thermal instabilities occur preferentially during the propagation of an NMDW through a conductor [1], it can be concluded that the growth of thermal instabilities is the most probable mechanism of strata formation.

Acknowledgements
The work was supported in part by the Russian Foundation for Basic Research (grant No. 14-08 00524).

References

[1] Oreshkin V I and Chaikovsky S A 2012 *Phys. Plasmas* **19** 022706
[2] Peterson K J, Sinars D B, Yu E P, Herrmann M C, Cuneo M E, Slutz S A, Smith I C, Atherton B W, Knudson M D and Nakhleh C 2012 *Phys. Plasmas* **19**, 092701
[3] Chaikovsky S A, Oreshkin V I, Mesyats G A, Ratakhin N A, Datsko I M and Kablambaev B A 2009 *Phys. Plasmas* **16** 042701
[4] Shneerson G A 1992 Fields and Transient Processes in High-Current Devices [in Russian], Energoatomizdat, Moscow
[5] Chaikovsky S A, Oreshkin V I, Datsko I M, Labetskaya N A and Ratakhin N A 2014 *Phys. Plasmas* **21** 042706

Journal of Physics: Conference Series **552** (2014) 012014 doi:10.1088/1742-6596/552/1/012014

Effect of electron extraction from a grid plasma cathode on the generation of emission plasma

V N Devyatkov[1,*] and N N Koval[1,2]

[1] Institute of High Current Electronics SB RAS, 2/3 Akademichesky Ave., Tomsk, 634055, Russian Federation
[2] National Research Tomsk State University, 36 Lenin Ave., Tomsk, 634050, Russian Federation

E-mail: vlad@opee.hcei.tsc.ru

Abstract. The paper describes the operating mode of a plasma electron source based on a low-pressure arc discharge with grid stabilization of the plasma emission boundary which provides a considerable (up to twofold) increase in discharge and beam currents at an Ar pressure in the vacuum chamber $p = 0.02–0.05$ Pa, accelerating voltages of up to $U = 10$ kV, and longitudinal magnetic field of up to $B_z = 0.1$ T. The discharge and beam currents are increased on electron extraction from the emission plasma through meshes of a fine metal grid due to the energy of a high-voltage power supply which ensures electron emission and acceleration. The electron emission from the plasma cathode and arrival of ions from the acceleration gap in the discharge changes the discharge plasma parameters near the emission grid, thus changing the potential of the emission grid electrode with respect to the discharge cathode. The load is not typical and changes the voltage polarity of the electrode gap connected to the discharge power supply, which is to be taken into account in its calculation and design. The effect of electron emission from the plasma cathode on the discharge system can not only change the discharge and beam current pulse shapes but can also lead to a breakdown of the acceleration gap and failure of semiconductor elements in the discharge power supply unit.

1. Introduction

Grid stabilization of the plasma emission boundary has rather long been used in plasma-cathode electron sources. The authors of [1] considered the generation of pulsed electron beams with high current density ($j = 1–10$ A/cm^2). The basic processes involved in electron emission from plasma were considered elsewhere [2–4]. The grid stabilization method, which is applied to provide stable electron emission from a plasma cathode, assumes the use of a grid with a mesh size comparable to the extent of the space charge layer separating the plasma from anode and emission grid electrode. Electrons, in this case, are emitted from a partially open plasma surface in the center of grid meshes and through a potential barrier at their edges. The operating mode of the cathode is stabilized because the Langmuir layer thickness separating the plasma from the grid increases with an arbitrary increase in electron emission current and the emitting plasma surface area decreases with an increase in plasma potential. The electron emission from the plasma can change not only the plasma potential but also other plasma parameters, resulting in unstable operation of the discharge (which generates the emission plasma), high-frequency oscillations [5, 6], change in the plasma concentration, and attendant change in the discharge current. Besides, the maximum parameters of plasma-cathode electron sources are much

Content from this work may be used under the terms of the Creative Commons Attribution 3.0 licence. Any further distribution of this work must maintain attribution to the author(s) and the title of the work, journal citation and DOI.
Published under licence by IOP Publishing Ltd

affected by the ion current from the anode (beam) plasma surface formed through ionization of working gas and gas desorbed from the electrodes (anode, collector, drift tubes) to which the beam current takes its path [7, 8]. This plasma serves as a plasma anode and defines in many respects the main characteristics of an electron source.

The paper considers the mode of electron extraction from a plasma cathode based on a low-pressure arc discharge with grid stabilization of the plasma emission boundary which, in the presence of a longitudinal magnetic field in the emission grid electrode region, provides a considerable increase in discharge current amplitude compared to the mode with no electron extraction from the plasma cathode. The experimentally observed increase in discharge current, which is due to the energy brought into the discharge power supply circuit by an accelerating voltage source, provides an increase in emission current from the plasma cathode. The arising positive feedback induces a simultaneous increase in discharge and emission currents, and this can result in failure of the plasma cathode operation due to uncontrollable changes of the electron beam parameters and in a breakdown of the acceleration gap. This mode of plasma cathode operation, associated with discharge power supply parameters and electron emission from the plasma through emission grid meshes, can be considered as one of the factors that limit the maximum characteristics of energy complexes comprising a plasma cathode, power supply units, and accelerating voltage source. The processes occurring in the emission grid electrode region should be taken into account in designing both plasma emitters and their power supplies. Understanding these processes will allow one to more accurately define the requirements on power supply parameters, disclose the causes for emergency operation of semiconductor elements which leads to their failure, and determine the ways of increasing the operation stability of grid-stabilized plasma cathodes and their maximum parameters.

2. Experimental setup

In the experiments, an electron source with a grid plasma cathode was used [9, 10]. Schematic of the electrode and discharge power supply circuits is shown in figure 1. In this system, an arc discharge cathode spot is initiated on magnesium cathode *1* by a pulse from the secondary winding of transformer *TR1* having a voltage amplitude of up to 10 kV and current amplitude I_{tr} = 10 A at a pulse duration τ = 30 μs. The main arc discharge (current I_d) is powered through isolating transformer *TR2*. A partial discharge of storage capacitor C_2 is used. The discharge current pulse is formed when switch S_1 is closed. The discharge current amplitude is defined by the voltage across capacitor C_2 and is limited by ballast resistor R_3. Also shown in the figure are elements D_1, C_1 and D_2, R_4 (protection elements of switch S_1) which, along with *TR2*, can influence the arc discharge current pulse shape. The pulse voltage U_d across the gap between the cathode and grid electrode *3* is controlled with a Rogowski coil from the current in resistor circuit R_1. The plasma cathode discharge system is partially placed in a divergent magnetic field of solenoid *7* which provides B_z = 0.1 T in the coil centre (in the emission grid region, B_g = 35 mT). Solenoid *6* creates a field of B = 25 mT in the region of collector *5*. The electron beam is extracted through an emission window of diameter 60 mm covered with a stainless steel grid with a mesh of 0.3 x 0.3 mm and is transported in drift tube *4* of diameter 80 mm to collector *5* located 250 mm away from grid electrode *3*.

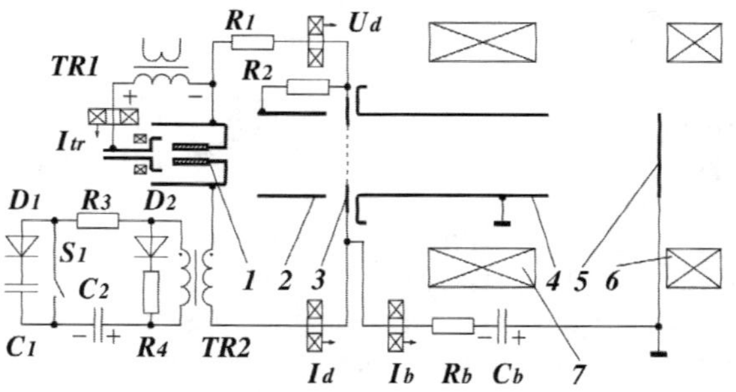

Figure 1. Electrode and power supply circuits of the plasma-cathode electron source: *1* – cathode; *2* – anode insert of diameter 80 mm; *3* – emission grid electrode; *4* – drift tube; *5* – collector; *6, 7* – solenoids. R_1 = 100 Ω, R_2 = 20 Ω, R_3 = 2.3 Ω, R_4 = 4.9 Ω, R_b = 1.25 Ω, C_1 = 1 μF, C_2 = 570 μF, C_b = 6 μF.

The electron source has two important peculiarities: operation with a plasma anode [11] and considerable ion current from the acceleration gap to emission electrode *3*. The ion flow governs the secondary ion-electron emission from the grid electrode surface with attendant increase in total current in the acceleration gap [8] and the ion injection from the acceleration gap into the plasma cathode discharge system.

3. Results and discussion

Figure 2 shows typical waveforms of the discharge current I_d and voltage U_d (between electrodes *1* and *3*) at an initial voltage U_{C2} = 400 V across capacitor C_2 and accelerating voltage U_{Cb} = 0. In this mode, almost the whole discharge current takes the path to emission electrode *3*. At p = 0.07 Pa, the discharge gap voltage U_d increased from 20 to 70 V as the discharge current I_d increased from 80 to 280 A. Figure 3 (a) shows a simplified discharge circuit. In this case, the load for the power supply is nonlinear load *RD* (discharge resistance) on which the voltage U_{rd} varies in the range 20–70 V. The current in the D_2R_4 circuit starts flowing only at the instant the discharge current pulse ends. The dependence of U_d on the operating pressure for the mode with no electron extraction from the plasma cathode is shown in figure 4 (curves *3*, *4*).

As an accelerating voltage U_{Cb} = 10 kV is applied and electrons are extracted from the plasma through the fine grid meshes, which stabilize the plasma emission boundary, the discharge current switches from the emission electrode to the acceleration gap and the voltage U_d changes its polarity (figure 4, curves *1*, *2*). In this case, the discharge circuit (figure 3 (b)) has, along with load *RD*, an additional voltage source $E > U_{rd}$ which increases the current I_d (from 152 to 218 A). Waveforms of the currents I_d, I_b and voltage $U_d = U_{rd} + E$ for this mode are presented in figure 5. The voltage source E is the potential difference between the anode discharge plasma and the emission grid electrode. The maximum voltage E in the experiments was $E \geq 1$ kV. This voltage was obtained with the discharge circuit shown in figure 6. Oscillogram 1 (figure 6) reveals an instant (no longer than ~ 2 μs) sharp increase in the current I_d due to a jump of the voltage E immediately before breakdown of the acceleration gap.

Despite the positive feedback, which simultaneously increases the discharge and emission currents due to the energy of the acceleration gap power supply, rather good pulse shape reproducibility of the discharge and emission currents is observed in the examined operating parameter range of the electron source. It is seen from the oscillograms that the voltage U_d (of positive polarity) increases with increasing the discharge current, magnetic field, and gas pressure. The discharge and emission currents cease to increase once the increase in U_d ceases. Besides, on cutoffs of the discharge current (with subsequent repeated discharge ignition and current recovery), a steep increase in U_d by 20–40 V is observed. A characteristic peculiarity is that U_d decreases by the end of the pulses and oscillations of U_d arise on the oscillogram (figure 5, curve *3*).

The operation of the electron source in the plasma anode mode defines a rather weak dependence of the discharge current on accelerating voltage. As the voltage U_b is decreased from 10 to 4 kV, the beam current and the voltage U_d decrease by no more than 5 %. A rather strong pressure dependence of U_d (figure 4) allows the assumption that the effect of emission from the plasma cathode on the discharge circuit is defined mainly by the accelerated ion flow arriving through the grid meshes from the acceleration gap in the discharge plasma.

The voltage U_d of reversed polarity (figure 4) is, in fact, equivalent to cutoff voltage applied to the emission grid of the plasma cathode. However, this cannot decrease the emission current under our conditions, because U_d changes precisely due to electron emission processes, but can probably be a factor that somewhat limits the increase in currents. On the other hand, the position of the plasma emission boundary in the grid meshes provides efficient electron emission from the plasma and most of the electron current not to the emission grid but to the acceleration gap through its meshes. The accelerating voltage source and the discharge power supply, in this case, are connected in series.

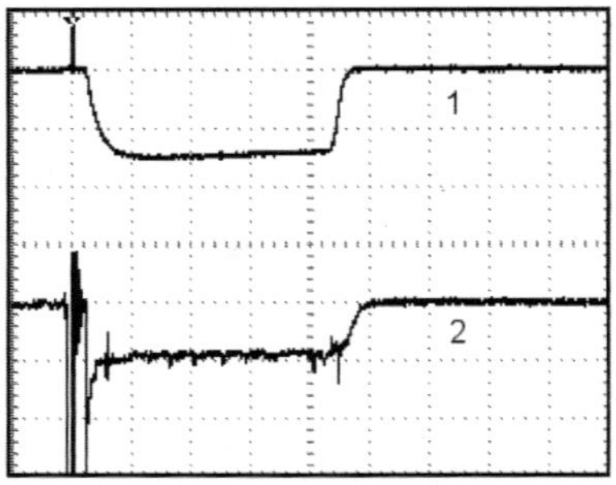

Figure 2. Typical oscillograms of the discharge current I_d (1) and voltage U_d (2). $B_g = 25$ mT, $p = 0.1$ Pa, $U_{Cb} = 0$. Scale: 100 A/div, 50 μs/div, 40 V/div.

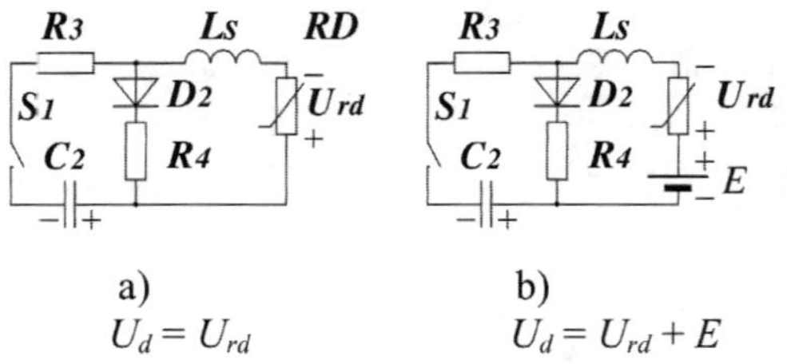

Figure 3. Simplified discharge power supply circuit and its equivalent load for the mode without (a) and with electron emission (b). L_S – inductance of transformer *TR2*.

A change in the grid electrode potential with respect to the discharge plasma can cause not only an increase in discharge current. An increase in plasma potential can lead to a breakdown of the plasma-anode (grid electrode) gap, resulting in cathode spots at the emission electrode surface or directly at the emission grid. This is confirmed by cathode spot erosion traces on the emission electrode surface and by occasional substantial short-time growth of I_b on the oscillograms of the currents I_d, I_b (figure 7) which does not, however, lead to a breakdown of the acceleration gap. It is seen that at the instant the beam current I_b increases, the discharge current I_d remains almost unchanged. It should be noted that if a cathode spot arises at the emission grid, the attendant steep rise of I_b is more likely to result in a breakdown of the acceleration gap, making this mode difficult to record. With the discharge circuit shown in figure 1 and $R_2 = 0$, we managed to obtain a stably reproducible mode in which it was possible to detect the incipience of a cathode spot at electrode *2* being, in this case, part of the emission electrode. Figure 8 shows a fragment of oscillograms (early in the pulse) for the currents I_d (1), I_b (2) and anode insert current (3). Analysis of these oscillograms allows the conclusion that at the instant the emission current reaches $I_b \approx 200$ A, a cathode spot arises at electrode *2* and an additional arc discharge (in addition to the main discharge) is ignited between electrode *2* and anode discharge plasma such that the arising additional electron current takes the path through the acceleration gap: the total emission current I_b increases along with the current of electrode *2*. The instant at which the emission current becomes higher than $I_b = 400$ A can probably be considered as the onset of breakdown of the acceleration gap.

The influence of electron emission on the generation of emission plasma can not only show up during the formation of the main beam current pulse but can also lead to another undesirable effect. With xenon (instead of argon) at high operating pressures, spurious repeated pulses of the discharge and electron beam currents are observed at times. Such pulses, having parameters different from the operating ones, are inadmissible for some applications of the electron source (e.g., for pulsed electron beam surface modification). The repeated pulses arose within ~ 100 μs after the end of the main discharge pulse at the instant storage capacitor C_2 in the discharge power supply started to be charged. When capacitor C_2 was charged, voltage of opposite polarity was applied to the electrodes of the plasma cathode discharge system which provoked the ignition of a discharge. Due to the effect of electron emission from the plasma cathode on the discharge circuit, repeated discharge current pulses and corresponding beam current pulses were formed. The parameters of the pulses unambiguously suggested that they were formed mainly due to the energy of the accelerating voltage source (the rise and fall times of the discharge current pulses were extended; their duration and amplitude differed from the parameters provided by the discharge power supply).

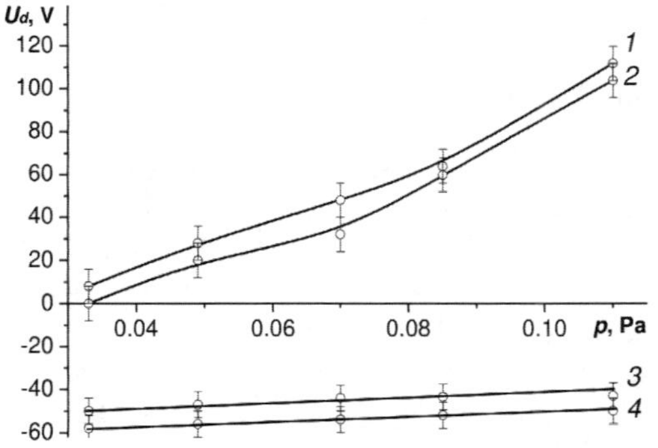

Figure 4. Voltage U_d vs pressure p in the emission grid region: $U_{Cb} = 10$ kV (*1,2*); $U_{Cb} = 0$ (*3,4*); $I_d = 250$ A (1,4); $I_d = 150$ A (*2,3*); $B_g = 25$ mT.

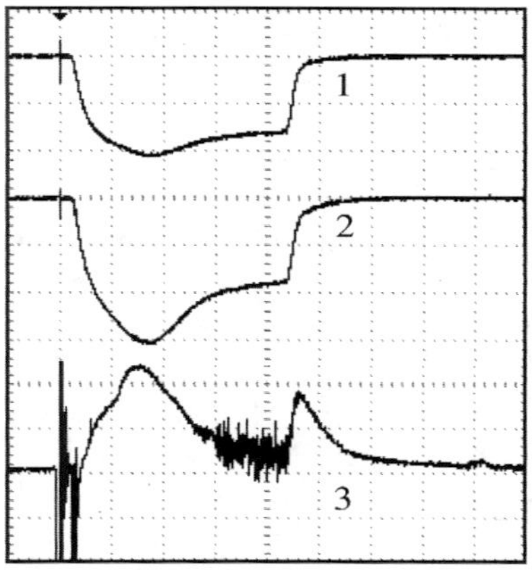

Figure 5. Typical waveforms of the currents I_d (1), I_b (2) and voltage U_d (3). $B_g = 25$ mT, $p = 0.1$ Pa, $U_{Cb} = 10$ kV.
Scale: 100 A/div, 50 μs/div, 40 V/div.

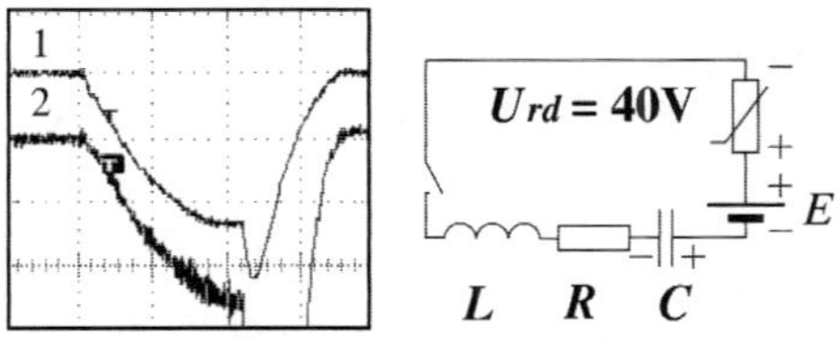

Figure 6. Waveforms of the currents I_d (1) and I_b (2) and discharge circuit in which $E \geq 1$ kV; $L = 20$ μH, $R = 0.7$ Ω, C = 50 μF ($U_C = 200$V). Scale: 100 A/div, 25 μs/div.

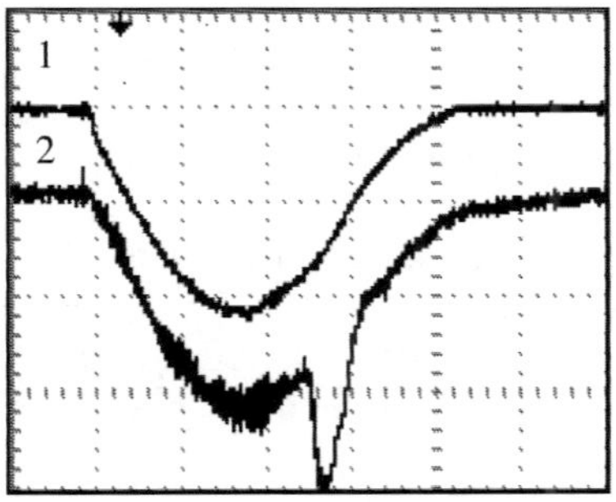

Figure 7. Waveforms of I_d (1) and I_b (2) for the discharge circuit in figure 6. Scale 100 A/div, 25 μs/div.

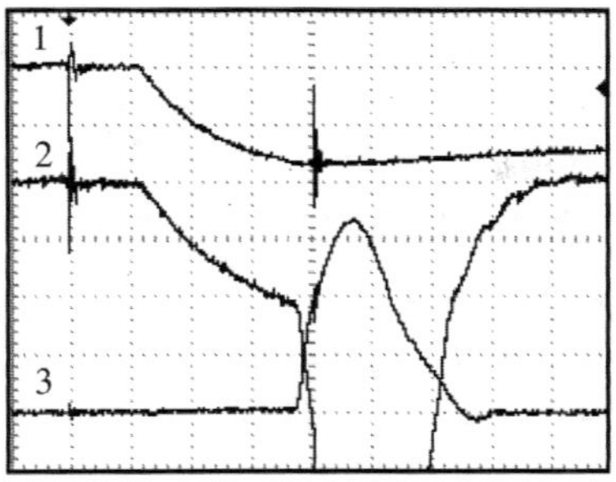

Figure 8. Fragment of waveforms of I_d (1), I_b (2), and current of electrode *2* (3) for the circuit in figure 1 at $R_2 = 0$. Scale: 100A/div, 10 μs/div.

4. Conclusion

The considered effects of electron emission from the plasma cathode on the generation of emission plasma often lead to uncontrollable changes in the electron beam parameters (pulse shape and current amplitude) and can be the factors that limit the maximum parameters of the electron source. At the same time, they can probably be used for increasing the efficiency of emission plasma generation due to the energy redistribution between the power supplies of the acceleration and discharge gaps.

The observed amplification of the discharge current depends on the voltage of the discharge power supply, its output resistance, and also on the grid geometry (mesh size), acceleration gap width, and high voltage applied to the gap. Low voltage and output resistance of the discharge power supply causes a substantial increase in discharge current. So, with seven concurrently operating arc discharge power supplies based on storage capacitors and ballast *LR* circuits at a charge voltage of 80 V (figure 6), the total discharge current increased from 120 to 300 A on applying the accelerating voltage. The current amplitude, in this case, failed to reach its maximum value due to breakdown of the acceleration gap. Stabilizing the discharge current requires a power supply with parameters approximating as close as possible those of a perfect current source. If it is required to have power supplies with high efficiency, they can probably be based on circuits with inductive energy stores, which efficiently stabilize load currents.

The described effect of electron emission on the generation of emission plasma results in two types of additional voltage present in the discharge power supply circuit: relatively low voltage in normal pulse formation mode and short high-amplitude voltage pulses arising with a steep increase in emission current before breakdown of the acceleration gap (figure 6). The generation of such pulses defines the necessity to use additional protective circuits in discharge power supplies. At the same time, the presence of protective elements in discharge power supplies can add to the increase in discharge current because these circuits decrease the output resistance of the power supplies.

This work was partially supported by program of the Presidium of RAS No. 12 and RFBR project No.13-08-98108.

References

[1] Koval N N, Kreindel Yu E, Schanin P M 1983 Generation of pulsed electron beams with uniform high current density distribution in systems with a grid plasma emitter *J Zh. Tekh. Fiz.* **53** Issue 9 1846

[2] Zharinov A V, Kovalenko Yu A, Roganov I S, Tyuryukanov P M 1986 Plasma electron emitter with grid stabilization I *J Zh. Tekh. Fiz.* **56** Issue 1 66

[3] Zharinov A V, Kovalenko Yu A, Roganov I S, Tyuryukanov P M 1986 Plasma electron emitter with grid stabilization II *J Zh. Tekh. Fiz.* **56** Issue 4 687

[4] Koval N N, Oks E M, Protasov Yu S, Semashko N N 2009 *Emission electronics* (Moscow, MGTU)

[5] Gavrilov N V, Emlin D R, Kamenetskikh A S 2008 High - efficiency emission from a grid-stabilized plasma cathode *J Zh. Tekh. Fiz.* **78**, Issue 10 59

[6] Gavrilov N V, Kamenetskikh A S 2013 Self-oscillating mode of electron beam generation in a source with a grid plasma emitter *J Zh. Tekh. Fiz.* **83** Issue 10 32.

[7] Astrelin V T, Kandaurov I V, Trunev Yu A 2014 Generation of a submillisecond beam of high current density in a plasma-cathode diode with emission from an open plasma boundary *J Zh. Tekh. Fiz.* **84** Issue 2 106

[8] Koval N N, Grigoryev S V, Devyatkov V N, Teresov A D, Schanin P M 2009 Effect of Intensified Emission During the Generation of a Submillisecond Low-Energy Electron Beam in a Plasma-Cathode Diode *IEEE Transactions on Plasma Science* **37** 1890

[9] Koval N N, Devyatkov V N, Grigoryev S V, Sochugov N S 2006 Plasma electron source «SOLO» *Proc.II Intern. Kreindel Workshop «Plasma emission electronics»* (Ulan-Ude: Russia) 79

[10] Grigoriev S V, Devjatkov V N, Koval N N and Teresov A D 2008 The automated installation for surface modification of metal and ceramic-metal materials and products by intensive pulse submillisecond electron beam *Proc. 9th Intern. Conf. on Modification of Materials with Particle Beams and Plasma Flows* (Tomsk: Russia) 19

[11] Devyatkov V N, Koval N N and Schanin P M 2001 Generation of high-current low-energy electron beams in systems with plasma emitters *J Russian Physics Journal* **44** No.9 937

Journal of Physics: Conference Series **552** (2014) 012015 doi:10.1088/1742-6596/552/1/012015

Effect of laser radiation wavelength on explosives initiation thresholds

R. Akhmetshin*, **A. Razin, V. Ovchinnikov, A. Skripin, V. Tsipilev, V. Oleshko, V. Zarko and A. Yakovlev**
Tomsk Polytechnic University

E-mail: kvant@tpu.ru

Abstract. In this paper the initiation thresholds were measured and kinetics of decomposition was studied for pressed powders of FTDO (furasan-1,2,3,4-tetrazin-1,3-dioxide) at pulsed laser action by different laser wavelengths: 266 nm (zone-to-zone absorption), 532 nm (transparency range), 1.064 nm (transparency range) and 10600 nm (phonon absorption) that corresponds to absorption region (266 nm), transparency range (532 nm; 1064 nm) and phonon absorption (10600 nm) of these explosives.

1. Introduction

Heated centers of various configurations form at laser initiation of explosives. Thus, varying the radiation wavelength results in laser energy localization in a thin subsurface layer of < 10−4 cm (spectral range of "zone-to-zone" absorption), or in "hot spots" of ~10−5 cm (spectral range of transparency), or in cylindrical volume with the height of 10−3 cm (spectral range of phonon absorption). This brings to the question about the reactive effectiveness of heated centers of different shapes which can be answered after measuring the initiation energy thresholds and kinetic parameters of the explosive decomposition process. Previous studies were performed for primary explosives [1]. It was found that specific heat in heated centers of different shapes makes almost equal value of ~2700 J/cm^3. These results suppose thermal heated center mechanism of heavy metal azides laser initiation [1]. This approach in secondary explosives studying has a significant interest for making of model representation of their initiation mechanisms, which is in its starting position. In this paper the given approach is used in PETN and FTDO initiation studying.

2. Experiments and methods

For the experimental works mount described in [2] was used. The base for this mount is YAG-laser LQ-929 with radiation on 1, 2 and 4 harmonics (wavelength 1064 nm, 532 nm and 266 nm at pulse duration of 10 ns), multimode beam structure with homogeneous central part. Mount includes pulse CO_2 laser (λ=10.6 μm) with pulse duration of 25 ns and energy ~ 0.1 J. This mount allows carrying out multiparameter process measurements at explosive decomposition of energetic materials.

As explosives powder samples energetic materials were used, such as PETN (tetranitropentaerytrite) and FTDO (furasan-1, 2, 3, 4-tetrazin-1, 3-dioxide), pressed in tablets with 3 mm diameter into the press mold. Maximum pressure was 5•108 N/m^2, weighed amount was 6±1 mg. It is known that secondary explosives initiation from open sample surface is hampered or impossible because of gas-dynamic unloading of heated center through available surface. To exclude this, experiments were carried out from covered samples surface with the covering pressure of FTDO =

Content from this work may be used under the terms of the Creative Commons Attribution 3.0 licence. Any further distribution of this work must maintain attribution to the author(s) and the title of the work, journal citation and DOI.
Published under licence by IOP Publishing Ltd

108 N/m2 and PETN = 3•108 n/m^2. Laser beam diameter made 1 mm and pulse duration made 10 ÷ 25 ns. In these conditions, threshold sensitivity of explosives initiation depends on the covering pressure. The experiments demonstrated optimal covering pressure. PETN optimal pressure was 3•108 n/m^2, whereas for FTDO this value was 108 n/m^2. At this covering pressure low-threshold mode of explosives initiation occurs.

Experiments determined samples initiation energy thresholds at different wavelengths within the range of 0.266-10.6 μm. Irradiation pulse from laser exposure zone [2] was determined and pressure pulse at samples initiation moment was registered. Initiation was accompanied by assembly destruction and strong bang. Sample rejection did not result in its structural changes and optical properties alteration [3]. Threshold level of laser exposure demonstrated pulse delay of explosive decomposition irradiation and delay of acoustic response with typical value of 4 ms for PETN and 1 ms for FTDO. Irradiation at 266 ns wavelength and 532 ns wavelength showed additional short irradiation peak, which permanent allocation corresponds to laser pulse allocation, what indicated luminescent character of this irradiation.

Threshold energy density and specific heat of micro heated centers are presented in Table 1. Average specific heat in zone-to-zone absorption ($\lambda_0 = 266$ nm) and phonon absorption ($\lambda_0 = 10600$ nm) was calculated by $\omega = H{\cdot}\mu/(1+ \mu{\cdot}\sqrt{\ \alpha{\cdot}\ \tau{\cdot}i}\ )$, where μ is explosive matrix absorption index (~10^5 cm^{-1} and ~10^{3}cm^{-1} for 266 nm and 10600 nm respectively), α is thermometric conductivity. At matrix transparency range ($\lambda_0 = 532$ nm and $\lambda_0 = 1064$ nm) because of very small values of μ (~10^{-1}cm^{-1} – 10^{-2}cm^{-1}), average specific heat is negligible. In this case it stems from the fact that laser pulse energy localizes at microinhomogeneities which exist in explosives matrix (absorption nano-sized inclusions, structural defects) with forming of thermal centers (heating, optical-induced breakdown) typically sized 10^{-5} cm. Thermal center heat was estimated by $\omega = 3{\cdot}H{\cdot}R{\cdot}2/(R+ \sqrt{\alpha{\cdot}\tau{\cdot}i}\ )^3$.

Table 1. Energy initiation thresholds (H) and specific energy heat of heated centers ω

		10 ns			25 ns
		266 nm	532 nm	1064 nm	10600 nm
PETN	H, mJ/cm^2	25±20	55±15	60±15	1800±20
	ω, J/cm^3	~1900	~1900	~2100	~1800
FTDO	H, mJ/cm^2	8±2	—	20±6	300±10
	ω, J/cm^3	~600	—	~700	~300

3. Results and discussions

Note that evaluated specific heat of different initiation centers (which correspond to different spectral ranges) for every explosive is almost equal when the initiation thresholds are significantly different. Critical value of specific heat for PETN comprised ~2.0 kJ/cm^3 and that for FTDO comprised 0.4 kJ/cm^3. It should be pointed out that laser initiation in phonon absorption is inherently thermal. It indicates thermal heated center character of PETN and FTDO initiation in different spectral ranges. Nevertheless, this brings to the question about the reactive effectiveness of heated centers with such specific heat, i.e. possibility of their development into explosives decomposition heated centers.

To clarify this statement, numerical simulation of PETN and FTDO ignition in zone-to-zone absorption and phonon absorption was made within the frames of thermal local initiation model, represented in [4]. Numerical simulation in transparency range was made by the method described in [5-6]. Simulation considers great alteration of absorption indices for these explosives at 266 nm, 532 nm, 1064 nm, and 10600 nm wavelengths. For that reason various light regime was made at given wavelengths in samples volume.

Thus, within transparency range of FTDO (1064 wavelength), matrix zone-to zone made it difficult to explain experimentally the resulted initiation thresholds. The thickness of absorption layer is 7.7 cm, i.e. heating temperature for exposure period of laser pulse can be easily evaluated

$$\Delta T = \frac{Q}{\frac{\pi \cdot d^2}{4} \cdot \left(h + \sqrt{\alpha \cdot \tau}\right) \cdot c \cdot \rho} = \frac{H}{\left(h + \sqrt{\alpha \cdot \tau}\right) \cdot c \cdot \rho} = \frac{w}{\left(1 + \frac{\sqrt{\alpha \cdot \tau}}{h}\right) \cdot c \cdot \rho} \tag{1}$$

According to the formula (1), heating temperature for 60 mJ thresholds is 0.003 K and for 40 mJ thresholds is 0.002 K.

In this case initiation mechanism can be explained by absorption microinclusions where radiation energy localization occurs. Thus, thermal local initiation model corresponds to the experiment well. Numerical simulation of light regime in FTDO matrix transparency range demonstrates 12 times more illumination enhancement in subsurface layer compared to surface illumination. Spectral beam is diminishing weakly while explosives basic layer is enhancing and it is estimated by absorption albedo. Any inclusion can be a heated center within the range of radiation penetration depth in the sample even up to 1 mm with its effective absorption crossing. As a result of inclusions ignition numerical simulation in active medium, initiation threshold was 40 mJ/cm^2 for FTDO and 80 mJ/cm^2 for PETN, what corresponds to experimental points of 20 mJ/cm^2 and 55 - 60 mJ/cm^2 by order of values. (Table 1).Effective absorption crossing for microinclusions particles with diameter 100 nm was calculated by Mie theory.

Edge of fundamental absorption at wavelength of CO2 laser (10600nm), sample refraction index becomes small, effect of full inner reflection stops contribute to light enhancement of subsurface explosive layer. Moreover, effective absorption crossing at this wavelength is approximately 4 times less than geometrical one for spherical inclusions. Thus, absorption particles 10-5 cm cannot be thermal centers of self-sustained decomposition, radiation is absorbed in sample volume with penetration depth 0.002 cm. As a result of numerical simulation of inclusions ignition at this wavelength with matrix absorption 500 cm^{-1} initiation threshold is 400 mJ/cm^2, what corresponds to experimental data (Table 1).

In the range of zone-to-zone absorption (wavelength 256 nm), relevant thickness of FTDO sample optical layer is determined by absorption index:

$$\tau = \frac{1}{\mu}, \tag{2}$$

Absorption index at this wavelength is considerably higher than dissipation index, which excludes volume light enhancement; all radiation energy localizes in volume of thin near surface layer. Light distribution on the sample volume can be described by Buger′ s law.

Edge of zone-to-zone absorption preconditions a very high absorption index, where the thickness of near surface layer localizes 1/e energy share of laser radiation will be several µm (1, 5 µm for absorption index 7000 cm^{-1} by the 2 formula). Heated center with plane geometry has two times less heating unloading than that with spherical thermal field. Moreover, plane subsurface layer at short impacts (15 ns) and poor thermal conductivity of covering pressure made this center one-side limited by adiabatic wall and reduce thermal unload by 2 times more. According to numerical simulation, because of small volume and high laser radiation energy absorption efficiency in this layer, threshold is 10 mJ that corresponds to the experiments too (Table 1).

Thus, calculations based on thermal microcenter initiation model give desirable results describing macro- and micro centers of various configurations. Besides, by the formula (1) it is obvious that in a sample transparency range, initiation is impossible because of matrix volume heating. In conclusion, it is necessary to point out that the given results are preliminary and need to be confirmed. Numerical simulation of FTDO thermo kinetic parameters demands special selection and validation because of their certain arbitrary character.

Acknowledgement
The work was supported by State assignment «Science» and State contract No 3.1329.2014

References

[1] Lisitsyn V, Tsipilev V, Damam J and Malis D 2011 *Comb. Explos. Shock Waves* **47** 5 106–115
[2] Korepanov V et al. 2006 *Comb. Explos. Shock Waves* **43** 1 106-119
[3] Aleksandrov Ye, Tsipilev V 1989 *Comb. Explos. Shock Waves* **26** 1 3-9
[4] Tsipilev V, Morozova E, Skripin A 2010 *Bull. of Tomsk Polytechnic University* **317** 4 149–155
[5] Aleksandrov Ye, Tsipilev V 1984 *Comb. Explos. Shock Waves* **20** 6 104-109
[6] Burkina R, Morozova E, Tsipilev V 2011 *Comb. Explos. Shock Waves* **48** 5 95-105

Journal of Physics: Conference Series **552** (2014) 012016 doi:10.1088/1742-6596/552/1/012016

Optimization of the vacuum insulator stack of the MIG pulsed power generator

G Khamzakhan[1] and S A Chaikovsky[2]
[1]Tomsk Polytechnic University, 30 Lenin Ave., Tomsk 634050, Russia

E-mail: goha_respect@mail.ru

[2]Institute of High Current Electronics, SB, RAS, 2/3 Akademichesky Ave., Tomsk 634055, Russia

E-mail: stas-chaikovsky@yandex.ru

Abstract. The MIG multi-purpose pulsed power machine is intended to generate voltage pulses of amplitude up to 6 MV with electron-beam loads and current pulses of amplitude up to 2.5 MA and rise time 100 ns with inductive loads like Z pinches. The MIG generator is capable of producing a peak power of 2.5 TW. Its water transmission line is separated from the vacuum line by an insulator stack. In the existing design of the insulator, some malfunctions have been detected. The most serious problems revealed are the vacuum surface flashover occurring before the current peaks and the deep discharge traces on the water–polyethylene interface of the two rings placed closer to the ground. A comprehensive numerical simulation of the electric field distribution in the insulator of the MIG generator has been performed. It has been found that the chief drawbacks are nonuniform voltage grading across the insulator rings and significant enhancement of the electric field at anode triple junctions. An improved design of the insulator stack has been developed. It is expected that the proposed modification that requires no rearrangement of either the water line or the load-containing vacuum chamber will provide higher electric strength of the insulator.

1. Introduction

Terawatt-class pulsed power generators are of great interest for use in basic research and applications dealing with extreme states of matter. In 1996, at the Institute of High Current Electronics (Tomsk, Russia), the MIG pulsed power machine [1] was built. The generator is designed to produce voltage pulses of amplitude up to 6 MV and current pulses of amplitude up to 2.5 MA with a peak pulse power of 2.5 TW. The generator currently operates by the following scheme: linear pulse transformer – water pulse-forming lines – vacuum transmission line – load. A capacitor bank capable of storing 400 kJ of energy is used as the primary energy store. The linear pulse transformer transfers the energy stored in the capacitor bank into the first pulse-forming line whose electrical length and impedance are, respectively, 75 ns and 1.3 Ω. When a single-channel water spark gap operates, part of the energy is transferred to the second pulse-forming line (26 ns, 0.65 Ω). On the operation of a nine-channel water spark gap switch, up to 120 kJ of energy is supplied, a power of 1.2–1.4 TW, through the 0.65-Ω transmission line to the (cone-shaped) vacuum transmission line and then to the load. The water-filled transmission line is separated from the vacuum line by an insulator stack of external diameter ~90 cm.

Content from this work may be used under the terms of the Creative Commons Attribution 3.0 licence. Any further distribution of this work must maintain attribution to the author(s) and the title of the work, journal citation and DOI.
Published under licence by IOP Publishing Ltd

The MIG generator is relatively compact among the machines of this power level. The dimensions (length, width, and height) of the pulse transformer with the primary energy store are 5.5×3.2×2.7 m^3. The length and diameter of the transmission lines are, respectively, 6 m and 2.3 m. The generator is located in a bunker-type premise of area 11×16 m^2. Using the MIG generator, various experiments have been performed to study intense electron beams [2–6], imploding Z-pinch plasmas [7–9], metal surfaces explosions in megagauss magnetic fields [10–12], and phase transformations of matter under high pulsed pressures [13]. When the generator operates into a load of inductance ~20 nH, the current amplitude reaches 2.5 MA within a rise time $\tau_{0.1-0.9} \approx 70$ ns.

The water-vacuum insulator stack is a key part of the MIG generator. The insulator provides the delivery of an electromagnetic pulse from the water transmission line to the load placed in a vacuum chamber. The insulator stack consists of five polyethylene rings of outer diameter 89.6 cm separated by stainless steel grading rings. A standard (+45°) configuration is used for the vacuum interface. The thickness of each polyethylene ring is 2.5 cm.

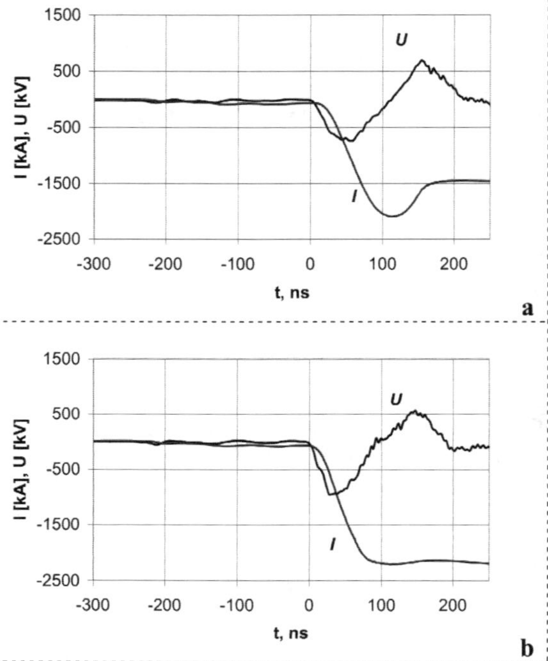

Figure 1. Load current and insulator voltage waveforms for shots without (a) and with (b) insulator flashover.

Figure 2. Top view of the insulator rings. The cathode is on the left.

The operation of the insulator stack is complicated at least by two adverse factors, which are possibly interrelated. The first one is illustrated by Fig. 1, which presents load current and insulator voltage traces recorded at different peak voltages. It can be seen that for the lower peak voltage (750 kV), the current trace has the typical (1 – cos) waveform until the time the reverse voltage peaks (see Fig. 1a). This time corresponds to the insulator flashover, as evidenced by abrupt flattening of the current waveform due to the crowbar effect. For the higher peak voltage (950 kV), the flashover time is closer to the time at which the current peaks. This indicates that the peak load current decreases due to the insulator flashover (Fig. 1b). The second adverse factor shows up as well-defined discharge traces on the water-facing surfaces of the polyethylene rings (Fig. 2). As can be seen in Fig. 2 (the insulator stack centerline is horizontal), only the two polyethylene rings nearest to the anode were damaged. Besides, these traces appeared only in the upper part of the insulator stack.

To estimate the flashover electric field, well-known relations can be used. The peak electric field at which the water-facing surface of the insulator is expected to flash, E_w, can be estimated by the formula [14]

$$E_w = K \cdot t_{0.63}^{-0.25} \cdot d^{-1/3} \text{ [MV/cm]}. \quad (1)$$

Here $K = 0.2$ for polyethylene, d is the insulator thickness [cm], and $t_{0.63}$ is the effective voltage pulse width [μs] defined as the full width of the applied voltage pulse at 63% of the peak voltage. Strictly speaking, this formula is valid only for a flashover in a uniform field. Therefore, to estimate the electric strength of an actual insulator, it is reasonable to compare E_w with an average electric field $E_{av} = U/d$, where U is the peak voltage across the insulator. Setting $U = 1$ MV, $t_{0.63} = 0.03$ μs, and $d = 12.5$ cm characteristic of the MIG insulator stack, we obtain $E_w \approx 210$ kV/cm, which is about 2.5 times the average electric field of 80 kV/cm typical of the MIG machine.

The vacuum flashover electric field can be estimated by the formula proposed by J. C. Martin (see, e.g., Ref. 15):

$$E_v = 175 \cdot t_{0.89}^{-1/6} \cdot A^{-1/10} \text{ [kV/cm]}, \quad (2)$$

where A is the insulator area [cm^2], $t_{0.89}$ is the effective voltage pulse width [μs] defined as the full width of the applied voltage pulse at 89% of the peak voltage. For our case, we have $A = 4959$ cm^2 and $t_{0.89} = 0.013$ μs, and formula (2) yields $E_v \approx 155$ kV/cm.

Thus, for both cases, the estimated flashover electric field is considerably greater than the average electric field, suggesting low flashover probability for the MIG insulator stack. Nevertheless, the existing insulator has often been observed to flash. The work presented in this paper was aimed at finding the reasons for the insulator flashovers and seeking ways to mitigate them.

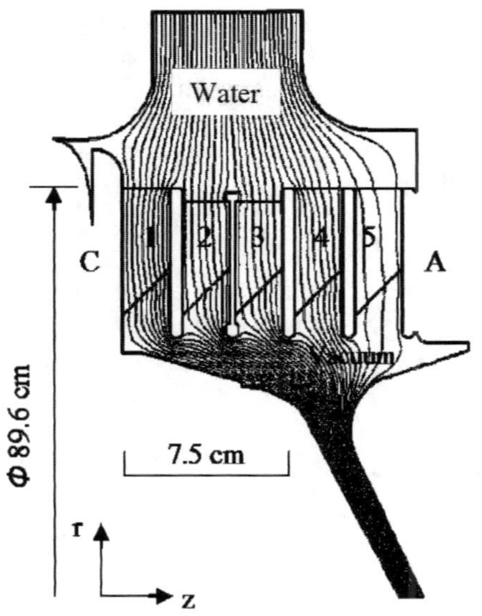

Figure 3. Calculated equipotential curves for the existing insulator. The insulator rings are marked with numbers. Cathode (C) is on the left. The z-axis in the figure indicates the horizontal axis of the MIG machine.

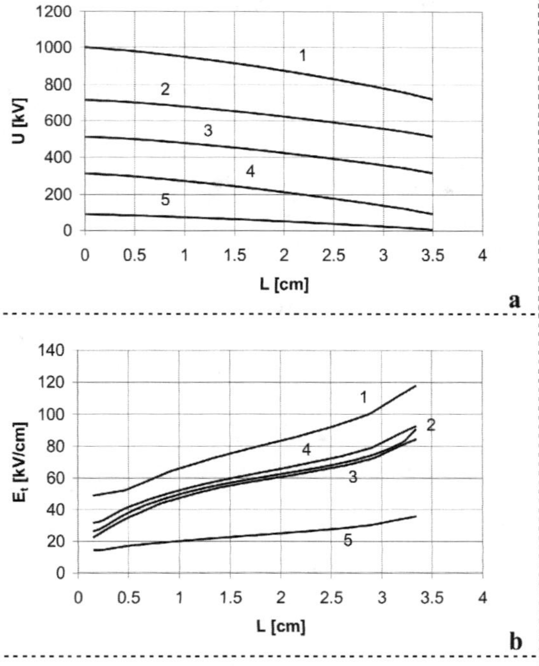

Figure 4. Calculated potential distribution along the vacuum surface of each ring (a) and distribution of the tangential electric field E_t at the vacuum surface of each ring (b) for the existing design of the insulator.

2. Numerical simulation of the electric field distribution at the insulator surface

We supposed that there are at least three possible reasons for the flashover: nonuniform voltage grading among the insulator rings, significant enhancement of the electric field at the anode triple junctions, and contaminants present on the insulator surfaces.

To check these suppositions, a two-dimensional electrostatic model was developed and corresponding numerical simulations were performed using the Elcut 5.1 software.

Figure 3 gives the equipotential curves calculated for the existing design of the insulator. The polyethylene rings are marked with numbers. The cathode (C) is situated on the left; it is at a potential of 1 MV, whereas the anode is grounded. The potential distribution along the vacuum surface of each ring is shown in Fig. 4a. It can be seen that ring 1 is at the highest voltage. The average electric field $E_{av} = U/d$ reaches 110 kV/cm. In contrast, the field E_{av} at ring 5 is not over 40 kV/cm. The calculated distribution of the tangential electric field E_t at the vacuum surface of each ring is shown in Fig. 4b. The cathode triple junction corresponds to $L = 0$, where L is the distance along the vacuum surface. The E_t distribution is evidently nonuniform. It is well known that the analytical solution of the Laplace equation for a 45° configuration gives an electric field infinitely increasing on approaching an anode triple junction. For our case, the electric field strength in the regions located about 2 mm away from the triple junctions should be considered not well defined because of the limitation on the computational spatial grid. The respective points were not included in the plots. It can be seen that E_t increases almost linearly, showing some enhancement near the anode triple junction. The maximum tangential electric field (120 kV/cm) near the anode triple junction has been predicted for ring 1. This is at least 1.5 times higher than the average electric field 80 kV/cm.

3. Numerical simulation of the electric field distribution for successive flashovers of the rings

Preliminary estimates have shown that the electric fields in the insulator are evidently insufficient for a total flashover to occur, but the flashover of ring 1 seems to be most probable. We have performed a numerical simulation assuming successive flashovers of the insulator rings. At the first step, ring 1 is considered to flash. This implies that the whole of the initial voltage will be applied to the other rings. At the next step, rings 1 and 2 are considered to flash. (The reason for the flashover of ring 2 is that when ring 1 flashes, the highest voltage appears to be applied to ring 2.) Thus, the whole of the voltage is switched to rings 3–5. So, the highest-voltage ring is considered to flash at each step. The results of this simulation are presented in Table 1.

Table 1. Simulation results for successive flashovers of the rings. Column 1 shows sequential numbers of flashing rings. Column 2 presents the voltage across each ring normalized to the voltage across ring 5. Columns 3 and 4 present maximum values of E_{vac} and E_{wp} for the whole of the insulator. Column 5 shows the ring's highest average electric field, estimated for each ring as $E_{av} = U/d$, where U is the voltage across the ring and d is its ring thickness.

Number of flashing rings	Normalized voltage grading among the rings	Maximum tangential electric field E_{vac} at the vacuum surface, kV/cm	Maximum tangential electric field E_{wp} at the water-facing surface, kV/cm	Maximum average electric field E_{av}, kV/cm
0	3/2.2/2.1/2.3/1	120	180	108
1	3.6/2.4/2.4/1	180	210	148
1–2	3.5/2.5/1	220	320	194
1–2–3	3.3/1	330	750	304
1–2–3–4	–	420	1000	400

It can be seen that successive flashovers of the rings make the voltage grading among the rings more and more nonuniform. As a result, even when only the first ring flashes, the maximum average electric field E_{av} increases by a factor of 1.4, reaching 148 kV/cm. Estimation of the flashover strength of the insulator stack for this situation seems to be complicated as the effective voltage pulse width is not well-defined. Nevertheless, it should be noted that the predicted average electric field is quite close to the vacuum surface flashover strength $E_v \approx 155$ kV/cm for the insulator stack. Therefore, it should be expected that the flashover of one ring will significantly increase the flashover probability for the whole of the insulator stack.

The data given in Table 1 also show that the maximum tangential electric field at the water-facing surface, E_{wp}, increases more sharply than the tangential electric field at the vacuum surface, E_{vac}. For rings 1–3 supposed to flash, the field E_{wp} is more than twice the field E_{vac}. This result can possibly explain the observed flashover traces along the water-facing surfaces of rings 4 and 5 (see Fig. 2).

4. Improved design of the water–vacuum insulator of the MIG generator

Comprehensive numerical simulations have been performed to find out the optimal design of the MIG water–vacuum insulator. In the course of the numerical optimization, the only requirement was set to be fulfilled. Namely, the modification of the design should not concern the water line and the vacuum chamber containing a load because of the high cost of these units. In the modified design, metal plugs were inserted in the polyethylene rings to reduce the electric field at anode triple junctions [16]. In addition, the insulator was shifted by several millimetres toward the cathode to provide a more uniform electric field distribution among the rings. A few minor changes have also been made. The results of the numerical modelling of the optimal insulator design are presented in Figs. 5 and 6.

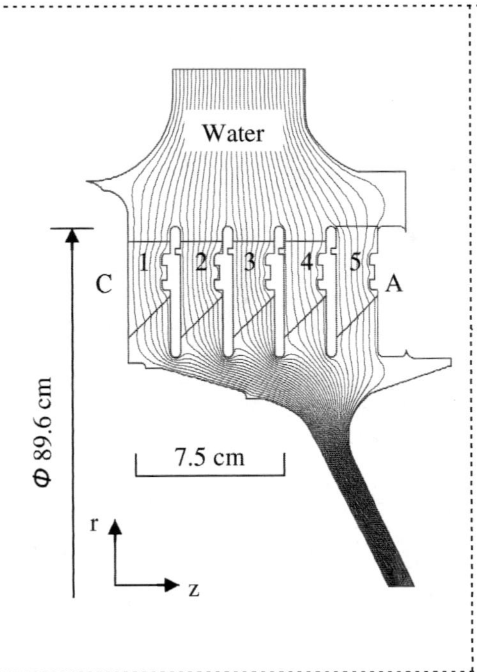

Figure 5. Calculated equipotential curves for the modified design of the insulator.

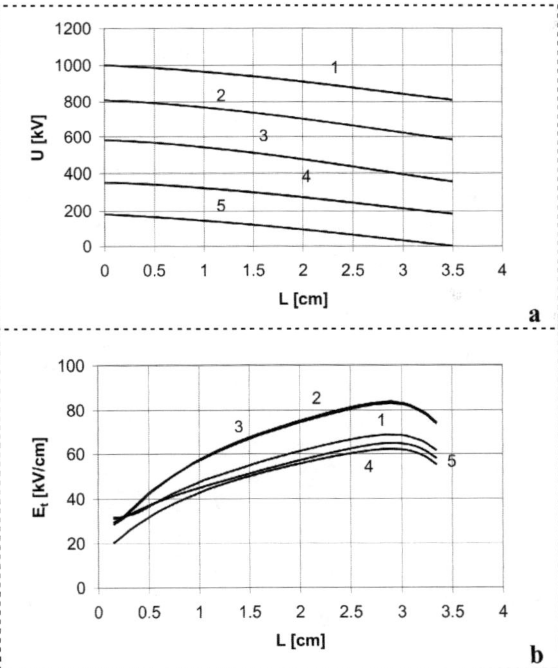

Figure 6. Calculated potential distribution along the vacuum surface of each ring (a) and distribution of the tangential electric field E_t at the vacuum surface of each ring (b) for the modified design of the insulator.

The improved design of the insulator provides a more uniform potential distribution along the insulator rings. The tangential electric field reaches the highest value (85 kV/cm) at rings 2 and 3, and it is only a little greater than the average electric field (80 kV/cm). A significant reduction in electric field is observed near the anode triple junctions. Therefore, the highest electric field for the modified design is no less than 1.4 times lower than the highest fields at the anode triple junctions calculated for the existing design. The proposed modification requires no rearrangement of either the water line or the load-containing vacuum chamber and should provide higher electric strength of the insulator stock.

5. Summary

A comprehensive numerical simulation of the electric field distribution has been performed for the water–vacuum insulator of the MIG pulsed power generator. The principal disadvantages of the existing design have been revealed. The simulations have confirmed that the nonunifrom voltage grading among the insulator rings and significant enhancement of the electric field at anode triple junctions could be responsible for the low electric strength of the insulator. A numerical modelling of successive flashovers of the insulator rings has revealed a more pronounced voltage grading among the rings. Also, it has been found that the tangential electric field at the water-facing surface increases sharper than the electric field at the vacuum surface. This seams to account for the observed flashover along the water–polyethylene interface of the two rings located closer to the anode. An improved design of the insulator stack has been developed. The proposed modification requires no rearrangement of either the water line or the load-containing vacuum chamber and should provide higher electric strength of the insulator stack.

Acknowlegments

The work was supported in part by Presidium of the Russian Academy of Sciences under a program of “Fundamental Problems of Pulsed Power” and by the Russian Foundation for Basic Research (grant No. 14-08-00524-a).

References

[1] Luchinskii A V, Ratakhin N A, Fedushchak V F and Shepelev A N 1997 *Russ. Phys. J.* **40** 1178.
[2] Petin V K, Shlyahtun S V, Oreshkin V I, Ratakhin N A 2008 *Zh. Tekh. Fiz.* **78** 103.
[3] Sorokin S A 2010 *IEEE Trans. Plasma Sci.* **38** 1723.
[4] Sorokin S A 2010 *Pis'ma Zh. Tekh. Fiz.* **36** 68.
[5] Sorokin S A 2012 *Radiography: Methods, Theory and Applications* (New York: Nova Science Publishers) p. 179.
[6] Kachalkov A A, Lavrinovich A V, Petin V K, Ratakhin N A, Shlyahtun S V, Vasil'ev K Yu, Kolesnikov S V, Serebriannikov M E 2012 *Vopr. At. Nauki Tekh.* **1** 107.
[7] Sorokin S A and Chaĭkovsky S A 2000 *Russ. Tech. Phys.* **45** 458.
[8] Sorokin S A 2009 *Zh. Tekh. Fiz.* **79** 43.
[9] Sorokin S A 2013 *Fiz. Plazmy* **39** 162.
[10] Chaikovsky S A, Oreshkin V I, Mesyats G A, Ratakhin N A, Datsko I M and Kablambaev B A 2009 *Phys. Plasmas* **16** 042701.
[11] Labetskaya N A, Oreshkin V I, Chaikovsky S A, Datsko I M, Sukovatitsyn Y A and Volkov E N 2012 *Proc. XXVth Int. Symp. on Discharges and Electrical Insulation in Vacuum, Tomsk, Russia* **2** 634.
[12] Chaikovsky S A, Oreshkin V I, Datsko I M, Labetskaya N A and Ratakhin N A 2014 *Phys. Plasmas* **21** 042706.
[13] Oreshkin V I, Chaikovsky S A, Labetskaya N A, Ivanov V F, Khishenko K V, Levashov P R, Kuskova N I, Rud' A D 2011 *Zh. Tekh. Fiz.* **82** 41.
[14] Ushakov V Ya 1985 *Elektricheskaya prochnost' zhidkostei i rabochaya napriazhennost' promezhutkov vysokovoltnykh impul'snyh ustroistv* (Novosibirsk, Nauka) p 114.
[15] Stygar W A *et al* 2004 *Phys. Rev. ST Accel. Beams* **7** 070401.
[16] Stygar W A *et al* 2005 *Phys. Rev. ST Accel. Beams* **8** 050401.

Journal of Physics: Conference Series **552** (2014) 012017 doi:10.1088/1742-6596/552/1/012017

Features of pulsed electron beam propagation with nanosecond duration in the air under the low (reduced) pressure

G Kholodnaya*, R Sazonov, D Ponomarev, G Remnev, A Vikanov
National Research Tomsk Polytechnic University, Lenin Avenue, 30, Tomsk, Russia

E-mail: galina_holodnaya@mail.ru

Abstract. This paper presents the research of pulsed electron beam propagation with nanosecond duration in the air under the low pressure (3-8 Torr) with the help of a calorimeter with a vacuum cutoff and a Faraday cup. Investigations have been conducted with the laboratory setup, including the TEU-500 electron accelerator and a drift tube. The major parameters of the accelerator are as follows: 450-500 keV electron energy; 9-12 кA ejected electron current; 60 ns half-amplitude pulse duration; 5 pps pulse repetition rate; and 250 J pulse energy.

1. Introduction

At present, the majority of experimental results in electron beam propagation are obtained in axially symmetric drift spaces with the use of metal and dielectric drift tubes with a diameter comparable to the cross-sectional dimension of the electron beam [1-3]. Numerous experiments connected with electron beams transfer in various gases (in the mode of self-focusing at the distance L numbered of some betatron lengths) show that the most effective electron beam propagation occurs in the area of low pressures p ~ 1 Torr. If area's pressure is lower than 1 Torr, effective pulsed electron beam propagation is hindered by the processes, connected with the specificity of current (magnetic) compensation, volume charge neutralization, and beam-plasma instabilities, changing the conditions of pulsed electron beam propagation (resistive firehose instability, lateral instability, etc.) [5-7].

Also, some theoretical investigations are devoted to the problem of electron beam propagation in gas environments (a computational study of low energy electron beam transfer along the drift tube, depending on the external magnetic field, experiment geometry, etc.) [9-11]. The authors of the paper [12] have obtained the results of the calculations of the magnetic field configuration and its size in the areas of forming, accelerating and transferring electrons in the electron-optical system of the electron plasma source. It is shown that the creation of a quasihomogeneous magnetic field in the electron-optical system of the electron plasma source allows improving beam focusing. In the given work the magnetic field source is a permanent magnet of a discharge chamber. This magnet is necessary for discharge initiation and combustion.

The paper [13] presents the results of a numerical simulation of non-steady processes in various schemes of charge neutralization for high-current electron beams (the calculations are obtained for hydrogen, nitrogen, and xenon plasma for various values of the external magnetic field and for its absence). The obtained results are in good agreement with the experimental data found for plasma-

Content from this work may be used under the terms of the Creative Commons Attribution 3.0 licence. Any further distribution of this work must maintain attribution to the author(s) and the title of the work, journal citation and DOI.
Published under licence by IOP Publishing Ltd

filled diodes (current increase of transferred pulsed electron beams occurs at the partial neutralization of their volume charge).

In the work [14] the authors have conducted theoretical research connected with the influence of current neutralization and reverse current distributor geometry on the transformation of the low energy high-current beam in the plasma channel. Experimental conditions (the geometry and material of the reverse current distributor) are close to those that will be used in the given paper, therefore, it is necessary to note the results obtained by the authors, although the injection of the pulsed electron beam in the work [14] has been carried out in plasma. The parameters of the system and beam used by the authors in their calculations: the energy of electrons is 27 keV; a beam radius in the transfer channel entrance is r_b=4.25 cm; beam current is I_b=20 kA; a drift tube radius is R=10.3 cm; a tube length from the cathode surface is 18.5 cm; the time of beam current front is 300 ns. In the calculations as a reverse current distributor was considered the following: 1) a symmetric current distributor (a drift tube); 2) two flat parallel bus-bars with the length L=4 cm and the width h=3..9 cm (current in each bus-bar I=10 kA); 3) four parallel studs with the length L=4 cm and the diameter d=0,5 cm (current in each stud 5 kA). As a result, the authors have found that the maximum change in the form of the beam circular cross-section can be observed in the systems with the current distributors in the form of narrow plates or studs, and on the target the beam cross-section is close to rectangular or square, respectively. Moreover, the deformation of the beam cross-section during beam transfer to the target is accompanied by the change of electron density distribution. The beam, homogeneous in the drift tube entrance, at its cross-section transformation becomes heterogeneous in density with its maximum in the centre. At the same time, heterogeneous initial electron distribution becomes flatter on the target and even closer to homogeneous in the central part of the beam at high current neutralization because of the change in the cross-section configuration [11].

Despite the number of investigation results in the sphere of transferring and focusing of electron beams in various environments, which exist nowadays in Russia as well as abroad, the area of research connected with the physics of interaction of high power pulsed beams of charged particles with a gas environment and condensed matter is still open. Especially, it is applied to the gas environment of high pressure, what can be potentially used in such technologies as: refining of dump uranium hexafluoride, associated gas processing, catalyst and photocatalyst synthesis, oil cracking, functional and structural composite nanomaterials obtaining, radiation biology, etc.

The authors have earlier conducted the study of the pulsed electron beam energy dissipation at injecting into electronegative gases of high pressure. This part of work is devoted to the problems of pulsed electron beam energy injection into gases and gas compositions, consisting of argon, hydrogen, oxygen, silicon tetrachloride [8].

The aim of the present paper is to research the factors of pulsed electron beam propagation with nanosecond duration with varied energy density in the air under the low pressure with the help of a calorimeter with a vacuum cutoff and a Faraday cup.

2. Experimental

The investigation of pulsed electron beam propagation in the air under the low pressure is conducted with the help of the TEU-500 accelerator [15]. A characteristic feature of the given accelerator construction is a matching autotransformer that provides matching of a low-resistance water double forming line with a high-resistance impedance of an explosive emission planar diode.

An electron beam in the given accelerator is formed due to autoelectron emission from the surface of a graphite cathode with the diameter of 45 mm. Electron beam ejection is carried out through the anode area, presented by a support grid and aluminum foil with the thickness of 140 micrometers. In the present work the authors used 2 types of anode grids: the 1-st type – a riffled grid with optical transparency of 70 %; the 2-nd type – a honeycomb type of the anode grid with optical transparency of 95 %. The anode –cathode gap for all the experiments is 13 mm.

The operation stability of the accelerator was monitored with the help of a Rogowski coil and a capacitive potential divider (parameter spread of current and voltage, registered by the sensors, did not exceed 5 %). To register current pulses a Tektronix TDS 2022V oscillograph was used. The current carried to a diode assembly was ~10-12 kA, accelerating voltage of electrons – 450-500 kV (Figure 1. oscillograph records of current and voltage), pulse duration at a half-height – 60 ns.

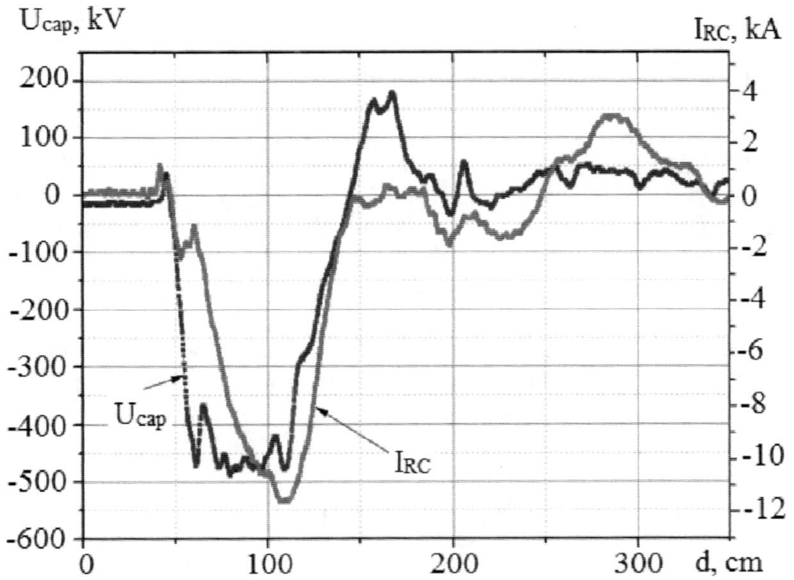

Figure 1. Average values of voltage (U_{cap}) and beam current (I_{RC}) carried to a diode assembly.

The investigation of pulsed electron beam propagation in the air under the low pressure was conducted with the help of the Faraday cup and the calorimeter with a vacuum cutoff. This vacuum cutoff of measuring equipment included the following: the collector of the Faraday cup (or calorimeter) was closed by the support grid (optical transparency - 70 %) and the aluminum foil of 140 micrometers. The Faraday cup (or calorimeter) was installed in the drift tube at a varied distance from the outlet hole of the diode chamber of the pulsed electron accelerator. The drift tube looked like a quartz tube with the diameter of 14 cm, the length of 38 cm and the walls thickness of 4 mm, which was pinned between two metal flanges with the help of two metal studs. Metal studs in the given geometry of the experiment played the role of the reverse current distributor. The pressure in the reactor was of 3-8 Torr for all the experiments, presented in this paper. Figure 2 demonstrates the scheme of the experiment.

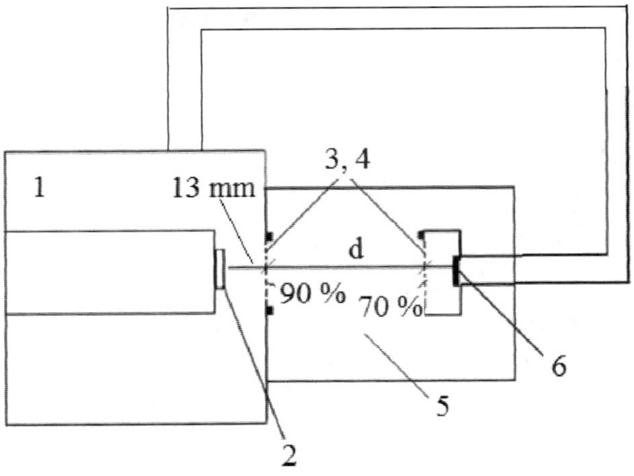

Figure 2. Scheme of the experiment: 1- diode chamber of the TEU - 500 accelerator; 2 – cathode; 3 – anode grid (3 mm) and aluminum foil (140 μm); 5 – drift tube; 6 – collector of the Faraday cup.

The experiment was conducted in the following way: the Faraday cup was installed at a long distance from the diode outlet hole of the accelerator (d=35 cm, further the distance was shortened at a pitch of 5 cm), drift tube volume was pumped down by a roughing pump, the electron beam was injected and entered the drift tube, and then, it reached the collector of the Faraday cup. As far as the pulsed electron accelerator works invariably, 5 redoubling each other experiments were conducted at each distance for statistics. In the series of experiments, when the calorimeter was used as a measuring device and was installed like the Faraday cup at the distances of d= 35, 30, 25, 20, 15, 10, 4,5 cm the output energy of the electron beam was measured. According to two parameters of the beam (energy and charge) the conclusion was made about the behaviour of the pulsed electron beam with nanosecond duration in the air under the low pressure.

3. Results and discussions

In the course of conducting the research the following results are obtained and presented in the given part of the report. Figure 3 shows the experimental investigations of the dependence of a total beam charge, having reached the collector of the Faraday cup, on the distance in the air under the low pressure.

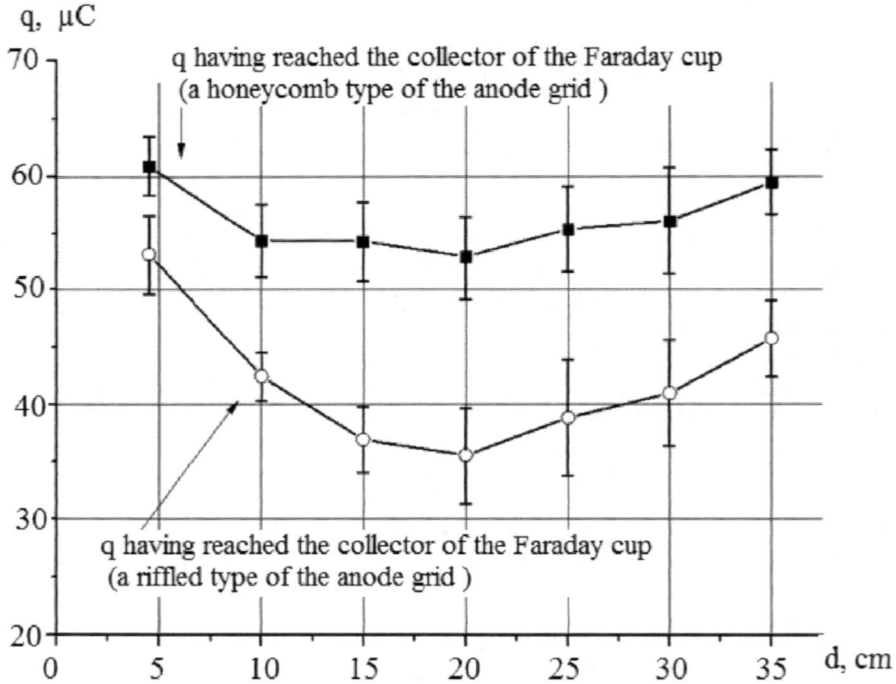

Figure 3. Dependence of a total beam charge, having reached the collector of the Faraday cup, on the distance in the air under the low pressure; a 5 pulses averaged model.

Figure 3 demonstrates that the dependence of the total beam charge, having reached the collector of the Faraday cup, on the distance in the air under the low pressure has a complicated character. When the collector of the Faraday cup is positioned at the distance from 5 to 20 cm from the outlet hole, the charge of the electron beam decreases. It should be noted that along the whole length of the reaction chamber the value of the beam charge, having reached the Faraday cup, in the case of using the anode grid with optical transparency of 95 % is higher than the similar value measured with the use of the anode grid with optical transparency of 70 %. This fact confirms bigger number of electrons injected in the drift tube. Having divided the charge value registered in the series of experiments with a honeycomb grid by the charge value registered in the second series of the experiments (a riffled grid) K=1.3895. It is compatible with the ratio of transparencies of these grids: 95/70=1.357 with an accuracy of 2.4 %. According to abovementioned and taking into the consideration the fact of a similar character between two beam charge dependencies, having reached the collector of the Faraday cup, on the distance further calorimetric measurements were conducted with the use of the anode grid with optical transparency of 95 %. Experimental set up was similar to the scheme presented in Figure 2. The energy of the electron beam, having reached the calorimeter's collector, was registered. In each series of experiments we calculated the value of the energy supplied to the diode from the data obtained with the help of the capacitive potential divider (accelerating voltage of electrons) and the Rogowski coil (I- accelerator total current). Data spread of energy delivered to the diode did not exceed 5 %. The graph of beam energy dependence on the distance in the air under the low pressure is presented in Figure 4.

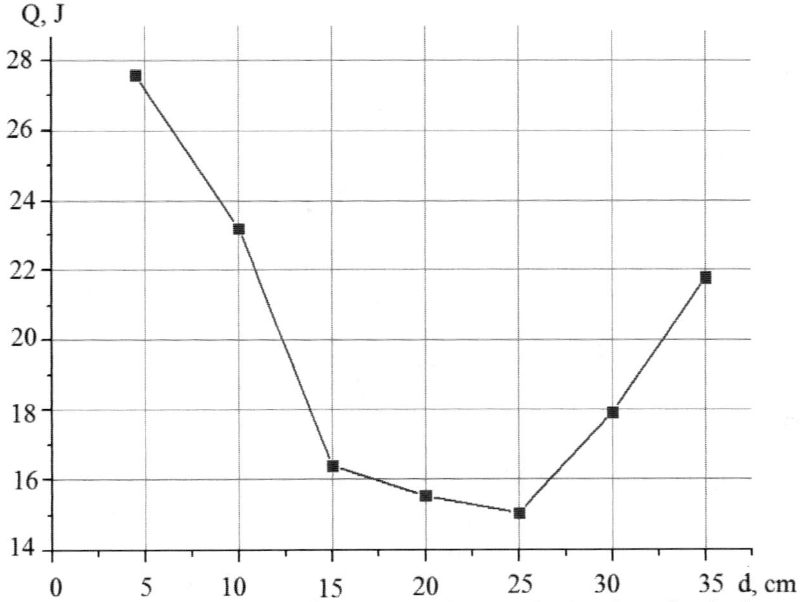

Figure 4. Dependence of ejected beam energy on the distance in the air under the low pressure (a 5 pulses averaged model).

Figure 4 shows that the dependence of ejected beam energy on the distance in the air under the low pressure is described by the dependence the character of which is similar to the dependence of the beam total charge, having reached Faraday cup's collector, on the distance (Figure 4), but with a more significant minimum. It is possible to make an assumption that at the distance from 0 to 20 cm the process occurs where the transfer of the beam without charge neutralization can be observed, otherwise, the value of the ion background, appearing at electron beam injecting in gas, is small. At insufficient charge neutralization the electron beam is pushed apart under the influence of Coulomb force, and only a small part of electrons reaches the collector of the Faraday cup. When the collector is positioned at the distance from 20 to 35 cm from the outlet hole, the value of the electron beam charge exceeds the value recorded at the distance of 20 cm. It can be explained by the following fact: the value of the ion background integral increases due to the increase of the distance where the beam is transferred. As a consequence, charge neutralization of the beam increases, and its transferring characteristics are improved.

4. Conclusion

Thus, experimental investigations of the pulsed electron beam propagation with nanosecond duration in the air under the low pressure are given. It can be seen from the observed graphs that the dependence of the energy and total charge (having reached the collector of the Faraday cup) on the length of electron beam run in the air under the low pressure does not have an exponential form. There is a critical distance in the drift tube where maximum beam dissipation occurs. After crossing this dissipation the beam focuses, what is proved by the increased value of the beam energy in the end of the drift tube. This behaviour of the pulsed electron beam may be influenced by many factors. For example, this behaviour may be determined by the experiment geometry. Moreover, in the series of experiments the dielectric materials for the drift tube and metal studs, playing the role of the reverse current distributor, have been used. In this area of pressures the processes, connected with the specificity of the current (magnetic) compensation value, neutralization of beam volume charge as well as beam-plasma instabilities, also influence on the conditions of pulsed electron beam propagation. Further experimental research of the authors will be devoted to the exposure of all the main factors, preventing effective energy injection of the pulsed electron beam into the gases and gas compositions not only in the given area of pressures, but also under the high pressure.

References

[1] Rudakov L I, Smirnov V P and Spektor A M 1972 *JETP Letters* **15** 9 p 540–543.
[2] Khodataev K V 1973 *JETP Letters* 18 3 p 184–186.
[3] Ryabchikov A I, Stepanov I B and Eremin S E 2010 *The Bulletin of Tomsk Polytechnic University* **316** 4 p 90–93.
[4] Jordan S, Ben-Amar Baranga A, Benford G and et al 1985 *J.Phys. Fluids* **28** 2 p 366–382.
[5] Glayshev M V and Nikulin M G 1997 *J. Tech. Phys* **67** p 94–98.
[6] Wang L, Ning W, Fu M, Wu C and Jia S 2013 *J. Plas. Phys.* **79** 5 p 683–689.
[7] Ivanov S N, Lisenkov V V and Shpak V G 2010 *J. Phys. D: Applied Physics* **43** 31 p 315–204.
[8] Sazonov R V, Novoselov Yu N, Kholodnaya G E, Kochkorov B S, Ponomarev D V, Remnev G E, Yalovets A P, Isakova Yu I 2010 Features of dissipation of a powerful pulsed electron beam in high pressure gases *15th Asian Conference on Electrical Discharge (Xian, China 7-10 November 2010).*
[9] Koval T V and Le X Z 2013 *Russian Physics Journal* **56** 1/2 p 103–108.
[10] Ozur G E, Grigoryev V P, Karlik K V, Koval T V and Le X Z 2011 *J. Tech. Phys* **81** 9 p 100–105.
[11] Glazychev L V, Norman G E and Sorokin G A 1989 *JETP Letters* 15 1 p 3–7.
[12] Kornilov S Yu and Rempe N.G 2012 *J. Tech. Phys* **82** 2 p 79–84.
[13] Agafonov A V 2010 *J. Partc. and Nuclei, Letters* **7** 163 p 793–797.
[14] Koval T V and Le X Z 2012 Bulletin of the Tomsk Polytechnic University **320** 2 p 43–47.
[15] Remnev G E, Furman E G, Pushkarev A I, Karpuzov S B, Kondrat'ev N A and Goncharov D V 2004 *J. Instr. Exp, Tech* **47** 3 p 394–398.

Journal of Physics: Conference Series **552** (2014) 012018 doi:10.1088/1742-6596/552/1/012018

Annular electron beam with virtual cathode in a coaxial diode with magnetic insulation

A A Grishkov[1], **I K Kurkan**[1, 2, *] **and I V Pegel**[1,2]

[1] Institute of High Current Electronics, 2/3 Akademichesky Ave., Tomsk, 634055, Russia
[2] National Research Tomsk Polytechnic University, 30 Lenin Ave., 634050 Tomsk, Russia

E-mail: ikk@lfe.hcei.tsc.ru

Abstract. Accumulation of electrons in course of transportation of tubular high-current electron beam in a coaxial diode with magnetic insulation was studied in the situation of counterstream electron motion occurring due to the formation of a virtual cathode. Accumulation of electrons in these flows is accompanied by gradual increase in their relative energy spread and simultaneous decrease in the maximum kinetic energy. The process is described analytically and confirmed in numerical modeling. The theory is an extension of the known theory of coaxial magnetically insulated diode (based on the conservation of axial component of the generalized momentum) taking into account the spread of electron energies. The numerical modeling employed the KARAT electromagnetic PiC-code. Experiments were performed using high-current electron accelerator SINUS-7 (pulsewidth 50 ns) in the range of diode voltages 350—800 kV, magnetic fields 8—24 kOe, electron beam currents 2—12 kA. A gradual increase of potential of the electron beam between the explosive-emission cathode and the virtual cathode was observed, which accompanies accumulation of circulating electrons as predicted in theory.

1. Introduction

Coaxial magnetically insulated diodes (CMIDs) with edge-type explosive-emission cathodes are used for production of tubular high current electron beams. Such diodes, immersed in strong magnetic field, produce electron beams with high enough density, stability and uniformity; they are applied in high power microwave tubes, electron beam ion sources/traps and other devices of electrophysics. Since the electron beam in a CMID is thin-walled, it possesses very low spread of particle kinetic energy.

The theory of coaxial magnetically insulated diode is well-studied and proven in numerous computer simulations and experiments. To date, the following problems have been solved: the limiting current of electron beam in a cylindrical channel [1], eventual states of electron beam [2], electron beam current in a CMID [3, 4]. Some approaches to description of electron beam states with longitudinal nonuniformities in such systems have been developed [5, 6]. Numerical modeling based on Particle-in-Cell (PiC) method enabled studying non-stationary states of electron beams with a virtual cathode (VC) [7]. A state of electron beam where the relativistic factor of electrons is lower than that for the limiting current (the compressed state) was demonstrated in axisymmetric and 3D simulations. At that time, this state was predicted in theory, but it was not experimentally investigated [8—11]. Success of numerical modeling stimulated certain improvements of analytical

Content from this work may be used under the terms of the Creative Commons Attribution 3.0 licence. Any further distribution of this work must maintain attribution to the author(s) and the title of the work, journal citation and DOI.
Published under licence by IOP Publishing Ltd

models of beams with a virtual cathode [15-17] and their experimental verification [18, 19]. It was the time-dependent PiC-modeling that allowed demonstration of situations where a spread of kinetic energy develops and grows in an initially monoenergetic electron beam. For the first time this was apparently observed in [7] and later similar effects were reported in [8, 17, 18]. The emergence of the energy spread, significantly affecting the observed phenomena, and accumulation of additional charge have been observed experimentally in the study of motion of the VC in sectioned channels [15].

The purpose of the present study was to demonstrate experimentally the accumulation of charge in an electron beam with a virtual cathode in a coaxial diode with magnetic insulation, and compare the parameters of the observed effect with the theory and numerical modeling. The modeling was made in 2D and 3D using KARAT electromagnetic PiC code [19]. The experimental setup was based on SINUS-7 high-current electron accelerator [20].

2. Theory and numerical modeling

In this section, let us consider a situation leading to development of kinetic energy spread in an initially monoenergetic hollow electron beam produced in a CMID comprising a virtual cathode.

Let the diode is extended with a channel having a step-wise widening sufficient to formation of VC, causing reflection of certain fraction of electrons (figure 1). Such a beam transport system is typical to many pulsed power applications. It is known that variation of current incident on VC causes corresponding changes in the current reflected from VC, while the transit current remains nearly equal to the limiting current for the wide section [2]. The wider this section the lower the transit current behind the VC. This is the way how a VC "stabilizes" the transit current in a CMID. Another essential issue is oscillations in the reflected and transit currents. These oscillations cause certain spread in the kinetic energy of electrons in both the reflected and transit electron flows. The amplitude of the oscillation is maximum if the beam incident on the VC is monoenergetic, and it reduces greatly when the energy spread exceeds a definite level. In the latter case the phase trajectories of electrons form a wide strip and the VC works as separator passing electrons with higher kinetic energies and reflecting electrons of lower energies with no considerable oscillations in the reflected and transit currents.

Let us consider evolution of the system assuming that the voltage applied to the diode is constant. At zero moment of time, electron emission from the cathode edge starts with the current limited by electron space charge. The emitted particles accelerated to the energy $\gamma = \gamma_F$ (which corresponds to Fedosov's current [3]) fill the transport channel up to the wider section where the VC is formed. The beam reflected from the VC gains some energy spread and runs back to the cathode edge where its most part is reflected forward. At the cathode, its potential barrier works similar to the VC separating the electrons by their reflection point according to their kinetic energy. The particles with the energy higher than the cathode potential hit the cathode and are absorbed here. Electrons reflected from the cathode (accompanied by newly emitted ones) reach the VC where they are separated into reflected and transit ones, and so on. The multiply repeated process results in accumulation of electrons between the cathode and the VC while the potential of the electron flow in this section gradually increases in time and the relative energy spread rises. Figure 2 illustrates the evolution of the system phase plots obtained from KARAT modeling. In the simulation, the cathode radius was 1.6 cm, the narrow section radius was 2 cm and the wide section radius was 4 cm. The lengths of the sections were 15 cm and 25 cm, respectively. The axial magnetic fields strength was 10 T. The diode was fed with a TEM wave adjustable amplitude so that the diode voltage was held near 500 kV. In this simulation, it took about 200 ns for the kinetic energy of the beam to reduce to about 50 kV.

Journal of Physics: Conference Series **552** (2014) 012018 doi:10.1088/1742-6596/552/1/012018

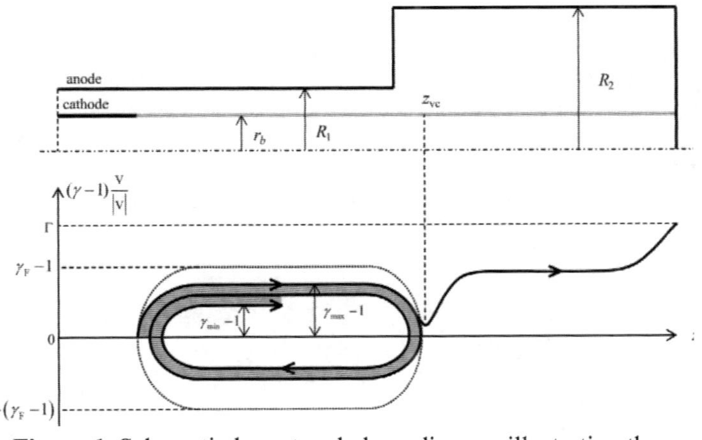

Figure 1. Schematic layout and phase diagram illustrating the analytical model.

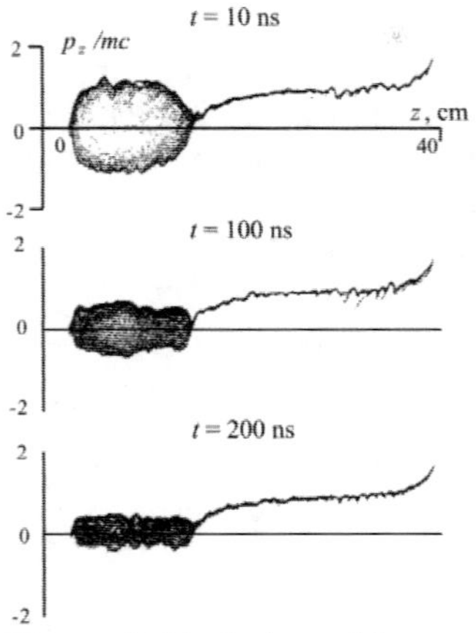

Figure 2. Evolution of the system phase plots (axisymmetric modeling using KARAT code).

Another series of numerical modeling was carried out to determine a feasible minimum for kinetic energy of electrons in a CMID with defined voltage level. It was demonstrated in a simulation involving a “counter-cathode” for particle reflection that the maximum energy of electrons could drop to as low as 1 keV while the diode voltage was sustained at about 500 kV [21]. The time for transition to this state was over 150 ns for the transport channel length 20 cm.

The theory considers a stationary tubular electron beam of radius r_b and infinitely thin wall guided by strong axial magnetic field in a uniform channel of radius R. The channel wall is at ground potential. The beam electrons can move in both positive and negative directions along z axis. The relativistic factor γ of electrons in the flow is assumed to be uniformly distributed from γ_{min} to γ_{max}. Using conservation of energy and conservation of the generalized (particle and field) momentum the following expression for the electron current was obtained:

$$I=\frac{I_0}{2\ln R/r_b}\cdot\frac{\Gamma-\gamma_{max}}{\gamma_{max}-\gamma_{min}}\left[\sqrt{\gamma_{max}^2-1}-\sqrt{\gamma_{min}^2-1}-\mathrm{arctg}\sqrt{\gamma_{max}^2-1}+\mathrm{arctg}\sqrt{\gamma_{min}^2-1}\right],$$

$$(\Gamma-1)^2=(\Gamma-\gamma_{max})^2+2\cdot(\Gamma-\gamma_{max})\left\{\frac{\gamma_{max}+\gamma_{min}}{2}-\frac{\ln\gamma_{max}/\gamma_{min}}{\gamma_{max}-\gamma_{min}}\right\},$$

here Γ is the relativistic factor corresponding to the diode voltage, and $I_0 = mc^3/e \approx 17$ kA.

The equations determine the feasible range for the value of energy spread $\Delta\gamma/(\Gamma-1)$ in the transported beam. This range is rather narrow: from zero to nearly 0.25 in the non-relativistic case, and it contracts as Γ rises. It is important that with any preset value of $\Delta\gamma/(\Gamma-1)$ within this range, the beam can exist in two states with different relative energy spreads and maximum kinetic energies. The exception is the “boundary” state with maximum value of $\Delta\gamma/(\Gamma-1)$, which is instable with the respect to small variation of this value. It means that near this “boundary” regime, any small longitudinal disturbances in the electron beam will result in coexistence of segments with slower and faster particle motion, as it was observed, for example, in [7, 17, 22].

Some more details of the above theory are given in [21]. Of course, the theory can describe a sequence of some steady states, but it is unable to describe any oscillations (e.g. produced by virtual cathode), which determine the real rate of the system evolution process.

3. Experiment

The experiment employed a high current electron accelerator SINUS-7 equipped with a coaxial magnetically insulated diode. The experimental arrangement is depicted in figure 3. The electron beam was transported in the area with nearly uniform magnetic field. The lengths of the narrow and wide channel sections were $L_1 = 150$ mm and $L_2 = 300$ mm, with the radii of $R_1 = 24$ mm and $R_2 = 41$ mm, respectively. The diode current was varied by adjusting the distance L_{AC} between the anode and the cathode from 80 mm to -20 mm. It is negative when the cathode penetrates the channel narrowing. The electrostatic potential of the electron beam in the narrow section was measured with a capacitive voltage divider situated at the middle of the section.

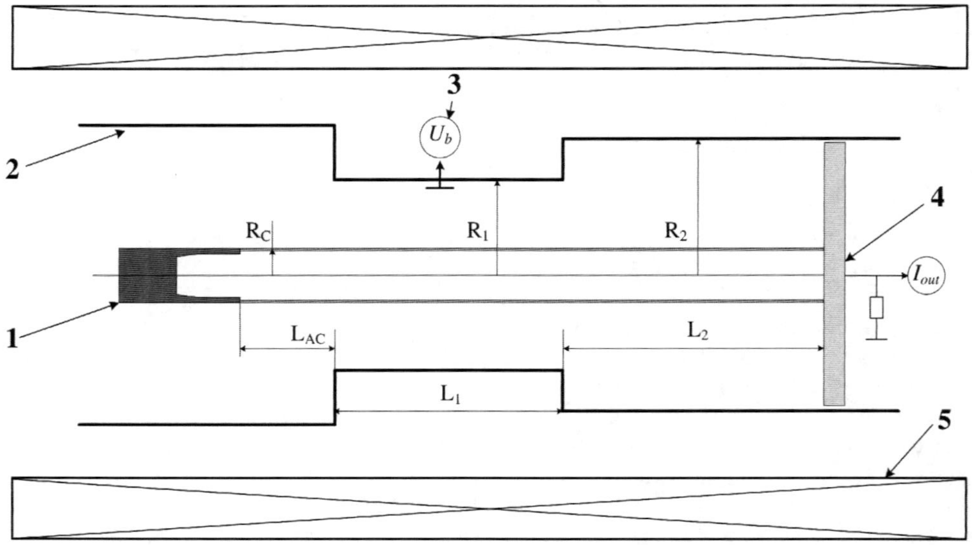

Figure 3. Experimental arrangement: 1 – cathode, 2 – anode, 3 – capacitive voltage divider, 4 – electron beam collector, 5 – magnetic coil.

The experimental parameters were varied over a wide range: cathode voltage 350—800 kV, magnetic field 8—24 kOe, and the diode current 2.0—12 kA. The voltage pulse duration was 50 ns (FWHM). The vacuum diode voltage and current were measured by means of capacitive voltage divider and low-inductance shunt, respectively. Both these sensors were situated in the oil-filled transmission line immediately near the vacuum diode. The diode current was also measured using a Rogowski coil mounted in the flange of the magnetic coil. The current of electrons at the collector was measured using a low-inductance, low-resistance shunt included in the collector circuit.

Electrons were emitted from an explosive-emission edge-type cylindrical graphite cathode with edge thickness 0.5 mm. Three cathodes with the radii of 20, 32, and 40 mm were tested. No assured indication to charge accumulation was obtained using a 20-mm cathode. For the 40-mm cathode, certain leakage of electron current to the walls of the transportation channel has been observed in lower magnetic fields.

Clear signs of charge accumulation with no radial current leakage were obtained with the use of the 32-mm cathode. The typical waveforms of diode voltage U_d, total diode current I_d, electron beam current I_b and electron beam potential (signal from capacitive divider, U_b) are in figure 4. Noticeable oscillations of the beam potential taking place near the pulse maximum are probably produced by the “strata” with lower and higher electron energies moving back and forth along the beam, as it was predicted by the theory and demonstrated in simulations [7, 17, 21, 22].

Knowing the waveform of the diode voltage U_d enables calculation of the beam potential, as though the beam is monoenergetic and there is no charge accumulation in the system. Some increase in the beam radius caused by expansion of the cathode plasma with typical rate $2{\cdot}10^6$ cm/s was taken into account. Comparing this calculated potential with the measured potential and with the simulated one allowed a conclusion whether there is a charge accumulation and how intense it is.

The waveforms of the potentials are presented in comparison in figure 5. The difference between the potential calculated under monoenergetic approximation and the measured potential is increasing with time making about 40% at the front of the pulse and exceeding 60% by the beginning of the fall.

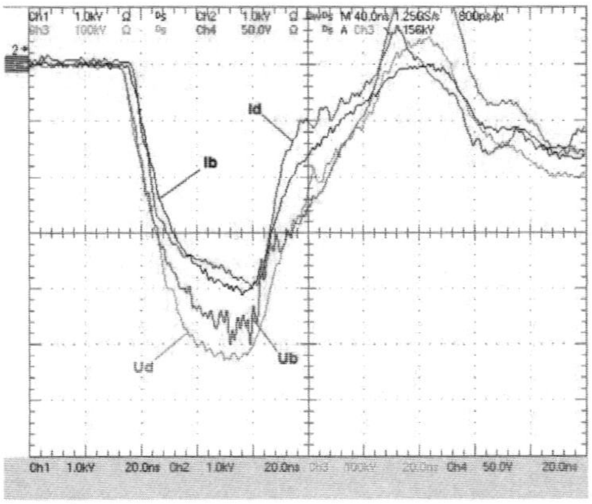

Figure 4. Waveforms of beam current I_b (Ch1), diode total current I_d (Ch2), diode voltage U_d (Ch3) and the beam potential U_b (Ch4).

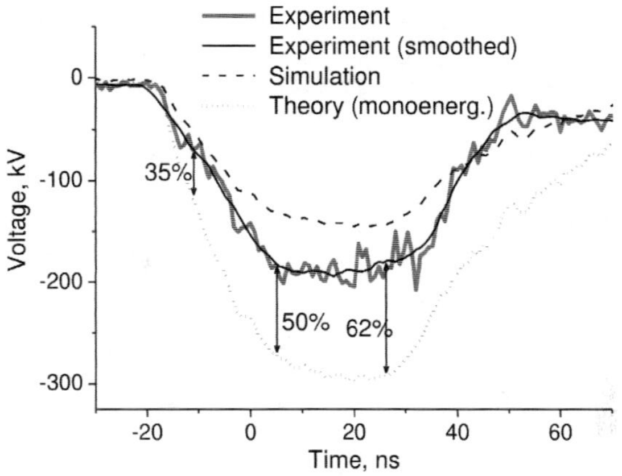

Figure 5. Waveforms of the beam potential with respect to the channel wall: experimental (original and smoothed), simulated with axisymmetric version of KARAT code, and calculated in the case of no energy spread of electrons

The increase in the beam potential within the pulse width is still notably less than the value predicted in theory for the case when the electron charge is completely accumulated. It is seen that the electron charge still continues accumulating at the end of the pulse. Note that in axisymmetric modeling the time for full accumulation was about 100 ns. Therefore, the width of the accelerator pulse is insufficient for observation of the complete process. Another factor that slows down the charge accumulation and the clearness of the final state of the beam is the three-dimensional instability of the beam leading to its filamentation. This effect is visual in simulation using 3D Cartesian version of KARAT code.

4. Conclusion

Thus, the process of electron charge accumulation in a coaxial magnetically insulated diode with virtual cathode was confirmed in experiment. The developing process is clearly visual even with relatively short pulse duration of the electron accelerator (50 ns). The result obtained is potentially important is such areas as generation of high-power microwaves in vircators, collective acceleration of ions by moving virtual cathodes, and operation of electron beam based ion traps.

Acknowledgments

The authors are grateful to Prof. Vladimir P. Tarakanov, the author of KARAT code.

The work was supported by RFBR grant 12-08-31447-mol-a.

References

[1] Bogdankevich L S and Rukhadze A A 1971 *Sov. Phys.Usp.* **14** 163

[2] Breizman B N and Ryutov D D 1974 *Nuclear Fusion* **14** 873

[3] Fedosov A I Litvinov E A Belomyttsev S Ya and Bugaev S P 1977 *Soviet Physics Journal* **10** 1367

[4] Belomyttsev S Ya Litvinov E A Mesyats G A and Fedosov A I 1981 *Sov. J. Plasma Phys.* **7**, 48

[5] Kovalev N F and Fuks M I 1980 *Preprint 13* (Gorki: Inst. Appl. Phys. USSR Acad. Sci.)

[6] Grishkov A A Ryzhov V V and Belomyttsev S Ya 2006 *Proceedings of the 14th Symposium on High-Current Electronics (Tomsk, Russia, 10-15 September 2006)* (*Izv. vuzov. Fizika,* vol 49 N11.Suppl.) ed B Kovalchuk and G Remnev (Tomsk, Russia: TSU) pp 42 -44

[7] Berezin Yu A Breizman B N and Vshivkov V A 1981 *Journal of Applied Mechanics and Technical Physics* **1** 1

[8] Antoshkin M Yu Grigor'ev V P Koval T V and Sablin N I 1992 *Radiotekh. Elektron.* **37** 1115

[9] Nikulin M G Stolbetsov S D Tarakanov V P Fedotov A V Shkvarunets A G 1992 *Radiotekh. Elektron.* **37** 1665

[10] Fedotov A V Ignatov A M and Tarakanov V P 1992 *Proc. of the 9th International Conference on High Power Particle Beams (Washington, DC, 22-25 May 1992) (Proc. of the 9th International Conference on High Power Particle Beams* vol 2*)* ed D Mosher and G Cooperstein (Springfield, VA: NTIS) pp 1373-1378

[11] Ignatov A M and Tarakanov V P 1994 *Phys. Plasmas* **1** 741

[12] Grishkov A A Belomyttsev S Ya Korovin S D and Ryzhov V V 2003 *Tech. Phys. Lett.* **29** 944

[13] Belomyttsev S Ya Grishkov A A Korovin S D and Ryzhov V V 2003 *Tech. Phys. Lett.* **29** 666

[14] Belomyttsev S Ya Grishkov A A Korovin S D and Ryzhov V V 2004 *Tech. Phys. Lett.* **30** 1060

[15] Belomyttsev S Ya Grishkov A A Kitsanov S A Korovin S D Polevin S D Ryzhov V V and Yachnyi A P 2005 *Tech. Phys. Lett.* **31** 982

[16] Belomyttsev S Ya Grishkov A A Kitsanov S A Kurkan I K Polevin S D Ryzhov V V and Tsygankov R V 2008 *Tech. Phys. Lett.* **34** 546

[17] Dubinov A E 1997 *Tech. Phys. Lett.* **23** 870

[18] Dubinov A E and Efimova I E 2001 *Tech. Phys.* **46** 723

[19] Tarakanov V P 1992 *User's manual for code Karat,* (Springfield, VA: Berkley Research Associate. Inc).

[20] Korovin S D and Rostov V V 1996 *Russian Phys. J.* **39** 1177

[21] Grishkov A A and Pegel I V 2013 Plasma Physics Reports 39 936

[22] Donets E D Donets E E Syresin E M Dubinov A У Makarov I V Sadovoy S A Saĭkov S K Tarakanov V P 2009 *Plasma Phys. Rep.* **35** 54

Current interruption in a low-pressure high-current pulsed discharge with hollow cathode

Y D Korolev[1,2,3,*], **N V Landl**[1,2], **V G Geyman**[1], **O B Frants**[1,2], **A V Bolotov**[1], **V O Nekhoroshev**[1,2], **V S Kasyanov**[1,2]
[1]Institute of High Current Electronics RAS, 634055 Tomsk, Russian Federation
[2]Tomsk State University, 634050 Tomsk, Russian Federation
[3]Tomsk Polytechnic University, 634050 Tomsk, Russian Federation

E-mail: korolev@lnp.hcei.tsc.ru

Abstract. This paper deals with the investigation of current interruption in the hollow-cathode discharge as applied to the electrode configuration of pseudospark switch. The phenomenon of interruption manifests itself at currents of several kiloamperes or higher and at a low gas pressure. It is demonstrated that the region of the hollow cathode is able to pass a high current density. The region that is responsible for the current interruption is the main gap of the switch, i.e., the positive column of discharge. The theoretical model of the positive column is proposed. The model implies that the gas ionization in the column is provided by the fast electrons which are accelerated in a double electric layer at the exit of cathode cavity. The estimates with a usage of the model are in agreement with the experimental data.

1. Introduction

This paper relates to investigations of the low-pressure pulsed glow discharges with hollow cathode. When we speak of the low pressure we imply that a neutral particle density is extremely small and electron free path for ionization exceeds a characteristic size of electrode system [1–4]. On the other hand, the neutral particles still play an essential role in forming gas-discharge plasma. Thus, the discharge regime keeps an intermediate position between the classical low-pressure discharge whose behavior is determined by electron avalanche ionization and the vacuum discharge that burns in cathode metal vapor [5–7].

There is a great variety of the gas-discharge devices functioning in the aforementioned regime. An illustrative example is the so-called pseudospark switch [1, 2, 8–13]. The electrode system of the switch typically consists of two cavities that communicate through axial bore holes. A bore hole diameter and a thickness of flat part of the cathode are comparable to the main gap spacing and amount to several millimeters. At a low gas pressure, the single electrons are not able to initiate the breakdown process in electrode system. A considerable prebreakdown electron flow is required for this purpose. Then a trigger unit, which is placed in the cathode cavity, is intended to provide a sufficient electron flow trough the bore hole into the main gap.

In some applications, the current densities in the pseudospark electrode geometry reach incredibly high values and a space of bore hole plays a role of hollow cathode [2, 14, 15]. A total discharge current from the cathode cavity can reach of several kiloamperes and higher. One of the phenomena, characteristic of the discharge with a high current and a deceased pressure, is the so-called current

Content from this work may be used under the terms of the Creative Commons Attribution 3.0 licence. Any further distribution of this work must maintain attribution to the author(s) and the title of the work, journal citation and DOI.
Published under licence by IOP Publishing Ltd

interruption or current quenching [15, 16]. This paper describes a model of current passage in the main gap of pseudospark switch that has been developed. Based on this model an interpretation of current interruption phenomenon is proposed.

2. Experimental arrangement and summary of experimental data

A schematic of experimental arrangement is shown in figure 1. The main interelectrode gap d = 4 mm is formed by the electrodes *1* and *2* with a diameter of central bore hole D = 4 mm. A thickness of the flat part of electrode *2* is h = 4 mm. The discharge is initiated due to the trigger system based on flashover [15, 16]. For instant of triggering a capacitor bank C_0 is charged to a voltage V_0. A discharge burning voltage V_d (i.e., the voltage between electrodes *1* and *2*) and the discharge current were recorded with a usage of resistive voltage dividers and a current shunt. Current measurements for the fast electron beam are provided by the Faraday cup *FC* located in the anode cavity.

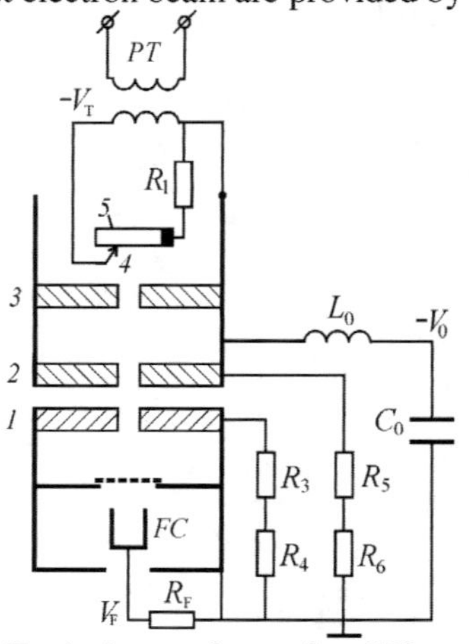

Figure 1. Schematic arrangement for investigations of the current interruption and generation of electron beam at the discharge axis in the pseudospark electrode system.
1–hollow anode, *2*–hollow cathode with the bore hole at the axis, *3*–baffle to separate the hollow cathode cavity from trigger system, *4*–pointer contact to provide a surface breakdown over semiconductor cylinder *5*. R_5/R_6 and R_3/R_4 are the resistive dividers for measuring a potential with respect to the ground. *FC*–Faraday cup for measuring electron beam current. R_F = 25 Ω, C_0 = 130 nF, L_0 = (27–300) nH.

Typical waveforms for different gas pressures are shown in figure 2. At the stage of lag time to breakdown, the voltage at the gap corresponds to V_0. Increase in the discharge current during development of breakdown is accompanied by the sharp decreasing in the discharge burning voltage. As shown in [2, 14–16], in the high-current stages the discharge burns in the regimes of the so-called dense glow and superdense glow.

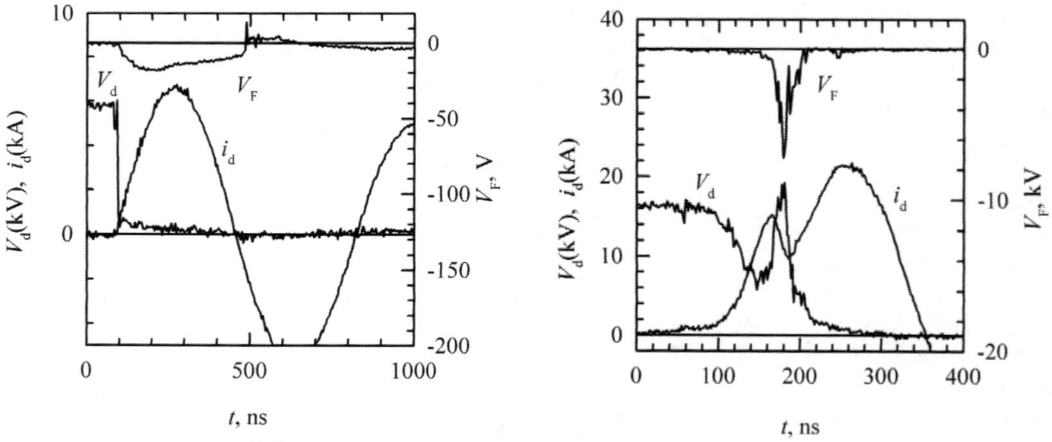

Figure 2. Current and voltage waveforms for discharge in xenon. C_0 = 130 nF. Current interruption is (*a*) absent and (*b*) available. (*a*) L_0 = 100 nH, $p \approx 2\cdot10^{-2}$ Torr. (b) L_0 = 27 nH, $p \approx 2\cdot10^{-3}$ Torr.

We can see that for a pressure $p \approx 2 \cdot 10^{-2}$ Torr the current waveform has a smooth shape, while decreasing a pressure to $p \approx 2 \cdot 10^{-3}$ Torr results in a sharp current interruption starting an instant t = 170 ns. Simultaneously, the inductive voltage kick is applied to interelectrode gap. A general tendency is that the current interruption manifests itself at a low pressure, for an enhanced discharge current and for short pulse duration.

In some papers, the effect of current interruption is associated with the processes inside the bore hole [17]. However, it seems [3, 15, 16] that the main physical processes responsible for the effect occur in the plasma column of the main gap. Then the description of the model of plasma column is presented below.

3. Description of the model and interpretations

Schematic of the discharge regions and the potential distributions for the temporal stage of the dense glow discharge is shown in figure 3. The space inside the hollow cathode C is filled with the negative glow plasma NG. Note that as applied to the pseudospark discharge a space of bore hole can play a role of the hollow cathode. The plasma is sustained due to gas ionization by the fast electrons which are emitted from a surface of hollow cathode and are accelerated in the cathode voltage drop region l_c.

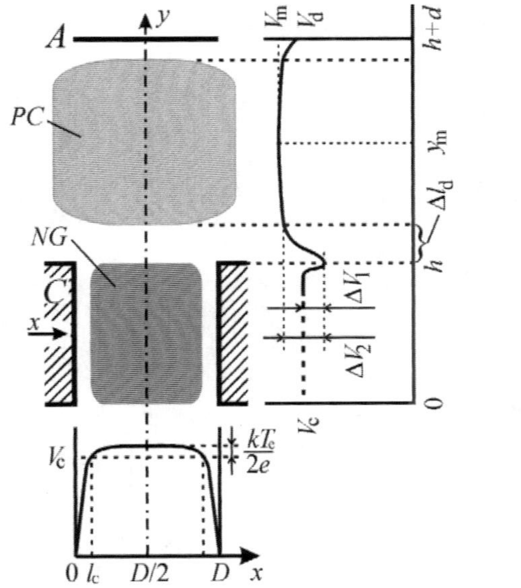

Figure 3. Schematic of the discharge regions and the potential distributions for electrode geometry of pseudospark discharge.

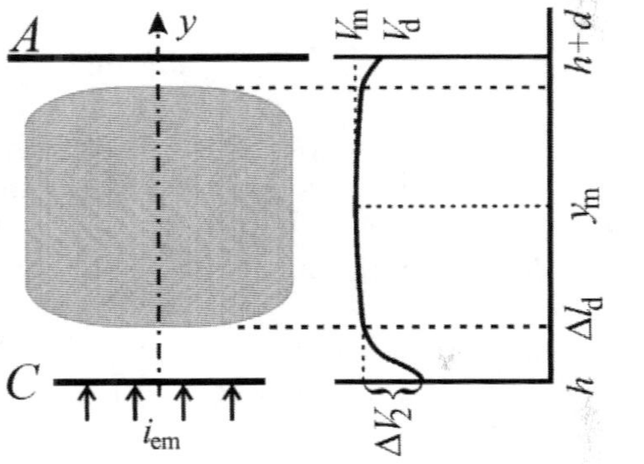

Figure 4. Schematic of the discharge regions and the potential distribution illustrating the model of current sustaining in the main gap.

At the exit of hollow cathode, a negative potential barrier ΔV_1 is available. The role of the barrier is to restrict an electron current emitted from the negative glow plasma in anode direction. The emission current has to be approximately equal to a total discharge current so that the height of potential barrier is established self consistently to satisfy this condition. The negative glow region represents a potential trap for the fast electrons, and an efficient plasma generation in this region is provided due to the hollow-cathode effect.

The main gap d is filled with the positive column plasma PC. The column is separated from the negative glow by a double electric layer Δl_d [15]. The electrons those are able to overcome the potential barrier ΔV_1 are accelerated by the voltage ΔV_2 and provide the gas ionization in the positive column. Due to a balance between the rate of ionization and the rate of losses of the charged particles, a steady state plasma density in the column is established ($n_i \approx n_e$). Conceptually, a role of potential drop ΔV_2 is the same as a role of the cathode voltage drop V_c in the hollow cathode. An electron current flows to the anode A from the positive column plasma. This current is controlled by a value of negative anode potential drop $\Delta V = (V_m - V_d)$.

To simplify the description of the positive column plasma we suppose that a certain emission current i_{em} enters into the column from a virtual cathode which is located in the plane $y = h$ (see figure 4). This current can be considered as an external current with respect to the main gap d.

Formally, the region of positive column also represents a potential trap for the plasma electrons. However, the effect of electron oscillations is expressed not so distinctively as that in the hollow cathode. A gas pressure in the discharge under consideration is rather low and some of the electrons, accelerated in the layer Δl_d, will go to the anode without ionizing collisions. These electrons, jointly with the plasma electrons, contribute to the total electron current at the anode. On the other hand, as shown in [2, 3], the plasma in the main gap can be generated if only a neutral particle density exceeds a certain critical value $n_a = n_{cr}$. Otherwise, the region between the cathode C and anode A will represent not a plasma gap with a high conductivity but a vacuum diode with low ion neutralization. Such a diode has a high resistance and is able to withstand a voltage of several tens of kilovolts without plasma generation. Similar regime is used in the sources of electron beams with plasma cathode [6].

For the stage of lag time to breakdown, n_{cr} is determined by the relation [3]:

$$n_{cr}\sigma_i d\left(\frac{M}{3m}\right)^{1/2} = \frac{d}{\lambda_i}\left(\frac{M}{3m}\right)^{1/2} = 1\,, \tag{1}$$

where σ_i is the average cross for ionization of the neutral particles by the electrons accelerated in the layer Δl_d, λ_i is the electron free path for ionization, M/m is the ratio of the ion mass to the mass of electron.

According to the model, in order that the positive column plasma be sustained, the neutral particle density has to exceed the critical value. On the other hand, as distinct to classical glow discharge or a high-pressure pulsed volume discharge [18–20], the neutral particle density is still rather low so that the number of ionizations produced by a single electron at the length of column $d/\lambda_i < 1$.

Establishing of a steady state plasma density in the column is provided as a result of balance between the rate of gas ionization and the rate of outflow of the charged particles to electrodes. The ions originated to the left of the point of potential maximum y_m travel to the cathode under the effect of potential difference $kT_e/2e$, thus providing the ion current i_{ic}. The ions originated to the right of the point y_m provide the ion current i_{ia} at the anode.

Ignoring the secondary processes at the cathode surface, we can write an expression for a total current at the cathode:

$$i = i(h) = i_{em} + i_{ic}. \tag{2}$$

The same current value should be provided at the anode. This current is added from the components entering to the following equation:

$$i = i_{eb} + i_{ep} - i_{ia} = i_{em}\left(1 - \frac{d}{\lambda_i}\right) + S\frac{1}{4}en_e v_e \exp\left(\frac{e\Delta V}{kT_e}\right) - i_{ia}\,, \tag{3}$$

where i_{eb} is the current of a fast electron beam (i.e., the current of the electrons which did not collide with the atoms while going to anode), i_{ep} is the current of the electrons from plasma that are able to overcome the potential barrier $e\Delta V$, S is the discharge area at the anode, $v_e = (8kT_e/\pi m)^{1/2}$ is the average thermal velocity of electrons.

In the further consideration we suppose that $i_{ia} \approx 0$ that is the point y_m is located near the anode. The positive column contains the plasma electrons and ions ($n_e \approx n_i$) and also the fast electrons of the electron beam whose density is n_{eb}. The rate of ionization can be written as

$$\Psi = \frac{1}{e}\frac{i_{\text{em}}}{S} n_{\text{a}}\sigma_{\text{i}} = \frac{j_{\text{em}}}{e} n_{\text{a}}\sigma_{\text{i}} . \tag{4}$$

Taking into account that an average time of ion outflow from the gap $T_{\text{i}} = d(M/kT_{\text{e}})^{1/2}$, we readily obtain the equation for plasma density in the positive column:

$$n_{\text{e}} = n_{\text{i}} = \Psi T_{\text{i}} = \frac{1}{e}\frac{i_{\text{em}}}{S} n_{\text{a}}\sigma_{\text{i}} d\left(\frac{M}{kT_{\text{e}}}\right)^{1/2} . \tag{5}$$

The region of positive column can be treated as an ionization chamber in which the originated ions go to the cathode. The same value of electron current has to flow to the anode via the potential barrier ΔV. Beside that, the fast electrons which appear in the column due to thermalization also have to leave the plasma column. Then for the current to the anode from plasma we obtain

$$i_{\text{ep}} = 2i_{\text{em}} n_{\text{a}}\sigma_{\text{i}} d . \tag{6}$$

A maximum possible current from plasma essentially exceeds i_{ep}. Then it would be concluded that the plasma is always able to carry the necessary electron current to anode. However, this situation holds true if only the gap d actually represents a plasma filled diode with the cathode layer Δl_{d}, but not a vacuum diode with a low ion neutralization. In other words, the plasma has to play important additional role that is to neutralize the excess space charge of electron beam. It means that the condition $n_{\text{e}} >> n_{\text{eb}}$ has to be fulfilled.

At the exit of the double electric layer, the fast electrons has a velocity $v_{\text{eb}} = (2e\Delta V_2/m)^{1/2}$, and then

$$n_{\text{eb}} = \frac{j_{\text{eb}}}{ev_{\text{eb}}} = \frac{1}{e}\frac{i_{\text{em}}(1 - n_{\text{a}}\sigma_{\text{i}} d)}{S}\left(\frac{m}{2e\Delta V_2}\right)^{1/2} . \tag{7}$$

To compare n_{e} and n_{eb} we can use the relation that allows interpreting different current regimes and the conditions of current interruption

$$\frac{n_{\text{e}}}{n_{\text{eb}}} = \frac{n_{\text{a}}\sigma_{\text{i}} d}{1 - n_{\text{a}}\sigma_{\text{i}} d}\left(\frac{M}{3m}\right)^{1/2}\left(\frac{6e\Delta V_2}{kT_{\text{e}}}\right)^{1/2} . \tag{8}$$

It is seen that with the condition $n_{\text{a}} \approx n_{\text{cr}}$ the ratio $n_{\text{e}}/n_{\text{eb}} \approx (6e\Delta V_2/kT_{\text{e}})^{1/2}$ essentially exceeds unity only for high values of voltage $\Delta V_2 = (10–20)$ kV. Such conditions are characteristic of the stage of lag time to breakdown. Transition to a high-current stage and low voltage ΔV_2 is inevitably accompanied by the current interruption. Note that the regime with a high voltage at the double electric layer is used in the electron sources with plasma cathode [6] where $n_{\text{a}} \le n_{\text{cr}}$. To provide the high-current discharge at a voltage ΔV_2 less than several hundred volts it is necessary to have $n_{\text{a}} >> n_{\text{cr}}$.

The regimes in which the current interruption is absent are determined not only by neutral particle density but also by total discharge current and current risetime, i.e., by the pulse duration. Figure 2*a* shows that for discharge in xenon at a pressure $p = 2\cdot10^{-2}$ Torr ($n_{\text{a}}/n_{\text{cr}} \approx 50$) the current waveform has a smooth shape. However, a decrease in pulse duration and an increase in current due to decreasing the inductance of electric lead to unstable regime of the current sustaining.

The illustrative example of current interruption is shown in figure 2*b*. Here, besides decreasing the pressure, the total discharge current has been increased to 20 kA. With such a current we have extremely high current density via bore hole, so that inside the hole the multicharged plasma is generated. Due to hollow cathode effect the plasma is able to be sustained and the bore hole region is able to pass the current. The effect of electron oscillations in the positive column is expressed not so distinctively as that in the cathode cavity. Correspondingly, a possibility of production of the multicharged ions in the positive column formally means that the ionization cross section in equation

(8) becomes much lower as compared to the regime of low current. In other words this is one more physical reason which encourages an increase in voltage ΔV_2 that is the current interruption.

Acknowledgments
The work was supported by Russian Scientific Foundation under the grant # 14-19-00139.

References

[1] Frank K and Christiansen J 1989 The fundamentals of the pseudospark and its applications *IEEE Trans. Plasma Sci.* **17** 748–53

[2] Korolev Y D and Frank K 1999 Discharge formation processes and glow-to-arc transition in pseudospark switch *IEEE Trans. Plasma Sci.* **27** 1525–37.

[3] Kozyrev A V, Korolev Y D, Rabotkin V G and Shemyakin I A 1993 Processes in the prebreakdown stage of a low-pressure discharge and the mechanism of discharge initiation in pseudospark switches *J. Appl. Phys.* **74** 5366–71.

[4] Lopatin I V, Schanin P M, Akhmadeev Y H, Kovalsky S S and Koval N N 2012 Self-sustained low pressure glow discharge with a hollow cathode at currents of tens of amperes *Plasma Physics Reports* **38** 583–89.

[5] Kondrat'eva N P, Koval N N, Korolev Y D et al 1999 A spectroscopic investigation of the near-cathode regions in a low-pressure arc *J. Phys. D: Appl. Phys.* **32** 699–705

[6] Koval N N, Oks E M, Schanin P M, Kreindel Y E and Gavrilov N V 1992 Broad beam electron sources with plasma cathodes *Nuclear Instr. Methods Phys. Res.* **A321** 417–28

[7] Feng J H, Meng S J, Fu Y C et al 2014 Spatiotemporal distribution of hydrogenous electrode vacuum arc discharge plasma *Acta Physica Sinica* **63** 145205

[8] Bochkov V D, Dyagilev V M, Ushich V G, Frants O B, Korolev Y D, Shemyakin I A and Frank K 2001 Sealed-off pseudospark switches for pulsed power applications (current status and prospects) *IEEE Trans. Plasma Sci.* **29** 802–8

[9] Hu J and Rovey J L 2013 Experimental investigation of time-resolved electron beam energy distributions generated in a transient hollow cathode discharge *J. Appl. Phys.* **114** 073301

[10] Larsson A 2012 Gas-discharge closing switches and their time jitter *IEEE Trans. Plasma Sci.* **40** 2431–42

[11] Zhao H T and Kirkici H 2012 Carbon-nanotube-triggered pseudospark switch *IEEE Trans. Plasma Sci.* **40** 2225–31

[12] Sozer E B, Gundersen M A and Jiang C Q 2012 Magnesium-based photocathodes for back-lighted thyratrons *IEEE Trans. Plasma Sci.* **40** 1753–58

[13] Meena B L, Rai S K, Tyagi M S, Pal U N, Kumar M and Sharma A K 2010 Characterization of high power pseudospark plasma switch (PSS) *J. Phys. Conference Series* **208** 012110

[14] Stetter M, Felsner P, Christiansen J, Frank K, Mehr T, Stenzenberger J and Tkotz R 1995 First experimental observation of the ignition of a "superdense glow" before the glow-to-arc transition in a pseudospark discharge *J. Appl. Phys.* **79**, 631–36

[15] Korolev Y D, Frants O B, Landl N V et al 2013 High-current stages in a low-pressure glow discharge with hollow cathode *IEEE Trans. Plasma Sci.* **41** 2087–96

[16] Korolev Y D, Frants O B, Geyman V G, Landl N V, Ivashov R V, Shemyakin I A, Bischoff R E and Frank K 2005 Temporal structure of the fast electron beam generated in the pseudospark discharge with external triggering *IEEE Trans. Plasma Sci.* **33** 1648–53

[17] Mehr T, Tkotz R, Stenzenberger J, Hintz G, Christiansen J, Felsner P, Frank K and Stetter M 1995 The bottleneck in pseudospark discharges *J. Appl. Phys.* **79** 625–30

[18] Kozhevnikov V Y, Kozyrev A V and Korolev Y D 2006 Drift model of the cathode region of a glow discharge *Plasma Phys. Rep.* **32** 949–59

[19] Korolev Y D, Mesyats G A and Yarosh A M 1987 Film etching by particles produced by a pulsed bulk discharge in CF_4 *High Energy Chem.* **21** 389–92

[20] Korolev Y D, Frants O B, Geyman V G, et al 2012 Transient processes during formation of a steady-state glow discharge in air *IEEE Trans. Plasma Sci.*, **40** 2951–60

Journal of Physics: Conference Series **552** (2014) 012020 doi:10.1088/1742-6596/552/1/012020

Diffusion of fast rising strong magnetic fields into conductors

N A Labetskaya[1,*], V I Oreshkin[1,2], S A Chaikovsky[1], I M Datsko[1], N I Kuskova[3] and A D Rud[4]

[1]Institute of High Current Electronics SB RAS, 2/3 Akademichesky Avenue, Tomsk, 634055, Russia

[2]National Research Tomsk Polytechnic University, 30 Lenin Avenue, Tomsk, 634050, Russia

[3]Institute of Pulse Processes and Technologies NASU, 43a Zhovtnevy Avenue, Nikolaev, 54018, Ukraine

[4]G.V. Kurdyumov Institute for Metal Physics NASU, 36 Academician Vernadsky Blvd., Kiev, 03680, Ukraine

E-mail: natalia@ovpe2.hcei.tsc.ru

Abstract. The basic processes occurring in a conductor exploding in a current skinning mode are the propagation of a nonlinear magnetic diffusion wave in the conductor and the formation of low-temperature plasma at its surface. An experimental study of the phenomenon of nonlinear magnetic diffusion into conductors in magnetic fields of induction rising at a rate up to $3 \cdot 10^9$ T/s was carried out on the MIG generator capable of producing a peak current up to 2.5 MA within a rise time of 100 ns. It has been found experimentally that the average velocity of a nonlinear magnetic diffusion wave in an aluminum conductor placed in a strong magnetic field (up to 300 T) rising at a high rate (on average, $3 \cdot 10^9$ T/s) is $(2.7 \div 3.3) \cdot 10^5$ cm/s. This is comparable to the velocity of sound in aluminum under normal conditions and reasonably agrees with predictions of numerical simulations.

1. Introduction

The electrical explosion of conductors (EEC) has been studied for a long time and has a number of practical applications. An EEC mode of interest in the generation of superstrong magnetic fields and in the energy transfer using vacuum transmission lines is the current skinning mode [1, 2, 3, 4]. For this mode, the time of energy delivery to the conductor is less than or comparable to the time of magnetic diffusion into it. The basic processes inherent in the current skinning mode are the propagation of a nonlinear magnetic diffusion wave (NMDW), the formation of low-temperature plasma at the conductor surface, and the development of thermal instabilities [5, 6].

Nonlinear magnetic diffusion features a speed of field penetration into a conductor anomalously high compared to conventional magnetic diffusion. The high diffusion speed is related to a decrease in conductivity of the metal due to its heating by electric current. Nonlinear diffusion can occur only in a rather strong magnetic field whose induction for the most used metals should be no less than 25÷45 T.

The nonlinear diffusion of a magnetic field into a conductor in a current skinning mode was investigated experimentally on the MIG pulse power generator [7]. The goal of the experiment was to determine the velocity of a nonlinear magnetic diffusion wave at a magnetic field induction rising up to 300 T at a rate of $3 \cdot 10^9$ T/s. To interpret the experimental results obtained, a numerical model was used, which described an NMDW propagating in a conductor [8].

Content from this work may be used under the terms of the Creative Commons Attribution 3.0 licence. Any further distribution of this work must maintain attribution to the author(s) and the title of the work, journal citation and DOI.
Published under licence by IOP Publishing Ltd

2. Numerical model

The process of nonlinear magnetic diffusion is described by Maxwell's equations written in a quasistationary approximation (not taking into account displacement currents) and complemented with the Ohm's and Joule–Lentz laws. In cylindrical coordinates, the equations take the form:

$$\frac{1}{r}\frac{\partial}{\partial r}(rB) = \mu_0 j\,; \quad \frac{\partial E}{\partial r} = \frac{\partial B}{\partial t} \tag{1}$$

$$E = j\delta \tag{2}$$

$$\frac{\partial \varepsilon_w}{\partial t} = j^2\delta \tag{3}$$

where B is the induction of the azimuthal magnetic field, E is the electric field strength, j is the axial current density, δ is the resistivity of the metal, ε_w is the energy density due to Joule heat release, and μ_0 is the permeability of vacuum. The boundary conditions for the system of equations (1)–(3) are specified as $B^{out}(t) = \frac{\mu_0}{2\pi}\frac{I(t)}{R}$ at the conductor outer boundary, where $I(t)$ is the current flowing in the conductor, R is the external radius of the conductor; and as $B^{\text{inner}}(t) = 0$, at the conductor inner boundary.

For the majority of metals, the resistivity increases with temperature T as:

$$\delta(T) = \delta_0\left(1 + \frac{\partial\delta}{\partial T}T\right) \tag{4}$$

where δ_0 is the resistivity under normal conditions and $\frac{\partial\delta}{\partial T}$ is the derivative of resistivity with respect to temperature. As $\varepsilon_w = \rho c_V T$, where ρ is the density of the conductor material and c_V is the heat capacity at constant volume, the resistivity can be expressed in terms of thermal energy density as [9]:

$$\delta(\varepsilon_w) = \delta_0(1 + \beta\varepsilon_w) \tag{5}$$

where $\beta = \frac{1}{\rho c_V}\frac{\partial\delta}{\partial T}$.

For diffusive penetration of a magnetic field into a material, the thermal energy density is approximately equal to the magnetic energy density [9]; that is, we have $\varepsilon_w = \frac{B^2}{2\mu_0}$. The magnetic field of induction B_0 at which the resistivity of the medium is doubled, that is, the condition $\beta\varepsilon_w \approx \beta\frac{B_0^2}{2\mu_0} = 1$ is satisfied, is generally termed the *characteristic* field [9, 10], whose induction is given by [8]:

$$B_0 = \left(\frac{2\mu_0}{\beta}\right)^{1/2} \tag{6}$$

Formula (6) gives a lower threshold field, such that at higher fields the effects related to nonlinear diffusion must be taken into consideration. As mentioned above, the characteristic magnetic induction is 25÷45 T for the majority of metal.

It has already been noted that if a strong magnetic field is applied at the outer boundary of a conductor, a nonlinear magnetic diffusion wave propagates into the conductor. This can be seen from figure 1a, where the results of solving the system of equations (1)–(3), (5) for an aluminum tube of external diameter 3 mm and wall thickness 250 μm are presented. The calculations were performed for the case where a magnetic field corresponding to a current pulse of the MIG generator recorded in an experimental shot was applied at the outer boundary of the conductor. The current pulse waveform is given in figure 1b (curve *I*). Figure 1a shows that the current density peaks at the NMDW front and the resistivity increases behind the front. The arrival of the NMDW at the conductor inner surface is accompanied by occurrence of an electric field inside the tube (curve U^{inner} in figure 2b). As the NMDW arrives at the inner surface, the current density at this surface peaks and the time derivative of voltage has a pronounced maximum (curve $(dU/dt)^{\text{inner}}$ in figure 1b).

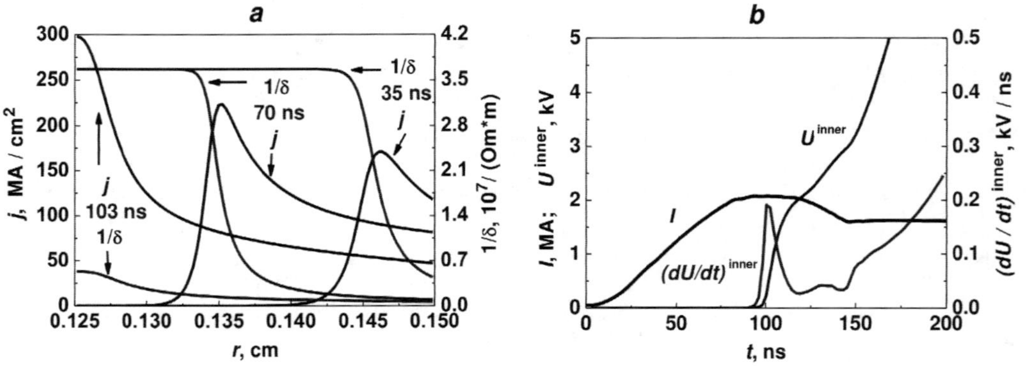

Figure 1. Distribution current density (*j*) and conductivity (1/δ) at different times during the propagation of an NMDW in an aluminum conductor (tube of external diameter 3 mm and wall thickness 250 μm) (*a*) and an experimental current waveform *I*, a predicted waveform of the voltage U^{inner} and its derivative $(dU/dt)^{\text{inner}}$ at the conductor inner surface (*b*).

Thus, the calculation results show that to determine experimentally the time at which the NMDW arrives at the conductor inner surface (to measure the velocity of the NMDW), it is necessary to record the waveform of the voltage at the inner surface and to determine the time at which the derivative $\left(\frac{dU}{dt}\right)^{\text{inner}}$ reaches a maximum.

3. Experimental results

The experiment was carried out on the MIG pulse power generator [7] at a current up to 2.5 MA with 100 ns rise time. Aluminum tubes of external diameter 3 mm (see figure 2a) and wall thickness from 100 to 400 μm were used for the generator load. To record the voltage waveform at the inner surface, a resistive voltage divider was used. The location of the divider connection to the load unit is shown schematically in figure 2. As the load unit was assembled with the tube, the divider cable was inserted through the open end of the tube on the low-voltage side, so that the cable conductor was in contact with the tube farther ("high-voltage") end and the cable braid was in contact with the tube inner surface on the low-voltage side. The measured voltage amplitude at the tube inner surface was up to 20 kV.

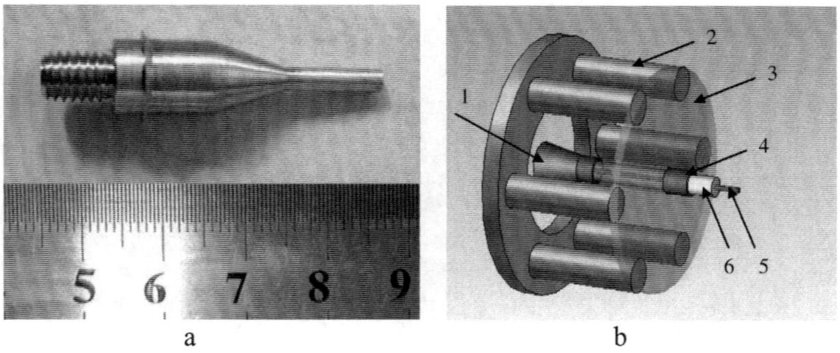

Figure 2. Photo of the generator load (*a*) and a schematic image of the voltage divider connection to the load unit of the MIG generator: 1 – conical cathode fitting piece to which the test tube is fastened, 2 – return current rods, 3 – anode disk, 4 – tube, 5 – divider cable conductor, and 6 – polythene insulator of the divider cable (*b*).

For the above generator parameters, the thickness of the skin layer at normal conductivity was, for all tubes, substantially smaller than the tube wall thickness, which was varied in the range 100÷400 μm. Thus, a current skinning mode, which should necessarily be realized to investigate magnetic diffusion into a conductor, was certainly provided. According to the experimental data, the characteristic field for aluminum (37 T) at the tube surface was reached within 18÷25 ns after the onset of current flow in the tube. Thus, for the given current waveform, magnetic diffusion into the conductor became nonlinear at the 18th÷25th nanosecond.

The experimental waveforms of the current through a tube *I*, of the voltage at the tube inner surface, U^{inner}, and of the derivative $(dU/dt)^{\text{inner}}$ obtained in one of the shots are presented in figure 3.

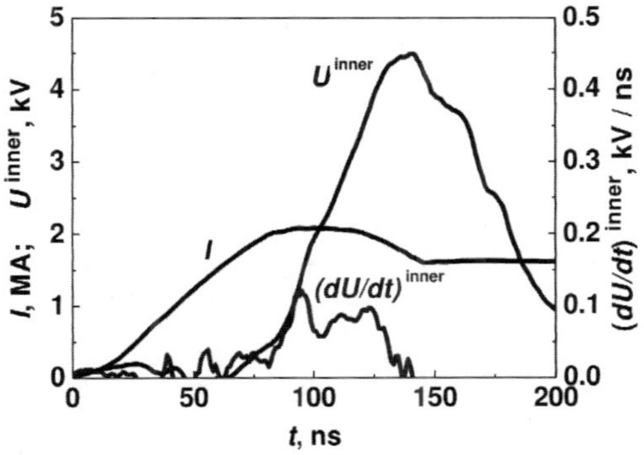

Figure 3. Experimental waveforms of current *I*, voltage U^{inner}, and derivative $(dU/dt)^{\text{inner}}$ at the inner surface of an aluminum tube of external diameter 3 mm and wall thickness 250 μm.

These experimental waveforms were obtained for an aluminum tube of wall thickness 250 μm, the same as was used in the calculations discussed in the previous section. Comparison of the experimental and calculated curves of voltage and its derivative shows that the curves are in rather good agreement at the beginning of the pulse and then disagree. The disagreement seems to be related to that the plasma formation at the conductor surface observed in the experiment was not considered in the calculations. Both the experimental and the calculated curve of $(dU/dt)^{\text{inner}}$ have a maximum associated with the arrival of the NMDW at the inner surface of the tube.

Signals of the U^{inner} sensor were similarly processed for all tube wall thicknesses, and for each case the time of arrival of the NMDW at the tube inner surface was determined. Figure 4 presents the results of the processing: the measured time of NMDW arrival at the tube inner surface and the NMDW velocity are presented as functions of the tube wall thickness in figure 4a and figure 4b, respectively. The NMDW velocity was determined as the ratio of the tube wall thickness to the difference between the time of NMDW arrival at the tube inner surface and the time at which the magnetic field at the outer surface became equal to B_0.

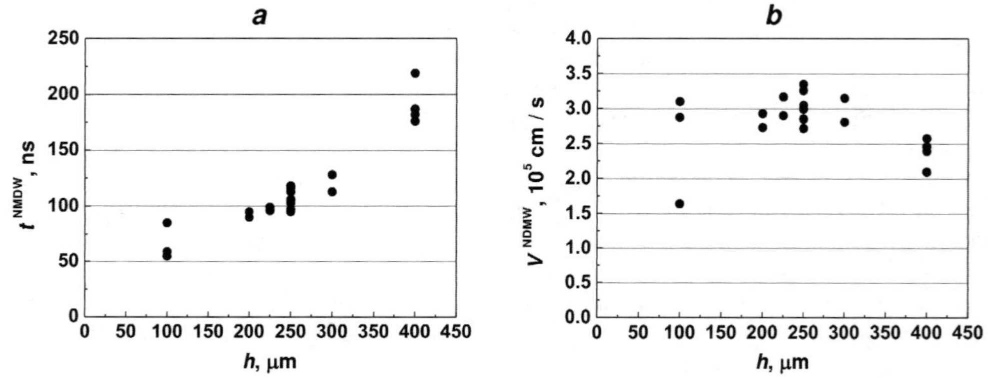

Figure 4. Experimental values of the time of NMDW arrival at the inner surface of an aluminum tube (*a*) and of the NMDW velocity at this time (*b*) versus tube wall thickness *h*.

As can be seen from figure 4b, for aluminum conductors placed in magnetic fields of induction up to 300 T rising at an average rate of $3\cdot10^9$ T/s, the NMDW velocity is $(2.7\div3.3)\cdot10^5$ cm/s, which is comparable to the velocity of sound in aluminum under normal conditions, and the NMDW velocity weakly depends on the conductor thickness.

4. Conclusion

According to the predictions of numerical simulations, as the NMDW arrives at the inner surface of an aluminum tube and the current density at this surface peaks, the time derivative of the electric field strength has a pronounced maximum. Measurements of the voltage at the tube inner surface also show the presence of a pronounced maximum in the voltage derivative. This makes it possible to measure the velocity of propagation of the NMDW more precisely, as the voltage increases rather smoothly.

Experimentally, for aluminum conductors placed in strong magnetic fields (up to 300 T with an average induction rise rate of $3\cdot10^9$ T/s) it has been found that the average velocity of the NMDW is $(2.7\div3.3)\cdot10^5$ cm/s. This is comparable to the velocity of sound in aluminum under normal conditions and reasonably agrees with the predictions of the numerical model.

Acknowledgements
The work was supported in part by the joint project of SB RAS – NAS of Ukraine (No. 03-09-12).

References

[1] Bichenkov E I and Voitenko A E 1969 *Zh. Prikl. Mekh. Matem. Fiz.*, No. 6 21–26
[2] Shneerson G A 1973 *Zh. Tekh. Fiz.* **18** (2) 419
[3] Chaikovsky S A, Oreshkin V I, Mesyats G A, Ratakhin N A, Datsko I M, and Kablambaev B A

2009 *Phys. Plasmas* **16** 042701

[4] Chaikovsky S A, Oreshkin V I, Datsko I M, Labetskaya N A and Ratakhin N A 2014 *Phys. Plasmas* **21** 042706

[5] Valuev A A, Dikhter I Ya, and Zeigarnik V A 1978 *Zh. Tekh. Fiz.* **48**, 2088

[6] Oreshkin V I 2008 *Phys. Plasmas* **15** 092103

[7] Luchinsky A V, Ratakhin N A, Feduschak V F, and Shepelev A N 1997 *Izv. Vyssh. Uchebn. Zaved., Fiz.* **40** (12) 67

[8] Oreshkin V I and Chaikovsky S A 2012 *Phys. Plasmas* **19** 022706

[9] Knoepfel H 1972 Pulsed high magnetic fields [in Russian], Mir, Moscow

[10] Shneerson G A 1992 Fields and transient processes in high-current devices [in Russian], Energoatomizdat, Moscow

Numerical simulation and analysis of energy loss in a nanosecond spark gap switch

I V Lavrinovich[1,2,*], **V I Oreshkin**[1,2]
[1]Institute of High Current Electronics SB RAS, 2/3 Akademichesky Avenue, Tomsk, 634055, Russia
[2]National Research Tomsk Polytechnic University, 30 Lenin Avenue, Tomsk, 634050, Russia

E-mail: lavrivan@mail.ru

Abstract. A system of differential equations for the RLC circuit of a capacitor-switch assembly was derived being supplemented with an equation for the spark resistance of the switch in accordance with the Braginsky model. The parameters that affect the solutions of equations for the circuit with parallel or series connection of several capacitor-switch assemblies to a common inductive load were determined. Based on numerical solution of the system of equations, a dependence of the energy E_S released in the spark within the first half-period on the discharge circuit and switch parameters was found.

1. Introduction

The spark gap switches are much used for switching high-current pulses over a wide (from subnanoseconds to microseconds) time range [1-4]. This type of switches finds application in different electrophysical experiments: for generation of electron and ion beams [5-7]; excitation of high-power microwaves [8,9]; design of high-intensity X-ray sources [7,10–12]; research in gas discharges [13–15], etc. The challenge of the last years is to create setups capable of transferring megajoule energy to a load in ~100 ns [16], which is required for inertial thermonuclear fusion. In particular, high-current pulse generators serve as drivers in the circuits in which a thermonuclear target is excited with Z-pinches [17–20]. The generation of nanosecond high-power pulses requires multi-module systems in which the energy from each module is transferred through a spark gap switch either to an intermediate energy store [21,22] or to a transmission line [16,23,24]. An important problem in designing this type of setups is to provide efficient use of the energy of the primary store [25] and this involves, among other things, minimization of the energy lost in the switch.

The operation of high-current switches is described on the basis of gas discharge theories. However, despite many works devoted to this problem [13,26,27], the now available discharge theories are still incomplete and contradictory. Nevertheless, the main conclusions following from analysis of rather simple discharge mechanisms are of practical value and are used in designing high-current switches [2].

The operation of a gas-filled spark gap switch results in a current channel of resistance equal to

$$R_s = \frac{l_s}{\pi a^2 \sigma}, \tag{1}$$

Content from this work may be used under the terms of the Creative Commons Attribution 3.0 licence. Any further distribution of this work must maintain attribution to the author(s) and the title of the work, journal citation and DOI.
Published under licence by IOP Publishing Ltd

where l_s is the channel length; σ is the channel conductivity; a is the channel radius. The time dependence of the channel radius and conductivity is determined with the use of various models.

For nanosecond spark gap switches, the best agreement with experiment is ensured by the Braginsky model [28] which describes the development of a spark channel in gas at high pressures and moderate currents. The model assumes that in the gas there arises a rather narrow current-conducting channel in which Joule heat is released resulting in a higher gas pressure and in a shock wave in the gas. It is assumed that the heat from the channel is removed by the radiation absorbed in the shock wave region; in this shock wave region, gas ionization takes place. In the self-similar solution derived for this case, the pressure as well as the temperature and density are constant along the channel radius and the velocity is proportional to the radius. The energy balance equation in this solution has the form:

$$\frac{I^2}{\sigma} = \frac{\pi^2 \xi}{4} \rho \left(\frac{\mathrm{d}a^2}{\mathrm{d}t} \right)^3 , \tag{2}$$

where ρ is the gas density ahead of the shock wave; ξ is a dimensionless quantity which is generally a function of time.

It is shown that in the range of currents I in which the model is applicable, the temperature of the expanding channel varies slowly and can thus be taken roughly constant [28]. Hence, the channel conductivity and the dimensionless function ξ are found constant. According to [28], the values of these functions for a discharge in air are $\sigma = 2 \cdot 10^{14}\ \mathrm{s}^{-1}$; $\xi = 4.5$.

An expression for the channel resistance in the Braginsky model can be derived from (1) in view of (2). This expression has the form:

$$R_\mathrm{s}(t) = \left(\frac{\xi \rho}{4 \pi \sigma^2} \right)^{1/3} \frac{l_\mathrm{s}}{\int_0^t I^{2/3} \mathrm{d}t'} . \tag{3}$$

The differential form of expression (3) is as follows:

$$\frac{1}{R_\mathrm{s}^2} \frac{\mathrm{d}R_s}{\mathrm{d}t} = -\frac{1}{l_\mathrm{s}} \left(\frac{4 \pi \sigma^2}{\xi \rho} \right)^{1/3} I^{2/3} . \tag{3a}$$

The initial condition for equation (3a) is $R_s(t=0) \to \infty$.

In the paper presented, we perform numerical study of equation (3a), which is based on the Braginsky model, and generator circuit equations for inductive load. Based on the numerical solutions, expressions which determine the energy released in the switch are derived.

2. Dimensionless circuit equations

Let the generator under study be an RLC circuit consisting of N parallel-connected modules (Figure 1). The model includes RLC circuit equations and equation (3a) describing the switch resistance. The RLC circuit equations for each module have the form (in the SI system):

$$L_0 \frac{\mathrm{d}I}{\mathrm{d}t} = U - U_\mathrm{L} - IR_\mathrm{s} ; \quad \frac{\mathrm{d}U}{\mathrm{d}t} = -\frac{I}{C_0} , \tag{4}$$

where I and U are respectively the current in a single circuit module (current through the switch) and the voltage across its capacitor bank; U_L is the voltage across the load, C_0 is the capacitance of the capacitor bank of the module, L_0 is the inductance of the module which comprises the inductances of the capacitor bank, switch, and transmission line.

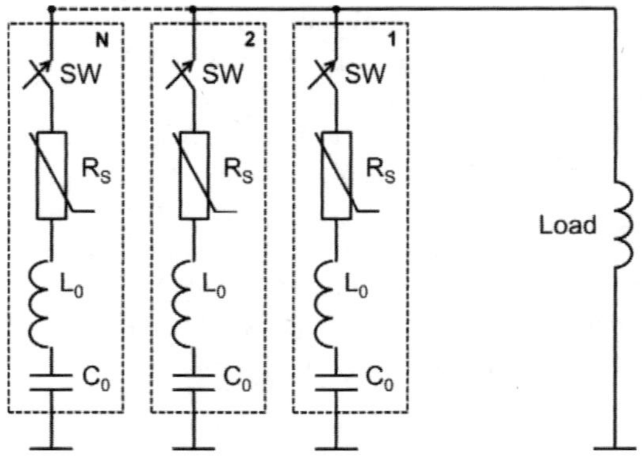

Figure 1. RLC circuit.

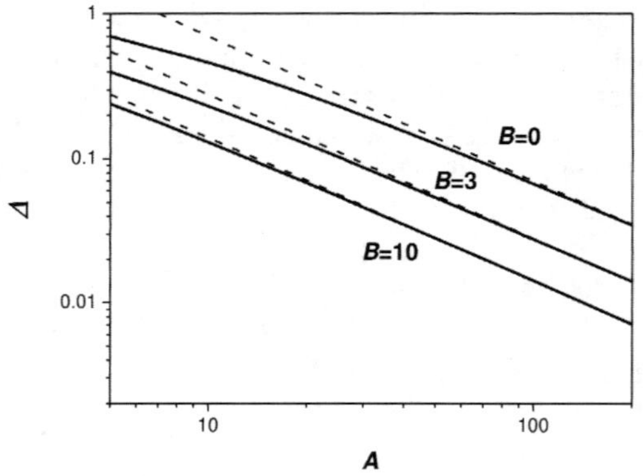

Figure 2. Dimensionless energy loss in the switch (first half-period) vs. parameter A.

The initial conditions of equation (3a) are supplemented with two more initial conditions $I\,(t = 0) = 0$, $U\,(t = 0) = U_0$, where U_0 is the voltage to which the capacitor bank is charged.

Let us rewrite equations (3a, 4) in the dimensionless form:

$$\frac{\mathrm{d}i}{\mathrm{d}\tau} = \frac{\pi}{2}\left(u - u_{\mathrm{L}} - ir_{\mathrm{s}}\right); \quad \frac{\mathrm{d}u}{\mathrm{d}\tau} = -\frac{\pi}{2}i\,; \quad \frac{\mathrm{d}r_{\mathrm{s}}}{\mathrm{d}\tau} = -Ar_{\mathrm{s}}^2 i^{2/3}. \tag{5}$$

Here, $\tau = t/t_0$ is the dimensionless time; $i = I/I_0$ is the dimensionless current; $u = U/U_0$ is the dimensionless voltage; $u_L = U_L/U_0$ is the dimensionless voltage across the load; $r_s = R_s/Z_0$ is the dimensionless switch resistance; $Z_0 = \sqrt{L_0/C_0}$ is the wave impedance of a single circuit module; $A = \frac{1}{l_s}\left(\frac{8\pi\sigma^2}{\xi\rho}L_0^2\mathrm{E}_0\right)^{1/3}$ is a dimensionless parameter; $\mathrm{E}_0 = \frac{C_0U_0^2}{2}$ is the energy stored in the capacitor bank of the module. The time and current scales are determined by the expressions:

$$t_0 = \frac{\pi}{2}\sqrt{L_0C_0}\,,\; I_0 = \frac{U_0}{Z_0}. \tag{6}$$

The initial conditions for the system of equations (5) are the following: $i(\tau = 0) = 0$; $u(\tau = 0) = 1$; and $r_s(\tau = 0) = 100$. The latter initial condition implies that early in the switching, the switch resistance is equal to 100 wave impedances of the module Z_0. Note that even at $r_s(\tau = 0) > 10$, this initial condition hardly affected the processes occurring in the oscillatory circuit.

3. Parallel connection of generator modules

Let us consider the operation of the generator onto an inductive load – inductance L_L. In this case, the voltage across the load is $U_L = L_L N\frac{dI}{dt}$, where N is the number of modules in the current generator. Then, the system of equations (5) takes the form:

$$\left(1 + B\right)\frac{\mathrm{d}i}{\mathrm{d}\tau} = \frac{\pi}{2}\left(u - ir_{\mathrm{s}}\right); \quad \frac{\mathrm{d}u}{\mathrm{d}\tau} = -\frac{\pi}{2}i\,; \quad \frac{\mathrm{d}r_{\mathrm{s}}}{\mathrm{d}\tau} = -Ar_{\mathrm{s}}^2 i^{2/3}, \tag{7}$$

where $B = NL_L / L_0$ is the dimensionless parameter numerically equal to the ratio between the load inductance multiplied by the number of parallel modules and the inductance of one generator module. Hence, the system of equations (7) is two-parametric. Let us consider solution of the system of equations (7) at different values of the parameters A and B.

The energy released in the switch with respect to the energy stored in the capacitor bank of one stage is defined by the expression:

$$\delta(\tau) = \frac{E_s}{E_0} = \frac{\int_0^t I^2 R_s(t')dt'}{\frac{C_0 U_0^2}{2}} = \pi \int_0^\tau i^2 r_s(\tau')d\tau' . \quad (8)$$

Figure 2 shows the dependence of $\Delta = \delta(\tau_h)$ on the parameter A at different values of the parameter B. For determination of Δ, the upper limit of integration in (8) was τ_h taken as the time of completion of the first half-period when the current in the circuit becomes equal to zero. It is seen in figure 2 that the calculated curves of Δ (solid lines) are rather accurately approximated by the expression:

$$\Delta \approx \frac{7}{A(1+B)^{2/3}} . \quad (9)$$

The curves corresponding to approximation (9) are shown by dotted lines in figure 2. To the short circuit mode there corresponds the curve $B = 0$. As can be seen from the figure, approximation (9) is asymptotically valid in describing the behavior of Δ at $A \to \infty$.

Substitution of the value of the parameter A in (9) gives us the following expression for the relative energy loss in the switch in the case of inductive load:

$$\Delta \approx \left(\frac{\xi\rho}{8\pi\sigma^2 L_0^2 E_0} \right)^{1/3} \frac{7 l_s}{(1+B)^{2/3}} \quad (9a)$$

Thus the relative loss decreases with increasing the load inductance and number of modules.

Let us rewrite expression (9a) in practical units. We assume that the discharge gap is filled with air whose density at atmospheric pressure is $\rho_0 = 1.29\times10^{-3}$ g/cm^3, $\sigma = 200\ \Omega \times$ cm, $\xi = 4.5$ [28, 29]. Then,

$$\Delta \approx 5.8 \cdot 10^{-5} \left(\frac{p}{L_0^2 E_0} \right)^{1/3} \frac{l_s}{(1+B)^{2/3}} , \quad (10)$$

where p is the air pressure in atmospheres, inductance is in henrys, energy is in joules, and length is centimeters.

The absolute loss in the switch is defined by the expression:

$$E_s = E_0 \Delta \approx 7 l_s \left(\frac{\xi\rho}{8\pi\sigma^2} \right)^{1/3} \left(\frac{E_0}{L_0 + NL_L} \right)^{2/3} . \quad (11)$$

Note that the energy lost in the switch can be expressed in the form proposed by Martin, Seaman, and Jobe [29]:

$$E_s \approx 7l_s\left(\frac{\xi\rho}{32\pi\sigma^2}\right)^{1/3} I_m^{4/3}, \quad (12)$$

where $I_m = \dfrac{U_0}{\sqrt{(1+B)\dfrac{L_0}{C_0}}}$ is the current close to the maximum circuit current.

4. Series connection of generator modules

Let us consider an RLC circuit consisting of the N number of series-connected modules. This connection is used, for example, in Arkadiev–Marx generators, see, e.g., Ref. 1. In this case, the RLC circuit equations have the form:

$$L_N\frac{dI}{dt} = U - U_L - NIR_s; \quad \frac{dU}{dt} = -\frac{I}{C_N}, \quad (13)$$

where $C_N = \dfrac{C_0}{N}$, C_0 is the capacitance of the capacitor bank of one module, $L_N = L_0 N$, L_0 is the inductance of the module. Division of equations (13) by N gives a system of equations which differs from (4) only in one term in the first equation; namely, instead of U_L in (13), we have U_L/N.

Therefore, all results obtained for parallel connection of generator also valid for series connection of the modules if the parameter B is defined as follows:

$$B = \frac{L_L}{NL_0} \quad (14)$$

Thus, with parallel connection, increasing the number of generator modules N decreases the energy lost in the switch, and with series connection, vice versa, increases this loss.

5. Conclusion

In the work, the spark channel in the generator discharge circuit is considered as a nonlinear element in which the energy loss in switching may amount to a large portion of the total stored energy and may thus greatly affect the efficiency of the whole generator. The main task of the work was to determine the parameters responsible for the energy loss in the switch and the degree of their influence.

Analysis of the operation of spark gap switches with the Braginsky model confirms the dependences of energy loss in a switch found earlier [2, 3, 29]: the increase in the relative loss Δ with an increase in gap width and gas density in the switch; the decrease in Δ with an increase in circuit inductance; and the decrease in Δ with an increase in capacitor stored energy.

The derived quantitative dependences of relative and absolute losses in the switch on the parameters of the switch and discharge circuit as well as on the number of parallel- or series-connected generator modules can be useful in designing high-current generators of the next generation[16-18]. The reduction of the energy loss to the number of modules, i.e., to the number of capacitor-switch assemblies, allows an optimal choice of the parameters and number of both to minimize the specific and absolute losses.

References

[1] Mesyats G A 2004 *Pulsed power and electronics* (Moscow: Nauka) p 704
[2] Kovalchuk B M, Kremnev V V and Potalitsyn Yu F 1979 *Nanosecond high-current switches* (Novosibirsk: Nauka) p 176
[3] Vitkovitsky I 1987 *High Power Switching* (New York: Van Nostrand Reinhold) p 304
[4] Mesyats G A, Yalandin M I 2005 *Russia Phys.-Usp.* **48** N.3 211

[5] Yushkov G Y, Vodopyanov A V, Nikolaev A G, Izotov I V, Savkin K P, Golubev S V, Oks E M 2013 *IEEE Trans. Plasma Sci.* **41** N.8 2081

[6] Anders A, Yushkov G Y 2002 *Appl. Phys. Lett.* **80** N.14 p 2457

[7] Petin V K, Shljakhtun S V, Oreshkin V I, Ratakhin N A 2008 *Technical Physics* **53** N.6 776

[8] Rostov V V, Yalandin M I and Mesyats G A 2008 *IEEE Trans. Plasma Sci.* **36** N.3 655

[9] Mesyats G A, Korovin S D, Gunin A V, Gubanov V P, Stepchenko A S, Grishin D M, Landl V F, Alekseenko P I 2003 *Laser and Particle Beams* **21** N.2 197

[10] Pereira N R, Davis J 1988 *J. Appl. Phys.* **64** R1

[11] Rousskikh A G, Zhigalin A S, Oreshkin V I, Chaikovsky S A, Labetskaya N A and Baksht R B 2011 *Phys. Plasmas* **18** 092707

[12] Jons B, Deeney C, Coverdale C A, LePell P D, McKeeney J L, Apruzese J P, Thornhill J W, Whitney K G, Clark R W, Velikovich A L, Davis J, Maron Y, Kantsyrev V, Safonova A, Oreshkin V I 2006 *J. Quant. Spectrosc. Radiative Transfer* **99** N.1 341

[13] Korolev Yu D, Mesjats G A 1991 *Physics of pulsed breakdown of gases* (Moscow: Nauka) p 225

[14] Mesyats G A, Korovin S D, Sharypov K A, Shpak V G, Shunailov S A and Yalandin M I 2006 *Technical Physics Letters* **32** Issue 1 18

[15] Oreshkin E V, Barengolts S A, Chaikovsky S A, Oginov A V, Shpakov K V, Bogachenkov V A 2012 *Phys. Plasmas* **19** 013108

[16] Stygar W A, Cuneo M E, Headley D I, Ives H C, Leeper R J, Mazarakis M G, Olson C L, Porter J L, Wagoner T C and Woodworth J R 2007 *Phys. Rev. ST Accel. Beams.* **10** 030401

[17] Smirnov V P 1991 *Plasma Phys. Control. Fus.* **33** N.13 1697

[18] Ryutov D D, Derzon M S, Matzen M K 2000 *Rev. Modern Phys.* **72** No.1 167

[19] Haines M G 2011 *Plasma Phys. Control. Fus.* **53** 093001

[20] Smirnov V P, Zakharov S V and Grabovskii E V 2005 *JETP Letters* **81** N. 9 442

[21] Grabovskii E V, Smirnov V P, Vorob'ov O Yu, Dyabilin K S, Lebedev M E, Ostrik A V, Fortov V E 1994 *JETP Lett.* **6** N.1 1

[22] Sanford T W L, Lemke R W, Mock R C, Chandler G A, Leeper R J, Ruiz C L, Peterson D L, Chrien R E, Idzorek G C, Watt R G and Chittenden J P 2002 *Phys. Plasmas.* **9** N.8 3573

[23] Mazarakis M G, Fowler W E, Kim A A, Sinebryukhov V A, Rogowski S T, Sharpe R A, McDaniel D H, Olson C L, Porter J L, Struve K W, Stygar W A and Woodworth J R 2009 *Phys. Rev. ST Accel. Beams* **12** 050401

[24] Kim A A, Mazarakis M G, Sinebryukhov V A, Kovalchuk B M, Visir V A, Volkov S N, Bayol F, Bastrikov A N, Durakov V G, Frolov S V, Alexeenko V M, McDaniel D H, Fowler W E, LeChien K, Olson C, Stygar W A, Struve K W, Porter J, Gilgenbach R M 2009 *Phys. Rev. ST Accel. Beams* **12** 050402

[25] Oreshkin V I 2013 *Phys. Plasmas* **20** 112505

[26] Loeb L B 1939 *Fundamental Processes of Electrical Discharge in Gases* (New York; J. Willey and Sons) p 426.

[27] Oreshkin E V, Barengolts S A, Chaikovsky S A, Oreshkin V I 2012 *Phys. Plasmas* **19** 043105

[28] Braginskii S I 1958 *JETP* **34** N. 6 1068

[29] Martin T H, Seamen J F, Jobe D O 1993 *Proc. 9th IEEE Inter. Pulsed. Power Conf., (Albuquerque, NM, USA)* pp 463–470

International Congress on Energy Fluxes and Radiation Effects (EFRE-2014) IOP Publishing
Journal of Physics: Conference Series **552** (2014) 012022 doi:10.1088/1742-6596/552/1/012022

Oversized interference switches of active resonant microwave compressors

V Avgustinovich[1], S Artemenko[1], V Kaminsky[1], S Novikov[1,*], Yu Yushkov[1], V Osipov[2]
[1]Tomsk Polytechnic University, 30 Lenina ave., Tomsk, 634050, Russia
[2] Institute of Electrophysics, Ural Branch RAS, 106 Amundsen str., Ekaterinburg, 620016, Russia

E-mail: nsa@tpu.ru

Abstract. Interference switches of resonant microwave compressors made of single mode waveguides provide high coupling factors after switching and rapid energy extraction. The disadvantage of these switches is low electrical strength which limits output power values. Oversized waveguides maintain higher electrical strength due to large dimensions of a cross section but mode transformation disturbs the regular operation of the switch. Operation of interference switches with a gaseous discharge gap as a switching element in oversized rectangular waveguides was studied experimentally. Conditions of their effective switching in active resonant microwave compressors were derived. It was shown that the stable microwave pulses of gigawatt power level in S-band and pulses of 0.1 GW pulse power in X-band can be produced in resonant microwave compressors with oversized interference switches. Possible switch designs are discussed.

1. Introduction

Microwave resonant compressors (MRC) produce pulses by accumulating energy in a resonant cavity during a relatively short time and subsequent rapid extraction of the energy into a load. The ultimate output power of a MRC is equal to a power of the travelling wave component in a resonant volume. So the output pulse power of MRC can reach high values, for example hundreds of megawatts in X-band and several gigawatts in S-band, if large volumes for energy storing are used. According to a method of a switching procedure MRCs come under the passive compressor group or the active compressor group. The energy extraction in a passive compressor is switched on by the phase inversion of an input feeding pulse at relatively low power level, the power gain does not exceed 9. In active compressors the switched power is comparable to the travelling wave power in a cavity and switch operation affects main output parameters such as the maximum power and efficiency. The switch design is the main problem in the course of every active compressor development when the switch invariably should introduce low losses and provide the stable rapid energy dumping from large volume cavities.

Interference switches designed as the H-tee are commonly used for rapid energy extraction. The coupling between a cavity and a load is controlled by the phase shift between the wave radiating from the cavity and the wave transferred from the side arm of the tee. The phase is inverted by the gaseous microwave switch having the trigatron type triggering located in the tee side arm. The limited cross section of the waveguide limits the electrical strength of the elements and the output

Content from this work may be used under the terms of the Creative Commons Attribution 3.0 licence. Any further distribution of this work must maintain attribution to the author(s) and the title of the work, journal citation and DOI.
Published under licence by IOP Publishing Ltd

power value. An increase of gas pressure leads to big energy losses in the discharge plasma and a decrease of efficiency. For example the output pulse power for switches located in S-band waveguides of the cross section of 72×34 mm^2 does not usually exceed 200 MW [1] and the efficiency is within the range of 0.2-0.3 at the losses in the switch of 2-3dB. The switching in oversized cavities involving mode transformation increases the limiting output power [2,3], supposedly up to the gigawatt level in X-band, but the design of the referred compressors was meant to provide output pulsewidth values of tens of nanoseconds and they had low power gain.

This report presents the study of switch designs based on oversized waveguides but intended for fast energy extraction and keeping high values of amplification.

2. Oversized rectangular waveguides in an interference switch

The transition of the cavity from the storage mode to extraction one occurs by use of an additional resonator where the switched wave travels [2]. This resonator is strongly coupled with an output line and its reflection factor from the direction of a storage cavity volume is expressed by:

$$\Gamma \approx -1 + \frac{2\beta}{1+\beta}\frac{1}{1+\frac{\delta f^2}{\delta f_l^2}} - j\frac{2\beta}{(1+\beta)}\frac{\delta f}{\delta f_l}\frac{1}{1+\frac{\delta f^2}{\delta f_l^2}}, \tag{1}$$

where δf – difference between operational and resonant frequencies, δf_l – width of resonance curve, β – coupling with the storage volume. As follows from (1) the reflection factor of strongly coupled resonator at the resonant frequency is approximately equal 1. The reflection factor will be – 1 if the frequency will be changed beyond the resonance curve but this means already the phase inversion and switching over to the energy extraction. The reflection factor plotted against the relative frequency change for different coupling values is presented in figure 1.

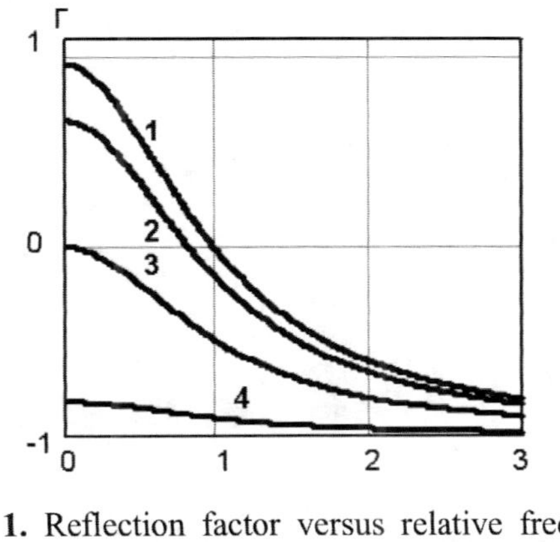

Figure 1. Reflection factor versus relative frequency deviation at following values of coupling: 1 - β = 50; 2 - β = 5; 3 - β = 1; 4 - β = 0.1.

It was assumed the strong coupling between the output line and the switching resonator and the necessary frequency change are enough for developing an effective interference switch no matter what an operational wave mode is used.

Switching resonators made of single mode waveguides are usually connected to storage volumes by T-junctions or by bridge junctions and demonstrate high switching efficiency. Although the cross section limited by the cut-off frequency value limits the electric strength and the output power.

One possible solution of increasing a waveguide cross section is the use of rectangular waveguides with H_{10} operational wave mode [4]. Corresponding output elements of the tee or bridge design keep the switching parameters of single mode analogues, does not require the special mode transformation into the primary mode of an output waveguide and make the maximum output peak power greater by several-fold. These oversized switches, besides usual requirements to the level of multimode transformation, raise some specific conditions to be provided for effective operation of the switch and the compressor. In order to provide switching over to extraction the phase of wave reflected from the tee arm should be changed by about 180^0 along with the change of the frequency beyond the resonance curve by value of $\delta f \approx nf/Q_a$, where $n \geq 3$ and Q_a - quality factor of the tee arm. Although it was found the dimensions of switching arm should be less some value determined by the ratio of the volume parameter transient time to the double time of wave travelling along the storage cavity *T*.

The expression for the limit of the oversized waveguide narrow wall size at the given arm length of L_{arm} and the size of the broad wall *a* was derived:

$$b_{\max} < \frac{z_0 Q_{arm} l^3}{90naL_{arm} \lg\left(\frac{2l}{r}\right)} = \frac{z_0 Q_{arm} \left(0.2Lv_{pl}/v_g\right)^3}{90naL_{arm} \lg\left(\frac{0.4Lv_{pl}/v_g}{r}\right)}, \qquad (2)$$

where v_g – wave group velocity, v_{pl} – plasma propagation velocity, $L_{arm} \approx Tv_g/2$, l – length of the plasma spark channel, r – cross section radius of the plasma channel. As is clear from (2) the size is proportional to $f^{1.5}$. This means the switches are more efficient for higher frequencies of the microwave band when the relative increase of the wall size is higher at given *T*.

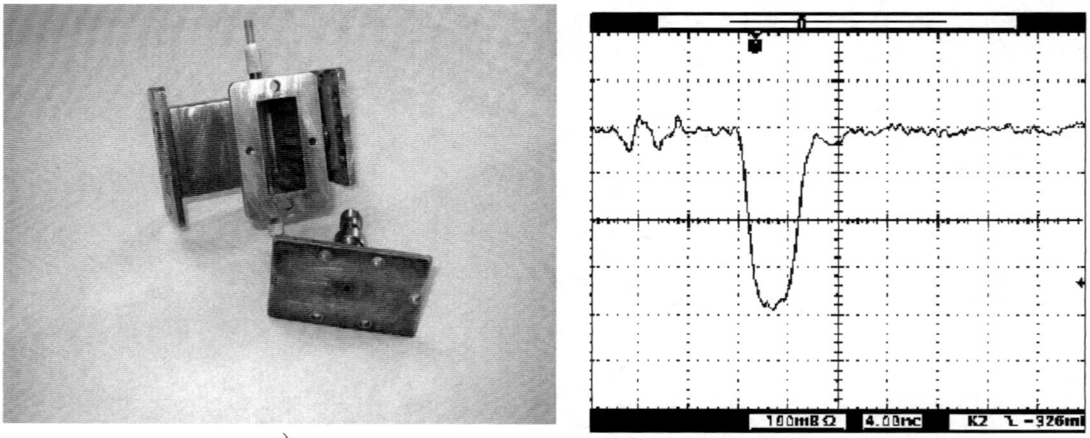

a) b)

Figure 2. Interference switch on the basis of oversized rectangular waveguide (a) and envelope of the formed microwave pulse. Exhibited parameters: pulse width of 3.5 ns, peak power of 2.8 MW, amplification of 17.5 dB.

Experimental study was made in X-band and the tee waveguides had the cross section of 58×25 mm^2. External view of the tee is shown in figure 2a. The longitudinal section of the tee is identical to the single mode one. So when dimensions are precise and operational mode H_{01} is not converted the switch of this type operates similarly to the common switch as there are no physical causes impeding that. It was proved by measuring the transition attenuation in the close state of the switch. The attenuation was 41±2 dB in the frequency range 8800-9500 MHz and that corresponds to the attenuation of a common tee. The tee with H_{01}-mode came to the open state by a slight change of the short-circuited arm parameters. It was verified as well when the switch was installed in the compressor. The calculated double time interval of wave travelling along the storage cavity was 4 ns at the quality factor of 1.6×10^4. That gave the calculated power gain of about 20 dB. The measured gain value was 17.5 dB at the pulse width of 3.5 ns. The primary exciting microwave source was the pulse magnetron generator with the power of 50 kW and so the output pulse power reached 2.8 MW. The switch was triggered by illumination of the discharge gap by an electric discharge spark or by a light beam of the nitrogen laser. The switch discharge was formed in argon at atmospheric pressure in a waveguide volume or in a quartz tube located in the area of maximum electric field and aligned along the electric flux. The typical output pulse envelope is shown in figure 2b. As the figure shows the energy extraction time is equal to the double time of wave travelling along the cavity that is similarly to operation parameters of a singlemode switch.

3. Pack of synchronized switches

Several synchronized switches connected to a single cavity can enable the energy extraction time close to the double time of wave travelling along the cavity volume [5]. The extraction time decreased with increasing the number of switches but the spread of switch triggering was compared to the output pulsewidth recorded at a single switch output. High-level synchronization was reached when the switches were arranged closely to each other i.e. they formed a packet of parallel switches [6]. External view of the switch packet in a rectangular waveguide of the tee, the short-circuit is disconnected, is shown in figure 3a.

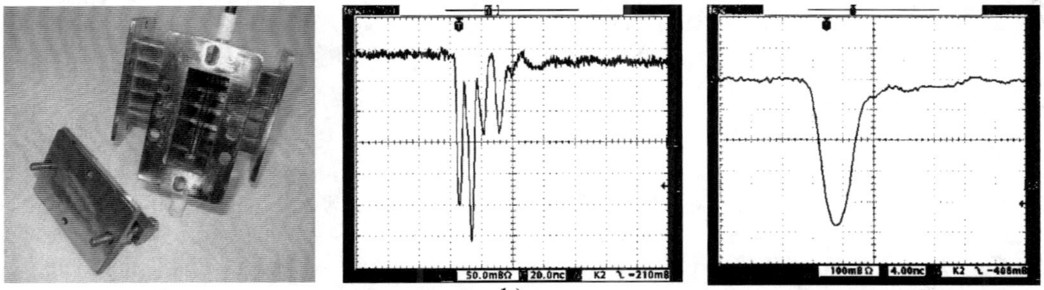

a) b) c)

Figure 3. Packet of switches with mutual side arm. a) layout of the tee; b) output pulse envelope when separate side arms were uncoupled; c) output pulse envelope with switching in mutual side arm.

The key problem of synchronous power extraction is maintenance of identical switching conditions in each gap. First, the electric field strength values in the gaps should be equal and this condition depends on the cavity geometrical arrangement, the type of the wave working mode, the waveguide interior surface quality, the coupling between switches and the cavity. For the proposed design the H_{01} mode of a prismatic cavity is most acceptable. The E-field strength along the larger wall is constant and this contributes to equality of field strength values in the switches. The wave modes H_{01} and H_{10} are easily transformed into each other and so combining the energy extracted through all switches does not become difficult.

The second issue is the quantity of switches in the packet which affects the compressor limiting power. The effective switching corresponds to the switching time less than T and if t_f is a characteristic time of the discharge development then the maximum number of switches is given by $n < T/t_f$. For example, in X-band the acceptable number is in the range 2…5 at $T = 5ns$ and $t_f \approx 2ns$ that is the number is not so great. For higher frequencies the usage of the packet may be efficient as $t_f < 1ns$ and, keeping the same T value, the allowable switch number is larger.

The larger number of switches requires hard locking of discharge triggering. The mutual switching area as a joint oscillation system was supposed to be realized as a mutual switching tee arm. The switching process should be controlled by a single discharge and there should not be a need for special synchronization of switches. But the resonant frequency of the arm should be withdrawn beyond the resonant curve during the time interval about *T*. It may be executed by the discharge having the spark channel length much less than the total length of channels when switching proceeds in uncoupled arms. The wave travelling time along the mutual arm is less than the discharge formation time so the phase inversion of waves reflected by the arm is synchronous and so the switches of the packet come to open state also synchronously. The maximum number of switches is estimated by $n < L/b$, where *b*- size of the narrow wall of the tee waveguide.

Experimental tests in X-band proved that the strong coupling between the switching arms is the requirement for synchronous operation. The coupling in the packet was introduced by different ways e.g. by holes in adjoining switch walls, by slots between the short-circuit and flat ends of side arms and by integrating the side arms into a mutual side arm in the form of an oversized regular waveguide section. The level of synchronism rose with increase of the coupling between the arms. The packet with uncoupled arms is responsible for unsteady process of energy extraction characterized by the spread of output pulses as shown in figure 3b. The simultaneous energy extraction during the double time of wave travelling is reached with the strong coupling. The power gain of the compressor with the oversized storage cavity and the mutual side arm of 58x25 mm^2 cross section was 16.5dB at the pulsewidth of 3.5 ns and the corresponding peak power of 2.2MW. According to estimate the energy extraction through a packet with 5…7 switches could form pulses with the peak power of 0.1GW in X-band and the peak power 1GW to 2GW in S-band. The packet of interference switches may combine parallel resonant cavities fed by different sources into a single high power compression system.

4. Conclusion

The microwave interference switch designed as the H-tee of oversized rectangular waveguides with the wave operational H_{01}-mode was experimentally studied. The results proved the possibility of energy extraction from storage cavities by this type of a switch. The packet of single mode switches was developed with a view to provide their synchronous operation for energy extraction. The synchronous operation was demonstrated, the necessary conditions were determined and the number of switches in the packet was estimated. Limiting power values attainable with these switches in X-band and S-band are given.

Acknowledgments

The work was partially supported by The Ministry of Education and Science of Russian Federation within the program "Nauka", project N 1814.

References

[1] Alvarez R A, Byrne D P and Johnson R M 1986 *Rev. Sci. Instrum.* **57** 2476
[2] Gorbachev A, Ivanov O and Isaev V 1998 *Tech. Phys. Letters* **24** 791
[3] Vikharev A, Gorbachev A, Ivanov O, Isaev V, Koldanov V, Kuzikov S, Hirshfield J and Gold S 2008 *Radiophysics and Quantum Electronics* **51** 597
[4] Avgustinovich V, Artemenko S and Zhukov A 2013 *Tech. Phys. Letters* **39** 492
[5] Avgustinovich V, Artemenko S, Chumerin P and Yushkov Yu 2000 *J. Tech. Phys. Letters* **45** 1475
[6] Avgustinovich V, Artemenko S and Arteev M 2013 *Tech. Phys. Letters* **39** 1014

International Congress on Energy Fluxes and Radiation Effects (EFRE-2014) IOP Publishing
Journal of Physics: Conference Series **552** (2014) 012023 doi:10.1088/1742-6596/552/1/012023

An avalanche of runaway electrons formed in an air discharge

E. V. Oreshkin,[1,2,*] S. A. Barengolts,[1,2] S. A. Chaikovsky,[1,3] and V. I. Oreshkin[3,4]

[1]P. N. Lebedev Physical Institute, RAS, Leninsky pr. 53, Moscow, 119333, Russia
[2]A. M. Prokhorov General Physics Institute, RAS, Vavilova 38, Moscow, 119333,Russia
[3]Institute of High Current Electronics SB RAS, 2/3 Academichesky ave., Tomsk, 634055, Russia
[4]Tomsk Polytechnic University, Tomsk Tomsk Polytechnic University, pr. Lenina 30, Tomsk, 634050 Russia

E-mail: oreshkinev@scalpnet.ru

Abstract. A numerical simulation of a beam of runaway electrons formed from an individual emission zone on a cathode has been performed for discharges in air of atmospheric pressure. The model is based on solving numerically two-dimensional equations of motion for the electrons and allows one to describe the dynamics of the fast electrons injected from the surface of the emission zone. In calculations it was supposed that the electric field at the surface of the emission zone is enhanced due to which conditions are realized for the electrons injected from the surface to switch into the mode of continuous acceleration.

1. Introduction

Runaway electrons (RE) were discovered in atmospheric pressure discharges in the late 1960th [1]. An RE beam passing through a gas initiates, due to avalanche multiplication of fast electrons, breakdown, which has been termed runaway electron breakdown (REB) [2, 3]. It is supposed that REBs take place in lightning discharges. The possibility of existence of REBs gave impetus to both theoretical [2-4] and experimental research [5,6] of RE avalanches. An RE avalanche was observed in an atmospheric pressure air discharge when the average electric field in the gap, E_{av}, was much greater than the dc breakdown electric field, E_{br} [6]. These experiments have shown that the pulsed RE current in discharges with $E_{av} >> E_{br}$ has the following structure. Its first portion of duration some tens of picoseconds, showing a strongly pronounced current peak, consists of high-energy runaway electrons. These are the electrons emitted from the cathode region where the field is enhanced due to the cathode geometry. Subsequently, with some delay, a second peak of RE current or a section of slowly decreasing current arises. This was accounted for by the formation of an avalanche of runaway electrons. The duration of the secondary electron beam was of the order of 100 ps and its current could be an order of magnitude greater than the current of the primary runaway electron beam. This took place if it was possible to avoid the decrease in electric field in the gap caused by the passage of the primary beam.

Laboratory investigations of REs in high pressure gas discharges are carried out both at microsecond rise times of the gap voltage [1,7,8] for $E_{av} < E_{br}$ and at subnanosecond rise times (that is, in overvolted electrode gaps) [5,6,9] for $E_{av} >> E_{br}$. In the latter case, the mechanism of formation of the runaway electron beam is the subject of discussion. The matter is that as the voltage across the gas

Content from this work may be used under the terms of the Creative Commons Attribution 3.0 licence. Any further distribution of this work must maintain attribution to the author(s) and the title of the work, journal citation and DOI.
Published under licence by IOP Publishing Ltd

diode increases, conditions arise for the runaway of the electrons emitted from the cathode microprotrusions [10] that have formed as a result of the ionization of the gas by the voltage prepulse and also of the electrons emitted from the boundary of the plasma cloud formed around cathode microprotrusions. The well conducting plasma cloud, which is a source of REs, is formed due to impact ionization of the gas by priming electrons emitted from the cathode.

The contribution of the electrons produced during the prepulse can be decreased by shortening it and by decreasing the voltage amplitude; the amount of field emission electrons is inappreciable in view of the small area of emission from cathode microirregularities. The emission current can increase abruptly only as a result of the transition of field emission to explosive emission. The electrons accelerated to high velocities in the high electric field region near microprotrusions [11] appear in a lower field region. If the density and energy of the electrons are high enough, this will lead to the formation of a virtual cathode [12] and, as a consequence, to disruption of the beam of the runaway electrons emitted by the cathode. Iin this work, we consider the emission zone formed by the plasma cloud surrounding an emitting microprotrusion to be a source of runaway electrons.

As mentioned above, in highly overvolted discharges, the source of electrons are individual emission zones (EZ) whose number in discharges with tubular cathodes can be more than ten [13]. Below we consider the formation of an RE beam for the case with a single emission zone. The size of the emission zone in atmospheric pressure air estimated by Babich et al. [1] and, in more detail, by Barengolts et al. [11] was obtained to be 50–150 μm.

2. Simulation of the evolution of a runaway electron beam in highly overvolted gas discharges

Suppose that an EZ is a plasma hemisphere of radius R_{ez}. In this case, a negative superficial charge is concentrated at the hemisphere surface which screens the external electric field. The number of electrons at the hemisphere surface (in the Debye layer) is given [14] by

$$N_d = \frac{E_{ez} R_{ez}^2}{2e}, \tag{1}$$

where E_{ez} is the electric field strength at the EZ surface. As conditions for runaway of electrons from the EZ surface are realized, electrons are injected with the current density [14]

$$j_{em} \approx \frac{3}{2\pi}\sqrt{\frac{e}{m\tilde{l}}}E_{ez}^{3/2}, \tag{2}$$

where $\tilde{l}$ is the mean free path of an electron (in elastic scattering by gas atoms). For $E_{ez} \approx E_{cr}$ (E_{cr} is the critical electric field strength, when all electrons switch into the mode of continuous acceleration) the current density of electrons injected from the surface of an EZ in air at atmospheric pressure is ~10^6 A/cm^2. The velocity of the injected electrons is determined by the drift velocity $v_D \approx \sqrt{\frac{e}{m}\tilde{l}E_{ez}}$. The characteristic time of electron injection from an EZ surface, determined by the effluxion of electrons from the Debye layer, is estimated [14] as $\tau_{em} \approx \frac{1}{6}\sqrt{\frac{m\tilde{l}}{eE_{ez}}}$.

We consider coarse particles each consisting of N electrons; the motion of a coarse particle is described a two-dimensional numerical model based on the Monte-Carlo method [5]. The basic equation of the model is an equation describing the variation of the momentum of an electron:

$$\frac{d\mathbf{p}}{dt} = e\mathbf{E} - \mathbf{F} - (\Delta\mathbf{p})_{el}, \tag{3}$$

where $\mathbf{p}$ is the electron momentum, $\mathbf{F}$ is the braking force, and $(\Delta\mathbf{p})_{el}$ is the variation of the electron momentum on elastic scattering. The elastic electron-atom collisions were taken into account using the

Monte-Carlo method [14]. Two components of the momentum were calculated: p_x, parallel to the vector **E** and p_y, perpendicular to the vector **E**.

The electric field strength in Eq. (3) is the sum of two terms: $\mathbf{E} = \mathbf{E}_{av} + \mathbf{E}_{ch}$, where $\mathbf{E}_{av}$ is the average electric field depending on time and $\mathbf{E}_{ch}$ is the electric field induced by the space charge. The space charge is generated by three species: the electrons injected from the EZ surface, the secondary electrons produced by fast particles, and ions. The equation of motion for ions is not solved. As their mobility is much less than that of electrons, they remain localized at the places of production of electron-ion pairs. The space charge field is assumed to be spherically symmetrical and it is found by solving numerically Maxwell's equations

$$div\mathbf{E}_{ch} = 4\pi e(n_e - n_i), \qquad (4)$$

where n_e is the total density of the electrons injected from the EZ surface and of the secondary electrons and n_i is the ion density. The boundary conditions to Eq. (4) specified at the EZ surface is $E_{ch}(R_{ez}) = \dfrac{2e(N_d - N_{em})}{R_{ez}^2}$, where N_{em} (t) is the number of electrons injected from the EZ surface within the time t; that is, after the injection, which lasts less than 1 ps, we have $E_{ch}(R_{ez}) = 0$.

3. Results of numerical simulations

The numerical simulation of the formation and acceleration of an RE beam was performed for air of atmospheric pressure and for conditions close to the conditions of the experiment [15]. The time dependence of the accelerating voltage amplitude was set by the relation

$$U(t) = U_0 \sin\frac{\pi}{2}\frac{t}{\tau_g},$$

where $U_0 = 160$ kV and $\tau_g = 100$ ps. In calculations, as well as in the experiment [15], the length of the electrode gap L was varied. The radius of the runaway electron EZ was set the same for all calculation runs: $R_{ez} = 0.01$ cm [1, 11]. As mentioned above, in experiments with highly overvolted gas discharges, the time of RE injection is determined by some critical electric field at the cathode. In the experiments described elsewhere [10], this critical field was determined from the time-of-flight characteristics of an electron in vacuum, and it turned out to be ~1500 kV/cm. However, in gas, unlike in vacuum, an electron undergoes many collisions with gas atoms and thus loses its energy and diverges from a rectilinear path. Both of these the factor lead to an increased time of flight of an electron in the electrode space and, hence, to an overestimated field strength [10]. Therefore, in the calculations discussed below, the injection of electrons from the EZ surface was assumed to begin as the electric field at the cathode edge, E_{ez}, reached 1000 kV/cm, the same value for all calculation runs. Accordingly, the time of the beginning of injection was calculated by the formula $U(t) = \dfrac{E_{ez}L}{\beta}$, where β = 10 is the field enhancement factor.

As in all calculation runs the values of E_{ez} and R_{ez} were the same, the number of electrons injected from the EZ surface, equal to the number of electrons in the Debye sphere (1), was the same: $N_d = 5{\cdot}10^8$. The velocity vector of the injected particles was directed normal to the EZ surface. The energy of the injected particles, depending on the drift velocity of electrons in a field equal to E_{ez}, was determined as $\varepsilon_k = \dfrac{mv_D^2}{2}(1 \pm 0.5) \approx 10 \pm 5$ eV. The spread in initial electron energies was set in a random manner in the range from 5 to 15 eV. The velocity vector of the injected particles was directed normal to the EZ surface. The initial positions of particles on the EZ surface (hemisphere of radius R_{ez}) were also set in a random manner.

The results of the calculations are presented in Figs. 1. Figure 1 gives the current of a runaway electron beam as a function of time for three runs with the electrode separation equal to 0.25, 1.0, and 1.25 cm. The vertical dashed lines mark the time of the beginning of injection of the electron beam from an EZ, lines 1 represent the voltage across the electrode gap as a function of time, lines 2 the total current of the RE beam, and lines 3 the current of the RE avalanche. It was assumed that the RE avalanche current was formed by electrons that had been produced more than 200 μm away from the EZ surface.

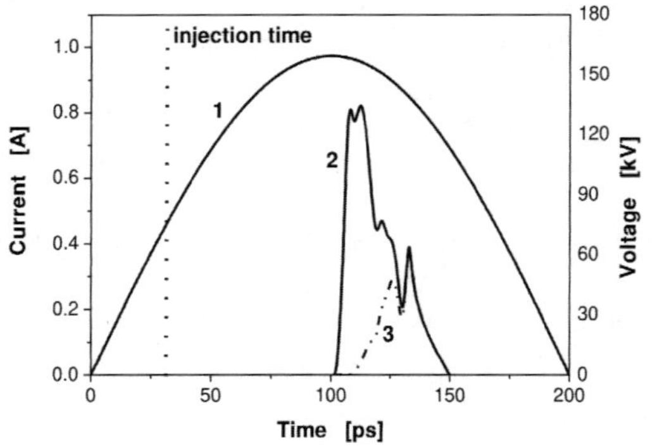

a)

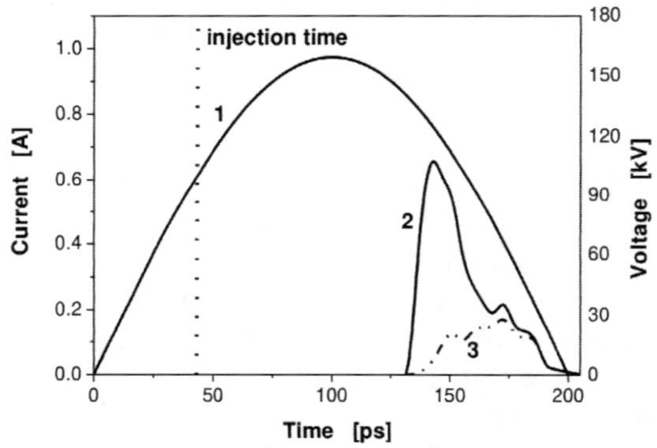

b)

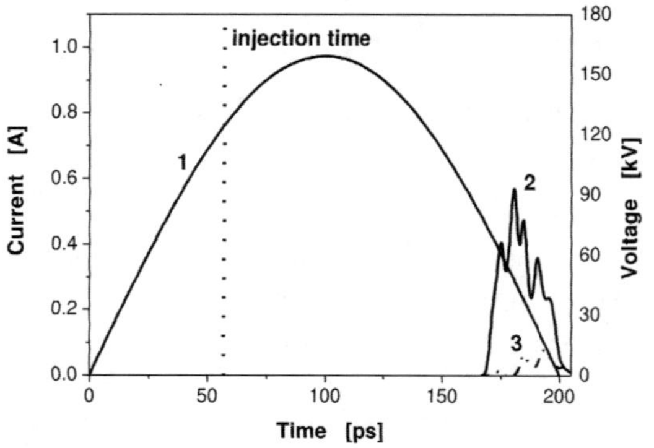

c)

Figure 1. Waveforms of gap voltage (curve *1*), total RE beam current (curve *2*), and RE avalanche current (curve *3*) calculated for an electrode separation of 0.5 (a), 1.0 (b), and 1.25 cm (c).

As can be seen from Fig. 1, in all calculation runs the RE current amplitude is close to 1 A with a current pulse FWHM of 12–25 ps and full duration of 50–100 ps. The peak in the beginning of the current pulse is caused by the electrons injected from the EZ surface and the extensive section by the avalanche electrons. Figure 1 clearly shows that avalanche processes play a great part in the formation of an RE beam at small lengths of the electrode gap, just as it was observed in the experiment [6]. Therefore, avalanche processes stronger affect the formation of an RE beam in shorter electrode gaps.

Acknowledgements
The work was supported in part by the Russian Foundation for Basic Research # 15-08-03284-a, Scholarship of the President of the Russian Federation grant # SP-1627.2013.1, Dynasty Foundation, program of the Presidium of RAS # 29P, grant support for young PhDs MK-1439.2014.2.

References

[1] Babich L P, Loiko T.V and Tsukerman V A, 1990 *Usp. Fiz. Nauk*, **60** (7), 49–82.
[2] Gurevich A V, Milikh G A and Rousel-Dupre R, 1992 *Phys. Lett. A*, **165**, 463.
[3] Gurevich A V and Zybin K P 2001 *Usp. Fiz. Nauk*, **171** (11b), 1178–1199.
[4] Oreshkin E V, et al. 2012 Characteristic length and enhancement time of a runaway electron avalanche in strong electric fields *Technical Physics Letters* **38.7** p 604-608.
[5] Tarasenko V F and Yakovlenko S I 2004 Usp. Fiz. Nauk, **174**, 953.
[6] Gurevich A V, Mesyats G A, Zybin K P, et al., 2011 *Phys. Lett. A*, **375**, 2845–2849 .
[7] Oreshkin E V, et al. 2011 Bremsstrahlung radiation from high-energy electrons in wide gas-filled gaps *Technical Physics Letters* **37.6** p 582-585.
[8] Oreshkin E V, et al 2012 Bremsstrahlung of fast electrons in long air gaps *Physics of Plasmas* **19.1** 013108.
[9] Levko D, Tarasenko V F, Krasik Ya E J. 2012 Appl. Phys., **112**, 073304.
[10] Yalandin M I, Mesyats G A, Reutova A G, et al 2011 *Pis'ma Zh. Tekh. Fiz.*, **37** (8), 56–54.
[11] Levko D, Yatom S, Vekselman V, Gleizer J Z, et al 2012 *J. Appl. Phys.*, **111**, 013304.
[12] Barengolts S A, Mesyats G A and Perelshtein E A 2000 *Zh.Eksper.Teor.Fiz.*, **91**, 1176.
[13] Tarasenko V F, Baksht E H, Burachenko A T et al. 2008 *Fiz. Plazmy*, **34**, 12.
[14] Oreshkin E V, et al. 2012 Simulation of the runaway electron beam formed in a discharge in air at atmospheric pressure *Physics of Plasmas* **19.4** 043105.
[15] Mesyats G A, Korovin S D, Sharypov K A, et al. 2006 *Pis'ma Zh. Tekh. Fiz.*, **32** (1), 35–44.

Study on nanosecond pulsed electron beam generation

D Ponomarev*, **G Kholodnaya, G Remnev, M Kaikanov, R Sazonov**
National Research Tomsk Polytechnic University, Lenin Avenue, 30, Tomsk, Russia

E-mail: ponomarev8105@mail.ru

Abstract. The paper presents the findings of an investigation on volt-ampere characteristics of the diode with explosive emission cathodes of different constructions (blade metal-dielectric (MD-cathode) and solid graphite cathodes) under the change of the anode-cathode gap in wide ranges. The investigations were carried out using the TEA-500 pulsed electron accelerator. The total current of the electron beam was measured using the Faraday cup (FC). A 0.5-mm foiled glass fiber laminate was used as an emitting edge of the cathode in the experimental study with the explosive emission blade MD-cathode. Based on the obtained results, the conclusion was made that the graphite cathode has the most effective efficiency factor.

1. Introduction

Radiation technologies are applied in many fields of science and technology: radiation treatment of medical articles, radiation doping of semiconductors, ion implantation, radiation treatment of the articles of electronic engineering, application of the ionizing radiation in environmental protection, radiation treatment of food products, etc.

The application of the powerful electron accelerators which generate the submicrosecond electron beams with the current amplitude up to several kA is rather promising [1]. Nowadays, different types of cathodes are developed and used. For example, the authors of [2-3] suggest using the planar cathodes from different metals (copper, tungsten, molybdenum) as electron sources. The works [4-5] present the investigation findings of the explosive emission cathodes made from carbon-based material. The disadvantages of such a type of the cathode are emission instability from a pulse to a pulse and emission distribution nonuniformity over the cathode area.

In [6], the diode with a multipoint explosive emission cathode located at external magnetic field. The duration of an optimal pulse is reached when $R > 15\ \Omega$. The authors emphasize that the increase in R up to several kΩ causes the increase in pulse duration by 15 %. In [7], the research results of the volt-ampere characteristics (VAC) of the planar diode with a multipoint explosive emission cathode under the agreement of the diode impedance with an output resistance of the nanosecond generator are shown. The diode impedance with needle multipoint cathodes and carbon-based cathode is obtained during the electron beam generation. The investigations are performed using the TEA-500 pulsed electron accelerator. In [8], the research results of the VAC of the planar diode with a graphite explosive emission cathode of the TEA-500 pulsed electron accelerator are presented.

The work of the diode during the electron beam generation can be divided into two stages. At the first stage, after the voltage is applied to the diode, the diode current is limited by the emissivity of the cathode. For the graphite cathode with a diameter of 45 mm, the duration of the first stage is 15-20 ns, and the increase is possible up to 30-40 ns when the voltage amplitude is decreased. The diode perveance at the first stage of the diode operation is much less than the perveance calculated according to the Child-Langmuir ratio. The TEA-500 pulsed electron accelerator used in [7-8] is also used in the current work. The purpose of the current work was the experimental study of VAC of the diode with the explosive emission cathodes of different construction (blade MD and solid graphite cathodes) under the change of the anode-cathode gap within wide ranges from 10 to 20 mm.

Content from this work may be used under the terms of the Creative Commons Attribution 3.0 licence. Any further distribution of this work must maintain attribution to the author(s) and the title of the work, journal citation and DOI.
Published under licence by IOP Publishing Ltd

2. Experiment

The works on the study of the generation of the pulsed nanosecond electron beam were performed on the TEA-500 accelerator. The main parameters of the accelerator: an accelerating voltage from 350 kV to 500 kV; an electron beam current from 6 kA to 11 kA; a half-amplitude pulse duration of 60 ns; a pulse energy to 200 J; a pulse repetition rate up to 5 pulses per second [9]. In the investigations, we used the diode with explosive emission blade MD-cathode [10]. A 0.5-mm foiled glass fiber laminate was used as an emitting edge of the cathode (figure 1). The effective diameter of the cathode is 62 mm, the height of MD-blades is 10 mm, and a number of blades is 11 pieces. The value of the accelerating gap changed within the interval from 12 to 22 mm. The application of metal-dielectric enables to initiate the explosive electron emission in triple points of the emitting edge of the cathode, that is it enables to increase the stability of the parameters of the electron beam during a large amount of pulses. To limit the impact of the edge effect on the generation of the electron beam in the construction of the cathodes, the shield is used.

Figure 1. A blade MD-cathode with a shielding electrode of the TEA-500 pulsed electron accelerator.

The application of the shield in the suggested geometry makes it possible to decrease the electron emission from the peripheral part of the cathode, that is varying the location of the shield enables to change the diode impedance under the constant values of the cathode-anode (C-A) gap. Consequently, the parameters of the generated pulsed nanosecond electron beam are changed.

3. Results and discussions

An optimal cathode-anode gap was chosen by determining the maximum energy applied to the explosive emission diode with a blade MD-cathode. Figure 2 presents the dependence of the energy on a C-A gap, the energy was calculated by 10 averaged pulses, the dispersion of the parameters did not exceed 7 %.

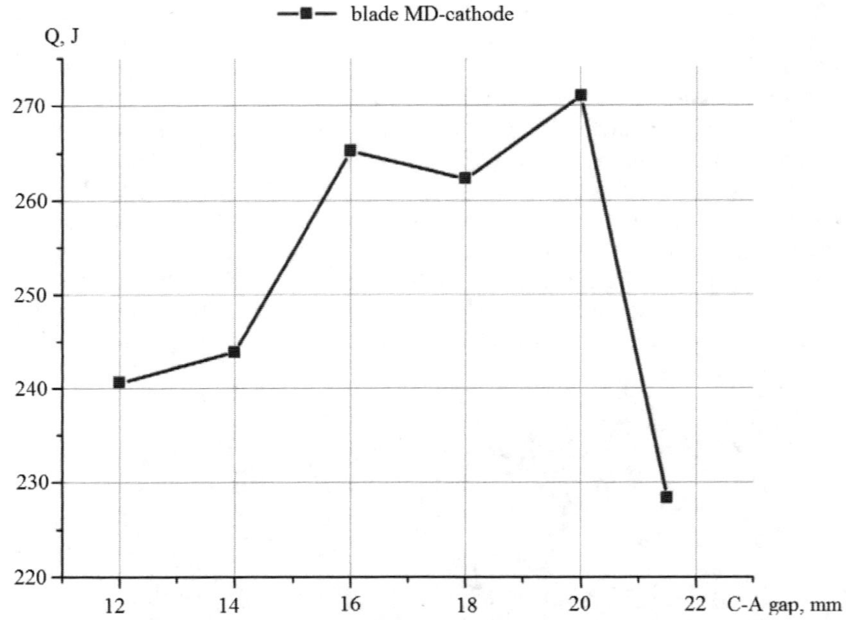

Figure 2. The dependence of the energy on the cathode-anode gap.

Figure 2 shows that the maximum energy applied to the explosive emission diode with a blade MD-cathode is achieved when the C-A gap is 20 mm. The oscillograms presented in figure 3 are obtained with this C-A gap:

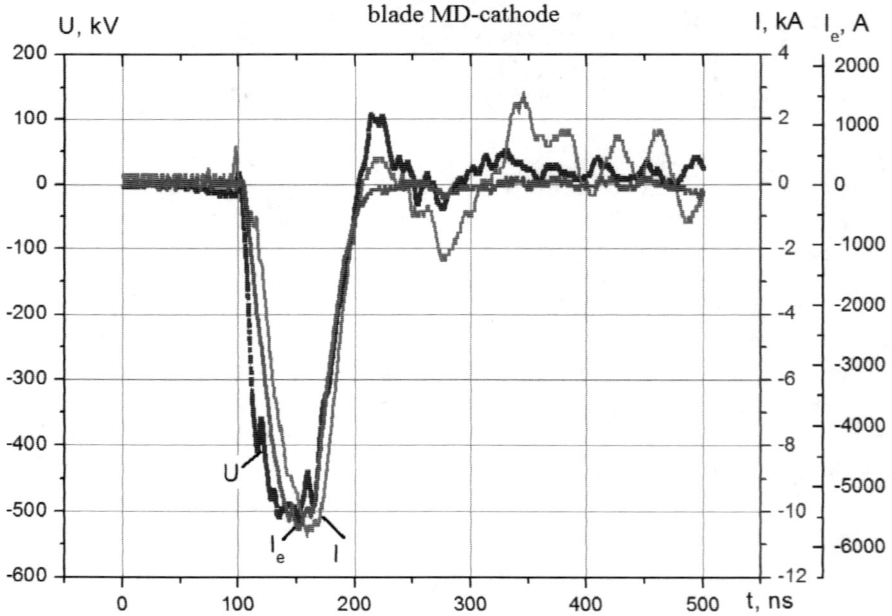

Figure 3. The voltage and current oscillograms which are typical to the 20-mm C-A gap. The MD cathode.

Journal of Physics: Conference Series **552** (2014) 012024 doi:10.1088/1742-6596/552/1/012024

The oscillograms show that the value of the electron beam current was ~5.5 kA. However, this value is lower compared to the case of using the graphite cathode with a diameter of 60 mm in the explosive emission diode. Figure 4 presents the oscillograms obtained at a C-A gap of 14 mm, and figure 5 shows the experimental impedances for the diodes with a blade MD cathode at a C-A gap of 20 mm and with a planar graphite cathode at a C-A gap of 14 mm.

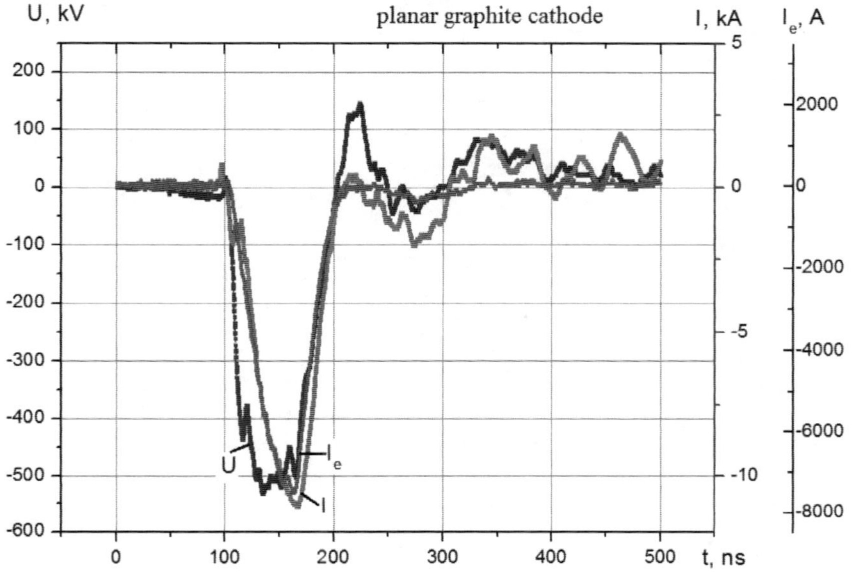

Figure 4. The voltage and current oscillograms which are typical to the 14-mm cathode-anode gap. The graphite cathode.

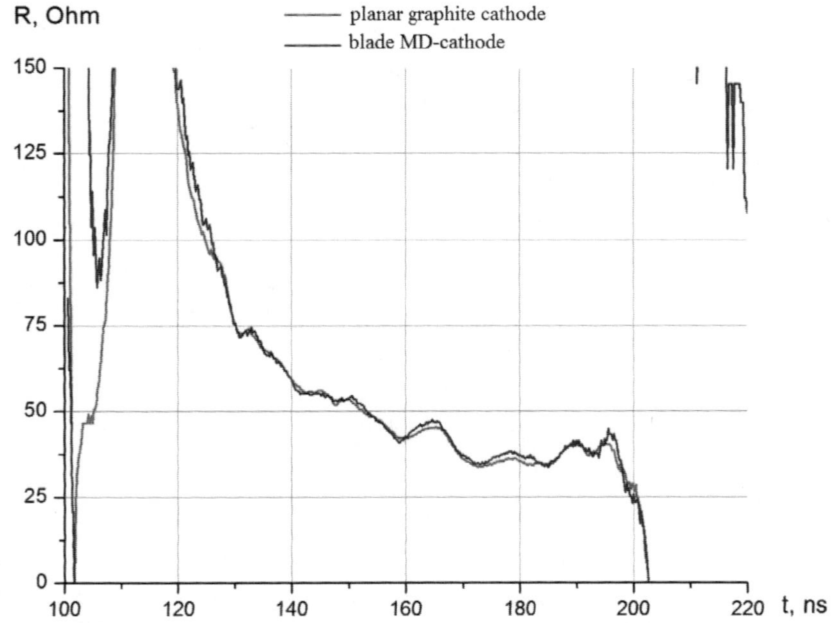

Figure 5. Comparison of the experimental impedances for different cathodes.

The decrease in the electron beam current is caused by the large C-A gap in the case of using the MD-cathode, and as a consequence, the major part of the electrons does not hit the output window (90 mm). To determine the impact degree of the electric field distribution on the efficiency of the electron beam ejection into the atmosphere, we carried out the experiments with a blade cathode with a shielding electrode moving to different V (figure 6). The value V=0 corresponds to the case when the shield is at the same level with the surface of the cathode. In the experiments, V was of the following values: 2; 4; 5; 7 mm (the cathode is dipped into the shield).

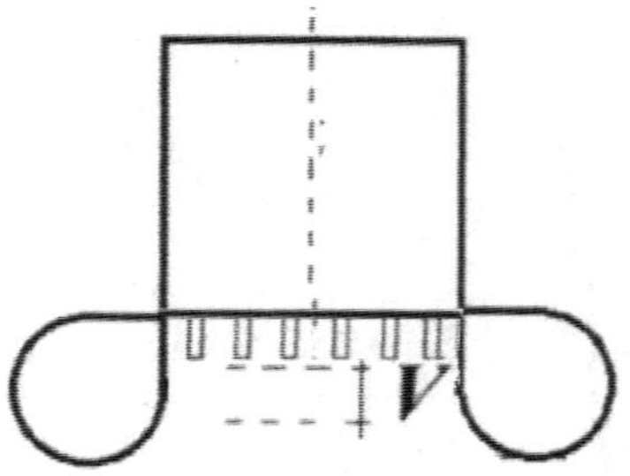

Figure 6. The escape of the shielding electrode from the cathode.

Figure 7 presents the oscillograms of the currents of the electron beam ejected into the atmosphere under different value of V.

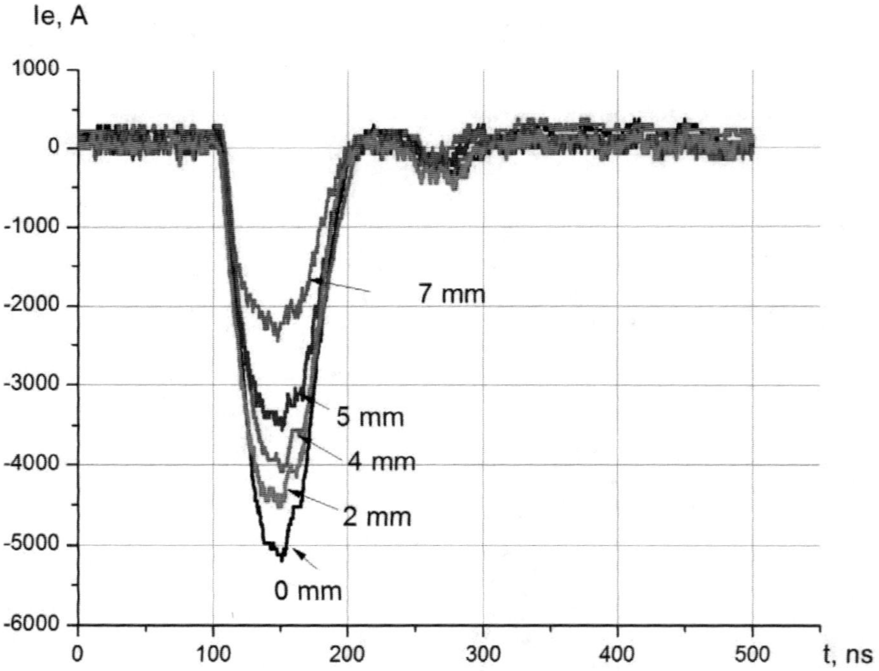

Figure 7. The electron beam current oscillograms under different value of V.

Figure 7 shows that the increase in the shielding electrode escape in relation to the cathode causes the decrease in the electron beam current, that is associated with the increase of the emitting surface (a shielding electrode), and consequently the smaller part of the electrons hit the output window.

4. Conclusion

The obtained results show that when the blade MD-cathode is used, the ejected current is less compared to the graphite cathode. The voltage and current variations after the main pulse for the blade MD cathode (figure 3) are higher than in the case of using the graphite cathode (figure 4), that indicates the limited matching of the accelerator operation mode. The application of the shielding electrode did not enable to increase the electron beam. The maximum value of the electron beam was ~ 4.5 kA, when the blade MD-cathode with the shielding electrode is used. The given value is 1.5 times less than the values of the electron current obtained when using the graphite cathode. For the given construction of the TEA-500 pulsed electron accelerator, which is used in plasma chemical processes, the graphite cathode is best applied. The further work in improving the cathode will be directed to the increase of the blades with the same geometrical size, that will enable to increase the emissive surface and thus to decrease the C-A gap.

References

[1] Bugaev S P, Litvinov E A, Mesyats G A and Proskurovskii D I 1975 *Usp. Fiz. Nauk* **115** 1 p 101-17.
[2] Mesyats G A, Ivanov S A, Komyak N I and Peliks E A 1983 *Powerful Nanosecond Pulses of X Radiation* (Moscow: Energoatomizdat).
[3] Abramyan E A, Alterkop B A and Kuleshov G D 1984 *Intense Electron Beams* (Moscow: Energoatomizdat).
[4] Selemir D V, Alekhin B V, Vatrunin E V et al. 1994 *Plasma Phys. Rep* **20** 621.
[5] Kostov K G, Nikolov N A at. al. 1992 *Appl. Phys. Lett* **60** 21 p 2548–52.
[6] Kovalev V G, Komarov O L, Pechersky O P, Savelyev Yu M, Tkachenko K I and Engelko V I 1990 *J. Tech. Phys.* **59** 1 p 133–139.
[7] Pushkarev A I, Novoselov Y N, Sazonov R V 2008 *J. Tech. Phys.* **53** 3 p 357–62.
[8] Isakova Yu I, Kholodnaya G E and Pushkarev A I 2011 Adv. High Energy Phys 2011 2011 ID 649828.
[9] Gunin A V, Landl V F, Korovin S D, Mesyats G A and Rostov V V 1999 *JETP Letters* **25** 25 p 84–94.
[10] Remnev G E, Furman E G, Pushkarev A I, Karpuzov S B, Kondratyev N A and Goncharov D V 2004 *Instr. Experiment. Tech.* **3** p 130–34.

Journal of Physics: Conference Series **552** (2014) 012025 doi:10.1088/1742-6596/552/1/012025

Features of silicon-containing coatings deposition from ablation plasma formed by a powerful ion beam

R Sazonov*, **G Kholodnaya, D Ponomarev, G Remnev, I Khailov**
National Research Tomsk Polytechnic University, Lenin Avenue, 30, Tomsk, Russia

E-mail: sazonr@mail.ru

Abstract. This paper presents the research of features of silicon-containing coatings deposition from ablation plasma, which is formed by a powerful ion beam at the influence on a microsized pressed powder of SiO_2. Experimental research have been conducted with a laboratory setup based on a TEMP-4M pulsed ion accelerator in a double-pulse forming mode; the first is negative (300-500 ns, 100-150 kV), and the second is positive (150 ns, 250-300 kV). A beam composition: C^+ ions (60-70 %) and protons, the ion current density on the target is 25±5 A/cm^2. An electron self-magnetically insulated diode has been used to generate the ion beam in the TEMP-4M accelerator. The properties of obtained silicon-containing films have been analyzed with the help of IR spectroscopy. A surface structure has been studied by the method of scanning electron microscopy.

1. Introduction

The durability of traditional materials used in the industrial power engineering in high-temperature corrosive environments is low. The resources on the increase of durable high-temperature stability and resistance to chemical and abrasive wear among various steels based on iron with alloy additives used in heat-power engineering have practically been depleted. It leads to the necessity to study the possibilities of new high-temperature corrosion-resistive materials application and high technologies of products manufacturing on their basis. Nowadays, the most prospective material is silicon-containing coatings, allowing to obtain desired combinations of properties, such as a high strength-to-weight and stiffness-to-weight ratio, heat resistance, wearability, high thermal conductivity and thermal properties, radiation durability, etc [1]. It is known that antireflection coatings (ARCs) became one of the basic components for mass-production manufacturing of solar batteries in 2006. The given coatings are obtained with the help of various methods (thermal evaporation, plasma-chemical deposition from a vapour phase, etc.). The first works on surfacing the silicon-containing coatings on a GaAs substrate were investigated and published in 1976 in the paper [2]. In this work the compounds of SiO_2 and Si_3N_4 were used as silicon-containing coatings. As a result, the authors obtained complex silicon-containing coatings, which had a high heat resistance. The paper [3] presents the method of obtaining SiO_2 films on semi-spherical kovar arbors of various sizes. SiO_2 was deposited on a kovar substrate by a method of magnetron sputtering. The thickness of a silicon-containing coating was 13 micrometers in the depth of 0.3 micrometers.

In the paper [4] monocrystalline titanium and silicone coatings (one-layer TiO_2, double-layer SiO_2/TiO_2, tri-layer $SiO_2/SiO_2 - TiO_2/TiO_2$) were obtained by a sol-gel method. The measured average optical reflection power (400-1000 nm) is about 9.3, 6.2, 3.2 % for one-layer, double-layer, and tri-layer coatings, respectively. In the work [5] the authors show that with the use of silicon-containing

Content from this work may be used under the terms of the Creative Commons Attribution 3.0 licence. Any further distribution of this work must maintain attribution to the author(s) and the title of the work, journal citation and DOI.
Published under licence by IOP Publishing Ltd

SiO_2 coatings it is possible to decrease the disturbances in photoresistors without using additional antireflection coatings. The authors of the paper [6] added hydrophobic SiO_2 nano-particles in polyurethane lacquer coatings by means of mechanical mixing and ultrasonic dispersion. As a result, they obtained nanocomposite coatings, the corrosion resistance of which was improved (the corrosion rate of a nanocomposite coating was, approximately, 20-70 % lower than the rate of the original polyurethane lacquer). Silicon-containing coatings are often combined with the coatings based on titanium dioxide [7-9]. Complex coatings $TiOxSiO_2$ (X – mole %) were obtained by the sol-gel method. In the paper [9] nano-crystalline films $TiOxSiO_2$ with a high photocatalytic activity were prepared on a glass substrate. The films were heat treated at 500° C for TiO_2 crystal growing. Obtained $TiOxSiO_2$ films were put into a methyl orange (MO) water solution (10 mgL^{-1}) and irradiated by ultraviolet (UV). As a result, it was shown that the TiO_xSiO_2 films possessed a high photocatalytic activity. The authors used the UV-spectrophotometric method to control by the methyl orange decrease the reconstruction of the main absorption peak at 464 nm.

The results showed that the complete corrosion was reached after 1.5 hours. Despite the fact that at present silicon-containing coatings are obtained, mostly, by the sol-gel method, it is necessary to point to the main disadvantages of this method. Firstly, this is a complex and multistage chemical process, and, secondly, it is necessary to use costly catalysts that must be extracted from the final product on completion the process. Therefore, the given paper presents the results of the investigations of silicon-containing coatings deposition from ablation plasma formed by a powerful ion beam.

2. Experimental

Experimental investigations were conducted at the TEMP-4M pulsed ion accelerator with a self-magnetically insulated diode with the following parameters: accelerating voltage – 200-250 kV, current on the target 150-200 A/cm^2, ion beam pulse duration (C^+ +H^+) 80 ns, the ion beam energy density on the target reached 4 J/cm^2. The pulsed ion accelerator consists of the following basic units: a pulse voltage generator, a double forming line, main and pre-TR gas dischargers, a vacuum strip diode, consisting of polarizing and grounded electrodes. The polarizing graphite electrode of the diode is connected through the pre-TR gas discharger with the internal electrode of the double forming line. The middle DBF electrode is connected with the pulse voltage generator [10].

The method of silicon-containing coatings deposition from ablation plasma formed by a powerful ion beam is as follows: the pulsed ion beam generated by the TEMP-4M accelerator falls on the target and dissipates it, making the ablation plasma. From this ablation plasma the deposition of the dissipated material on the substrate occurs. Figure 1(a) demonstrates the experimental scheme of the target, substrate and ion beam layout as well as the photo of the experiment (Figure 1(b)).

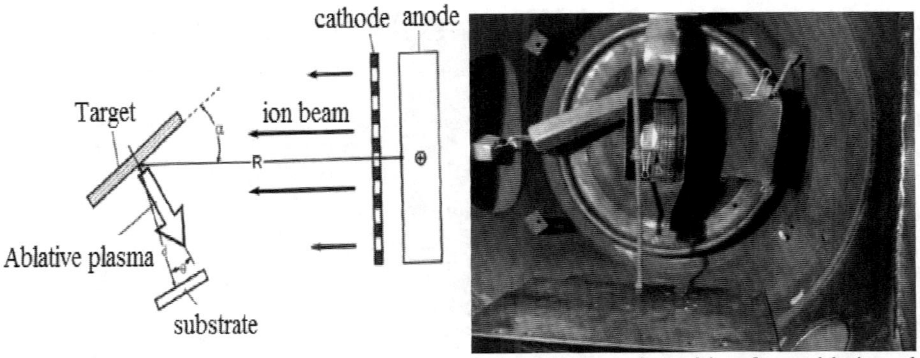

Figure 1. Scheme of the experiment on silicon-containing coatings deposition from ablation plasma formed by a powerful ion beam.

The target is installed inclined to the ion beam at the distance R from the anode of the diode. The dissipated surface of the target is inclined at an angle a=30–60° (the indication from the surface) towards the ion beam propagation. Normally to the target surface at the distance d from the target the substrate is installed parallel or at a small angle to the target surface (θ < 30°). In the research as a target the commercial samples of SiO_2 micron-sized particles are used. The SiO_2 powder was pressed in a tablet with a diameter of 2 cm and a thickness of 0,2 cm. To compare the coatings obtained from the powdery SiO_2, the authors suggested using common window glass as a target (GOST 111-65) with the size of 7*10 cm. A titanium foil with a thickness of 50 micrometers was chosen as a substrate (GOST 14958-69).

3. Results and Discussion

The properties of obtained silicon-containing films were analyzed with the help of IR-spectroscopy and scanning electron microscopy. Figure 2 shows photos of the original substrate as well as the target and substrates after the powerful ion beam irradiation.

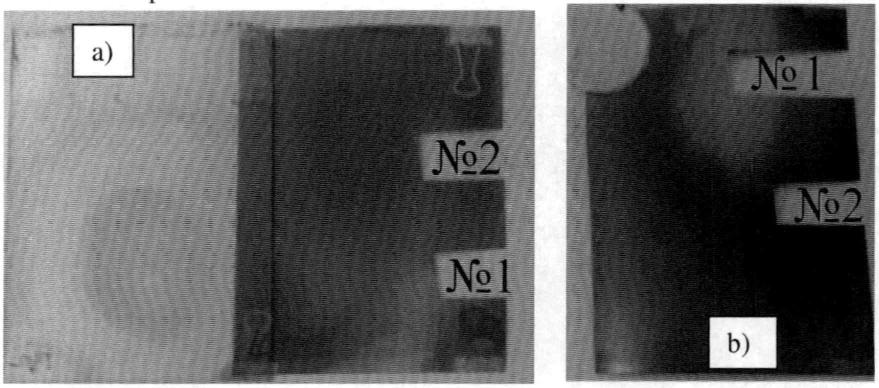

Figure 2. Photos of the target and the substrate: a) target – glass and substrate after the ion beam impact; b) target – the tablet of SiO_2 powder and the substrate after the ion beam impact.
**№1, №2 – cutout of the target part for analytical research of the obtained coating.*

Figure 2 shows the recorder of the powerful ion beam on the target, proving the glass dissipation under the influence of a concentrated energy flux (4 J/cm^2). The target material is deposited on the titanium substrate (it is possible to judge about it according to the change of the substrate colour after the ion beam impact – a bright blurred spot is observed). The same situation is observed in the case when the SiO_2 tablet is used as a target. But in this case the spot is more focused, and it has brighter colour. To conduct IR-spectroscopy 8 samples were prepared (Table 1).

Table 1. Types of samples used in IR-spectroscopy and scanning electron microscopy.

№	Comment
1	Common window glass (GOST111–65)
2	Titanium foil (GOST 14958–69)
3	Commercial sample of SiO_2 micron-sized particles
4	Pressed tablet from SiO_2 micron-sized particles
5	Part of the substrate having the maximum colour after glass irradiation (part №1, Figure 3a)
6	Peripheral area of a beam imprint on the substrate № 2, marked in Figure 3(a)
7	Part of the substrate having the maximum colour after the SiO2 tablet irradiation (part №1, Figure 3b)
8	Peripheral area of a beam imprint on the substrate № 2, marked in Figure 3(b)

Figure 3 presents the results of the IR-spectroscopy. The IR-spectroscopy is conducted with the IR-Fourier Nicolet 5700 spectrometer (a spectral dimension is 0,09 cm^{-1}). The spectrometer is equipped with attachments of the diffuse reflection and single-shot ATR for the analysis of liquid, solid and powdery materials. Figure 3 demonstrates typical absorption spectra of infrared irradiation of studied samples. Analyzing spectra (samples 2-7) according to water presence in its various conditions, it is possible to denote the following fact. An absorption band is 1500-1700 cm^{-1}, corresponding to deformation oscillations of OH-groups (sample 3). Frequencies in the area of 3200-3750 cm^{-1} correspond to OH-groups stretching vibrations. The appearance of this band is caused by the presence of adsorbed water in the samples.

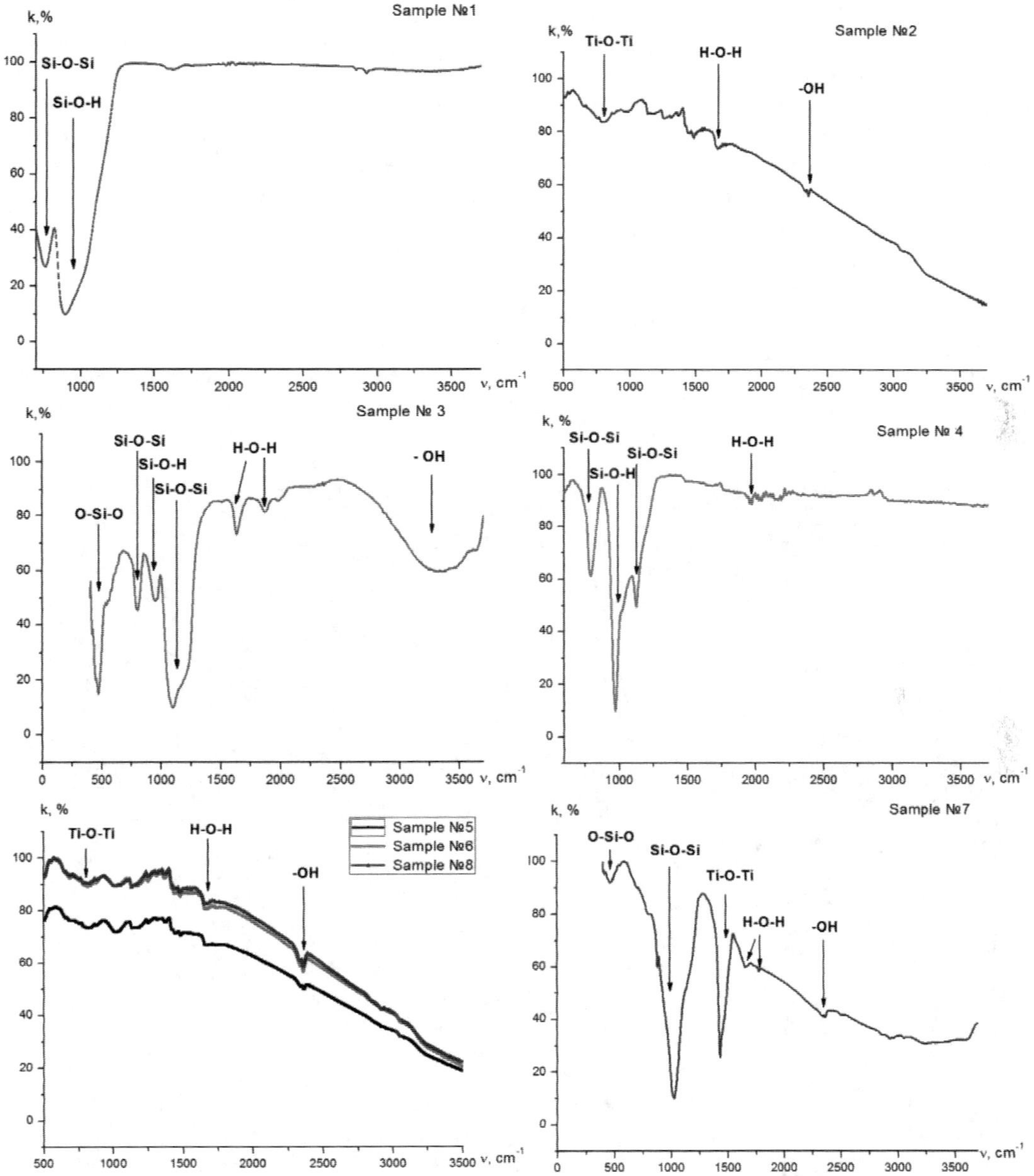

Figure 3. R- investigated samples spectra.

For common window glass (sample 1) we observe the collision of peaks, corresponding to the oscillations of bonds Si–O–Si and Si–O–H. For Sample 2 (a titanium substrate) are typical the peaks, corresponding to the oscillations of bonds Ti–O–Ti. On the absorption spectrum of infrared irradiation for the SiO_2 micron-sized powder (sample 3) are present the absorption bands of 480, 800, 1100 cm^{-1}.

These bands can be interpreted as asymmetric oscillations of external Si–O-bonds. The bond at 970 cm^{-1} can be associated with Si–O–H oscillations. The pulsed ion beam impact on the SiO_2 powder leads to the fact that on the IR-spectrum (sample 1) there is no frequency in the area of 3200-3750 cm^{-1} (OH-groups oscillations). However, in this case the absorption bands from 1100 to 1150 cm^{-1} are observed, and they can be interpreted as asymmetric oscillations of external Si–O-bonds [11-12]. The band in the area of 790 cm^{-1} can be interpreted by Si–O–Si bonds. IR-spectra for Samples 5, 6, 8 are identical to the IR-spectrum of the original titanium substrate. For sample 7 on the IR-spectrum two peaks are observed that deal with the oscillations of Si–O–Si (1020 cm^{-1}) and Ti–O–Ti (1400 cm^{-1}) bonds. Also, here we may observe absorption bands within the range of 1500-1700 cm^{-1} and the frequency in the area of 2000-2500 cm^{-1}, corresponding to the oscillations of OH-groups. Let us compare the IR-spectra of Samples 3 and 4. After the pulsed ion beam impact on the SiO_2 powder the number of peaks on the IR-spectrum has decreased (there is no frequency in the area of 3200-3750 cm^{-1}, corresponding to the oscillations of OH-groups), and the deformation of peaks, corresponding to the oscillations of Si–O–Si and Si–O–H bonds, has occurred (the peaks have become narrow). At the exposure of the pulsed ion beam on the tablet made from the SiO_2 powder the surface heating occurs, which, probably, facilitates the deformation of the mentioned peaks due to the changes in the phase composition of the surface layer. It is necessary to note the occurrence of the peak responsible for the oscillation of the Ti–O–Ti bond in Sample № 7. It can be explained by the chemical reaction of titanium oxidation either in plasma, or by interreacting of the sample with air oxygen at the further contact. Then, investigations were conducted with the help of the JEOL JSM-7500FA scanning electron microscope (Figure 4).

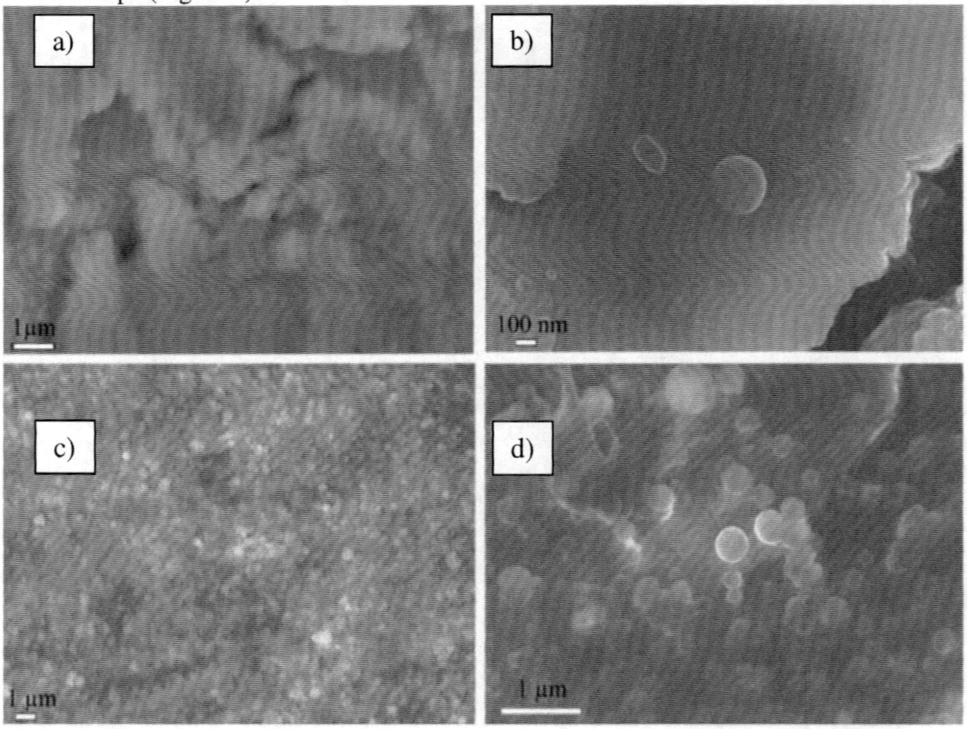

Figure 4. Microphoto of substrates' surface: a) sample № 5, b) sample № 6, c) sample № 7, d) sample № 8.

Figure 4 demonstrates the absence of a homogeneous coating on all the samples. However, on samples 6-8 SiO_2 is presented as circular particles with a diameter of 100-150 nm. When the SiO_2 powder is used as a target, the appearing coating in the centre of a plasma imprint is presented as an assembly of the particles with a regular shape. The coating is homogeneous, any other clusters of coarse or soldered particles are not met on it. It is difficult to say the same about the periphery of the plasma imprint, where the agglomerates, presenting a soldered joint of several particles, can be seen. Particles morphology for the given sample is presented by spheres with a rough surface (Figure 5). The size of these particles reaches several micrometers, while the size of the particles forming the coating in the centre of the plasma imprint reaches hundred of nanometers.

The elemental analysis of samples was conducted with the help of the energy dispersive X-ray spectroscopy method. Let us cite the results only for the titanium substrate and Sample № 7 as for the most informative example. The results of EDX-analysis are shown in Table 2.

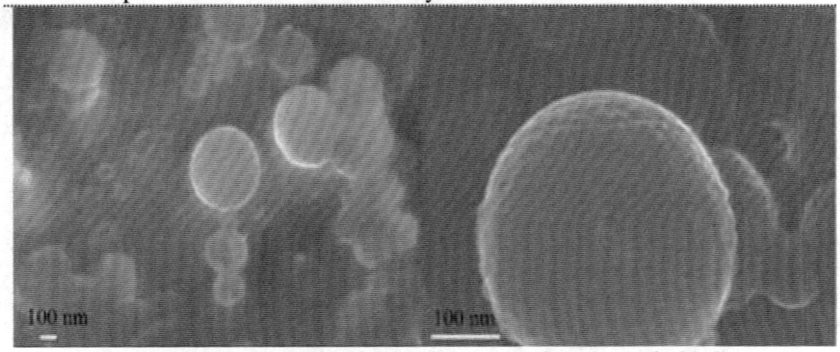

Figure 4. Microphoto of the substrate surface (sample № 8).

Table 2. Content ratio of basic elements and additives in the obtained silicon-containing coating (sample 7) and in the original substrate (sample 2) according to the data of EDX-spectra.

Sample 7		Sample 2	
Element	Wt.%	Element	Wt.%
Ti	22.07	Ti	84.34
Si	7.91	Si	0.02
O	64.04	O	-
N	-	N	15.48
Ni	0.06	Ni	0.16
Na, Mg, Al, K, Ca, Fe	5.98	-	-

It can be seen from the results of EDX spectroscopy that the coating obtained from ablation plasma formed by a powerful ion beam consists of 3 main elements: titanium, silicon and oxygen. The presence of such elements as sodium, potassium and calcium is explained by the presence of these elements in the SiO_2 powder. Magnesium, aluminum and ferrum were, hypothetically, inserted as a result of scattering of target holder structural components. In the original titanium substrate (sample 2) oxygen is lacking. After irradiation in Sample № 7 its weight content exceeds 50 % from the total weight content of the elements. Oxygen in the surface layer can occur in 5 the most probable states: SiO_x, TiO_x, adsorbed O_2, OH-groups and adsorbed H_2O. The IR-spectrometry of Sample № 7 proves the statement that the significant part of oxygen is contained in SiO_x and TiO_x oxides.

4. Conclusion

Therefore, in this work the silicon-containing coating from ablation plasma formed by a powerful ion beam was obtained. These coatings are not a solid solution, but they contain the deposited SiO_2 and TiO_2, having formed as a result of substrate surface oxidation. The coating morphology presents the surface, being covered intensively by the agglomerate of circular particles with a size of 100-150 nm. Obtained silicon-containing coatings may find their application in different fields of science and technology.

Acknowledgments

The research was conducted with the financial support of State task "Science", Ministry of Education and Science of the Russian Federation. Project number: 3.1301.2014.

References

[1] Lebedeva Yu E, Popovich N V and Orlova L A 2013 *Trudy VIAM* Access mode: http://viam-works.ru.
[2] Sealy B J and Ritchie J M 1976 *Thin Solid Films* **35** 1 p 127–130.
[3] Meyer S F and Hsieh E J 1979 *Thin Solid Films* **64** 3 p 383–388.
[4] Lien S Y, Wuu D S, Yeh W C and Liu J C 2006 *Solar Energy Materials and Solar Cells* **90** 16 p 2710–2719.
[5] Carvalho E J, Alves M A, Braga E S and Cescato L R 2006 *Microelectronics Journal* **37** 11 p 1265–1270.
[6] Zhang W, Wang X, Yao S and Wang H. 2006 *J. of Chem. Indust. and Eng. (China)* **57** 11 p . 2745–2749.
[7] Erdural B, Bolukbasi U and Karakas G 2014 *J. of Photochem. and Photobio. A: Chem.* **283** p 29–37.
[8] Liu H, Lv T, Li G, Dong X and Zhu Z. 2013 *J. Micro and Nano Letters* **8** 11 p 792–795.
[9] Rahmani E, Ahmadpour A and Zebarjad, M. 2011 *Chem. Engin. J.* **174** 2-3 p 709–713.
[10] Remnev G E et all 1991 *Vacuum* **42** 1/2 159-162.
[11] Zhang Y H and Reller A. 2002 *Mater. Sci. Eng* 323–326.
[12] Music S, Filipović-Vinceković N and Sekovanić L 2011 *Braz. J. Chem. Eng* **28** 1 p 89–94.

Assessment of thermodynamic parameters of plasma shock wave

O V Vasileva[1*], Yu N Isaev[1], A A Budko[1] and A I Filkov[2]
[1]National Research Tomsk Polytechnic University, 30 Lenina avenue, Tomsk, 634050, Russia
[2]National Research Tomsk State University, 36 Lenina avenue, Tomsk, 634050, Russia

E-mail: vasileva.o.v@mail.ru

Abstract. The work is devoted to the solution of the one-dimensional equation of hydraulic gas dynamics for the coaxial magneto plasma accelerator by means of Lax-Wendroff modified algorithm with optimum choice of the regularization parameter artificial viscosity. Replacement of the differential equations containing private derivatives is made by finite difference method. Optimum parameter of regularization artificial viscosity is added using the exact known decision of Soda problem. The developed algorithm of thermodynamic parameter calculation in a braking point is proved. Thermodynamic parameters of a shock wave in front of the plasma piston of the coaxial magneto plasma accelerator are calculated on the basis of the offered algorithm. Unstable high-frequency fluctuations are smoothed using modeling and that allows narrowing the ambiguity area. Results of calculation of gas dynamic parameters in a point of braking coincide with literary data. The chart 3 shows the dynamics of change of speed and thermodynamic parameters of a shock wave such as pressure, density and temperature just before the plasma piston.

1. Introduction

The precursor shock wave is formed before head part of plasma substance of the coaxial magneto plasma accelerator [1]. The assessment of thermodynamic parameters in a wake of the shock wave is presented in the article. Some simplifications were introduced – substance is conditionally called the blunted body or the piston, calculation was made for one dimension. Not perturbed gas-air will move on the piston with a piston speed if to pass to system of coordinates connected with the piston. This model is presented when modeling, as movement of two identical waves on a meeting each other [2-4].

2. Technique

The equations of conservation of mass, momentum and energy for the solution of one-dimensional nonstationary gas-dynamic tasks is written in divergent form were used:

$$\begin{cases} \dfrac{\partial \rho}{\partial t} + \dfrac{\partial (u\rho)}{\partial x} = 0 \\ \dfrac{\partial (u\rho)}{\partial t} + \dfrac{\partial (\rho u^2 + p)}{\partial x} = 0 \\ \dfrac{\partial \left(\rho(\varepsilon + \rho u^2/2)\right)}{\partial t} + \dfrac{\partial \left(u(\rho\varepsilon + \rho u^2/2 + p)\right)}{\partial x} = 0 \end{cases}, \qquad (1)$$

here ρ – gas density; p – gas pressure; u – velocity of gas; ε – internal energy of the gas; t, x – time and coordinate.

Content from this work may be used under the terms of the Creative Commons Attribution 3.0 licence. Any further distribution of this work must maintain attribution to the author(s) and the title of the work, journal citation and DOI.
Published under licence by IOP Publishing Ltd

The same equations (1) which have been written down in a vector form, convenient for numerical realization:

$$\left\{\begin{array}{l} \mathbf{s}=\begin{pmatrix}\rho\\ u\rho\\ \rho(\varepsilon+\rho u^2/2)\end{pmatrix},\\ \mathbf{f}(\mathbf{s})=\begin{pmatrix}u\rho\\ \rho u^2+p\\ u(\rho\varepsilon+\rho u^2/2+p)\end{pmatrix}=\begin{pmatrix}s_2\\ s_2^2/s_1+p\\ s_2(s_3+p)/s_1\end{pmatrix}\end{array}\right. \rightarrow \frac{\partial \mathbf{s}}{\partial t}+\frac{\partial \mathbf{f}(\mathbf{s})}{\partial x}=0, \qquad (2)$$

where **s** – vector of conservative variables; **f**(**s**) – flux vector.

Algorithm description:

1. Replacement of the differential equations containing private derivatives by finite differences (2).
2. Addition of optimum parameter of regularization artificial viscosity in the MathCAD wrapper.
3. Choice of the optimum artificial viscosity using the exact known decision (Soda problem).
4. Approbation of the developed algorithm of thermodynamic parameter calculation in a braking point (analytically).
5. Calculation of thermodynamic parameter change dynamic in front of plasma piston of the accelerator.

The modified of Lax-Wendroff algorithm used for the numerical decision of equations system which is that the equations containing private derivatives are replaced by finite differences [5]. Instability appears in finite differences, i.e. influence of high-frequency noise because of strong shock waves existence. The artificial viscosity was added. The optimum parameter of regularization (artificial viscosity) was picked up for ensuring regularization of the decision and suppression of its noise in the on-line mode.

3. Experimental part

The value of artificial viscosity was determined from the misfits-mismatch (figure 1, a). The exact known solution of Soda problem was compared to our algorithm if mismatches were not more than 10%, the regularization coefficient (artificial viscosity) was considered to be optimal, figure 1, b.

Values of pressure, density and temperature in environments were identical before collision of waves, and speeds of waves identical in size, but different in the direction (sign).

An inspection of work of algorithm was carried out on calculation of critical parameters – pressure, density and temperature braking p_T, ρ_T, T_T for given initial data of the undisturbed gas $p_0=1/\gamma$, $\rho_0=1$, $T_0=1$ (in relative units) and given Mach number M_0. Parameters of inhibition were calculated on known ratios [6, 7]. The Hugoniot-Rankine's adiabatic (shock wave) used to calculate the pressure, density and temperature of the undisturbed environment through the leap seal:

$$p_1=p_0\left(\frac{2\gamma}{\gamma+1}M_0^2-\frac{\gamma-1}{\gamma+1}\right),\ \rho_1=\rho_0\left(\frac{\gamma+1}{(\gamma-1)+2/M_0^2}\right). \qquad (3)$$

$$T_1=T_0\left(\frac{\gamma-1}{\gamma+1}\right)^2\left(\frac{2\gamma}{\gamma+1}-\frac{1}{M_0^2}\right)\left(\frac{2}{\gamma-1}+M_0^2\right). \qquad (4)$$

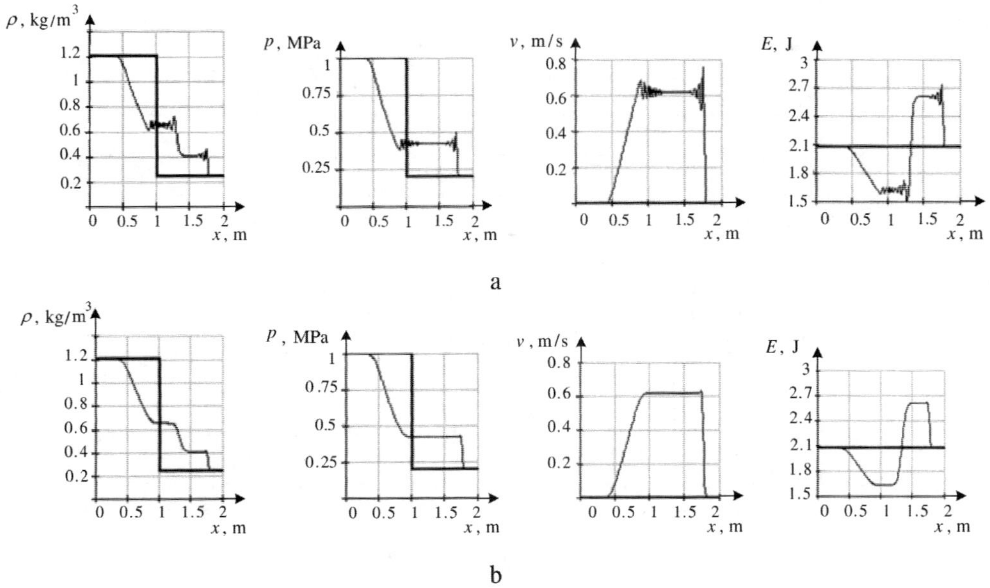

Figure 1. The restored thermodynamic parameters of a shock wave and exhaustion wave: a) with non-optimal viscosity μ=10^{-7}; b) with optimal viscosity μ=10^{-6}.

Formulas for engineering calculations use (3, 4):

$$\rho_1=\frac{1}{4}\left(5M_0^2-1\right)\rho_0,\ \rho_T=4\left(1+\frac{3}{M_0^2}\right)^{-1}\rho_0,T_T=\frac{5}{16}\left(1-\frac{1}{M_0^2}\right)\left(3+M_0^2\right)T_0. \quad (5)$$

The Poisson's adiabatic (rarefaction wave) used to calculate the parameters of inhibition:

$$p_T=\left(1+M_1^2\frac{\gamma-1}{2}\right)^{\frac{\gamma}{\gamma-1}}p_1,\ \rho_T=\left(1+M_1^2\frac{\gamma-1}{2}\right)^{\frac{\gamma-1}{\gamma}}\rho_1,T_T=\left(1+M_1^2\frac{\gamma-1}{2}\right)T_1, \quad (6)$$

where M_1 – Mach numbers after shock wave, are defined through Mach number in not indignant environment by expression: $M_1=\left(\frac{2+(\gamma-1)M_0^2}{2\gamma M_0^2-(\gamma-1)}\right)^{1/2}$.

For engineering calculations use (5, 6):

$$M_{1=}\left(\frac{3+M_0^2}{5M_0^2-1}\right)^{1/2},\ p_T=\left(1+\frac{M_1^2}{3}\right)^{5/3}p_1,\rho_T=\left(1+\frac{M_1^2}{3}\right)^{3/2}\rho_1,T_T=\left(1+\frac{M_1^2}{3}\right)T_1. \quad (7)$$

Initial data in relative units: $p_0=1/\gamma=0.6$, $\rho_0=1$, $T_0=1$.
Results of calculation of methods are given in the table.

Table. Results of comparison engineering (E) and program (P) calculations.

Mach number M_0	p_1	p_T	ρ_1	ρ_T	T_1	T_T	Calculation
1.5	1.536	2.280	1.714	2.172	1.495	1.750	E
	1.542	2.287	1.717	2.175	1.497	1.752	P
2	2.850	3.807	2.286	2.719	2.078	2.333	E
	2.914	3.893	2.306	2.744	2.106	2.364	P
3	6.600	8.204	3.000	3.418	3.667	4.000	E
	6.557	8.151	2.995	3.413	3.649	3.980	P
5	18.600	22.300	3.571	3.982	8.680	9.333	E
	18.522	22.206	3.570	3.980	8.647	9.298	P
10	74.850	88.397	3.883	4.291	32.123	34.333	E
	75.050	88.643	3.884	4.291	32.210	34.426	P

The distance between a shock wave and the end of the piston S is brought, using Lunev's formula and expression (7):

$$\frac{\rho_T}{\rho_0}=1/K_\rho=4\left(1+\frac{3}{M_0^2}\right)^{-1},\quad K_\rho=\frac{1}{4}\left(1+\frac{3}{M_0^2}\right), S=\sqrt{K_\rho}\left(1+0.6K_\rho\right). \tag{8}$$

The program gives good results according to table. By means of Lax-Wendroff numerical scheme and the entered artificial viscosity are calculated pressure, density and environment temperature just before the piston on values of speed of the shock wave, received of experiment (8). Experimental values of coordinate of a shock wave $L(t)$ (substance) and speed $v(t)$ are given in figure 2, a. Smoothed values $L(t)$ and $v(t)$ are given in figure 2, b. Values of coordinate of a shock wave approximated by splines and smoothed by "sliding average" filter, values of speed obtained by taking the derivative of a spline curve.

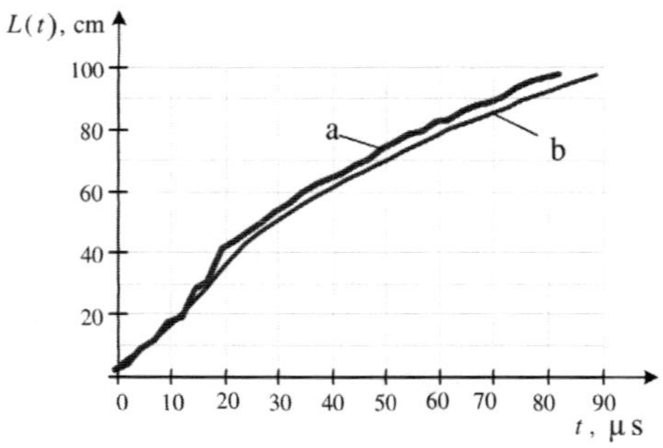

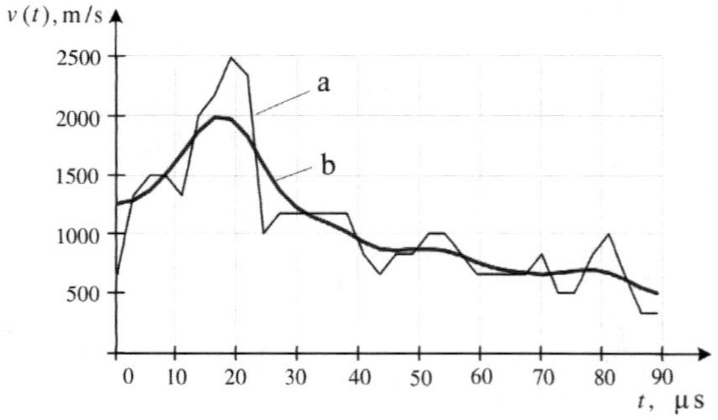

Figure 2. Values of coordinate $L(t)$ and speed of a shock wave $v(t)$ respectively: a) experimental values; b) smoothed values.

At calculation of thermodynamic parameters not indignant environment was considered as one-nuclear gas: constant of polytropic – $\gamma=5/3$, pressure – $p_1=105$ Pa, air-density – ρ_1 = 1.2 kg/m^3, the air temperature – 15°C, $t = T_1 = 273.15 + 15 = 288.15$ K, the speed of sound in not indignant environment – c = 340 m/s. Dynamics of change of thermodynamic parameters such as the shock wave velocity, pressure, density and temperature, directly in front of a plasma piston, shown in figure 3.

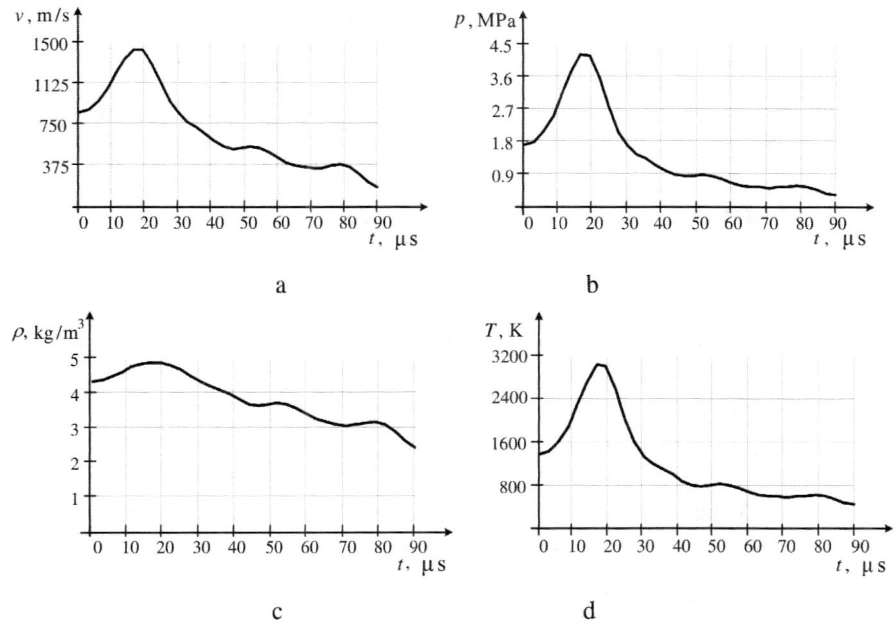

Figure 3. Dynamics of change of velocity and thermodynamic parameters: a) the shock wave velocity; b) pressure; c) density); d) temperature.

When plasma flies from an electrode trunk with a supersonic speed at a temperature of 1300 … 3000 K it represents the gas extending in the form of a stream which is called as an underexpanded stream [8]. The theory shows that when gas flies from Laval's nozzle [5] with a supersonic speed originally gas is accelerated, and then slowed down. This process is repeated periodically, but the intensity is less. It is observed on graph (figure 2, figure 3, a).

4. Results

Modeling of parameters of the plasma piston in of front a shock wave of the coaxial magneto plasma device is carried out on the basis of Lax-Wendroff modified algorithm with introduction of artificial viscosity of plasma. Unstable high-frequency fluctuations are smoothed using modeling and that allows narrowing the ambiguity area and to allocate only smooth decisions. Results of calculation of gasdynamic parameters in a point of braking are consistent with literature data.

References

[1] Sivkov A A, Saygash A Ya, Pak A A and Evdokimov A A 2009 *Nanotechnics* 2 (18) 38-44
[2] Kolesnikov P M 1971 *Electrodynamic acceleration of plasma* (Moscow: Atom publishing) 388
[3] Landau L D and Lifshits E M 1992 *Theoretical physics* vol 8 (Moscow: Science) p 664
[4] Morozov A I 2008 *Introduction in plasma dynamics* (Moscow: Physics mathematics literature) p 613
[5] Zaliznyak V E 2008 *Fundamentals of computing physics* part 1 (Moscow: Techno sphere) p 224
[6] Zeldovich Ya B and Rayzer Yu P 2008 *Physics of shock waves and high-temperature hydrodynamic phenomena* (Moscow: Physics mathematics literature) p 656
[7] Chernyak V G and Suetin P E 2006 *Mechanics of continuous environments* (Moscow: Physics mathematics literature) p 352
[8] Trubnikov B A 1991 *Theory of plasma* (Moscow: Energy atom publishing) p 464

Journal of Physics: Conference Series **552** (2014) 012027 doi:10.1088/1742-6596/552/1/012027

Experimental research of the fine foil explosion dynamics

A S Zhigalin[1,*], A G Rousskikh[1], V I Oreshkin[1,2], S A Chaikovsky[1], N A Ratakhin[1] and V V Kuznetsov[3]

[1]Institute of High Current Electronics SB RAS, 2/3 Academichesky ave., Tomsk, 634055, Russia
[2]National Research Tomsk Polytechnic University, pr. Lenina 30, Tomsk, 634050 Russia
[3]Kutateladze Institute of Thermophysics SB RAS, 1 Akad. Lavrent'eva ave., Novosibirsk, 630090, Russia

Zhigalin@ovpe2.hcei.tsc.ru

Abstract. The work is devoted to studying of substances properties at high specific deposit energy using double-frame pulsed backlighting system. The high specific deposit energy was reached at electrical conductor explosion (ECE). Fast mode of ECE was investigated. Fine foils of aluminum, cooper, titanium and nickel were used as conductors. Experiments were carried out on the experimental complex consisting of three current generators. The first generator WEG-1 was used for explosion of the fine conductors. This generator represents fast capacitor with capacity 250 nF, which was charged to voltage 10 to 30 kV. The investigated conductor was mounted in special holder and the foil contacts with the electrodes were soldered. Two other generators - radiographs XPG-1 and G2 with x-pinch load were used two frame X-ray backlighting imaging. The generators current pulses had amplitude 300 kA and rising time 180 ns with a low inductance load. Four crossed molybdenum wires with diameter of 25 μm were used to form an x-pinch. Using of the x-pinches soft x-ray radiation the images of exploded foil were registered with temporal resolution of 2 ns. The images were detected by a photo film located behind the filter. The x-ray imaging, together with the measurements of the current flowing through a conductor and voltage on the exploded conductor had allowed inferring of the energy deposited into the conductor, delay time of the bubbles formation relative to the moment of current-cutoff and the time dependence of the vapor bubbles quantity.

1. Introduction

Formation and disintegration of the metastable states are great of interest for fundamental and applied science. Metastable state of a substance occurs when the material is rapidly heated to temperatures higher than those of phase transition in conditions of suppression of nuclei of a new phase [1]. At a high level of metastability, phase separation occurs as nucleus or spinodal disintegration [2-5] whose properties still remain insufficiently studied at high density of energy generated or accumulated in the material. The phase separation mechanism is determined by kinetics of spontaneous structurization of material with a wide spectrum of applications in physics, astronomy, biology, material science and various technologies. This effect occurs in a wide class of problems of high energy density physics, mechanics, and power engineering, and is important for advance in technologies of nanopowder

Content from this work may be used under the terms of the Creative Commons Attribution 3.0 licence. Any further distribution of this work must maintain attribution to the author(s) and the title of the work, journal citation and DOI.
Published under licence by IOP Publishing Ltd

production by electrical metal wire explosion, methods for suppression of steam explosion as a critical stage of severe accident at nuclear power plants, critical and transient phenomena in multiphase flows, etc.

Disintegration of superheated metastable liquids is sufficiently studied for liquids with a temperature of boiling under normal conditions comparable with room temperature [1]. Such data are very scant for liquid metals whose boiling temperature is several thousand degrees. At the same time, they are important for studying such processes as explosive emission [6], where metal can stretch during explosion of explosion of micropoints under the action of electric forces [7], in production of nanopowders [8] during electrical wire explosion (EWE) [9,10], etc.

Formation of the metastable states could be investigated at the fine foil electrical explosion with the energy deposition comparable with sublimation energy. This paper describes the experimental results obtained at the electrical explosion of the aluminum, nickel, copper and titanium foils. The experimental purpose was to find out a relation between the energy deposition into the foil and the dynamics of the vapor bubble rising at the foil boiling. In order to get the relation a diagnostics set was developed that included precise current and voltage measurements combined with two frame soft x-ray imaging based on the x-pinches.

2. Experimental setup

The experiments had been performed using three synchronized generators. The generator WEG-2 [11] was used for foil electrical explosion. The generator consists of fast capacitor with capacity of 250 nF with a controlled high pressure gas switch. The charging voltage was varied in the range of 10-30 kV in order to match various foils parameters. The typical generator WEG-2 current pulse had a peak 8-15 kA with a rising time of 400 ns. Two frame soft x-ray backlighting system was based on two pulsed generators G-1 (XPG-1 [11-13]) and G-2 that could provide the current peak up to 300 kA with a rising time of 180 ns. Each generator had an x-pinch load that produced soft x-ray pulse. The generators' operating was approximately the same as described in [14]. Significant modification in the triggering scheme was performed. The primary trigatron type switches were replaced by thyratrons. As result the jitter between the generators current pulses and low voltage synchronization pulses (300 V, 100 ns) was reduced to ± 15 ns. The generators WEG-2, G-1, G-2 were synchronized by programmable three channel pulser DPG with independent controlled delays between the channels. The DPG provides pulses (300 V, 100 ns) with adjustable delays (0 – 1000 μs) with the time step of 25 ns.

The x-ray backlighting system arrangement is shown in Figure 1. X-pinches and the exploding foil were placed in line.

The x-rays from the x-pinch 1 passed through the foil and formed the backlighting image on the film 2. The x-rays from x-pinch 2 passed the foil from the opposite side forming the image on the film 1. In order to avoid film 1 exposure by x-pinch 1 a metal rods with diameter of 2 mm were placed between the x-pinches and the films. The rods were placed at the distance of 2 cm from the x-pinches that provided the rod shadow size on the film higher than the film size. Hence, the film 1 was not directly exposed by x-pinch 1 x-rays. The film 2 was protected from radiation of the x-pinch 2 in the same manner. The rod 2 shadow produced by x-pinch 1 on the film 2 (see Figure 1) had significantly lower size (10 mm) due to higher distance from x-ray source to the rod. That allowed enough wide foil region to be imaged (see Figure 2). The same is correct for the film 1. The magnification of the projection scheme was 10.8 that made it possible to get quality images. In order to exclude the

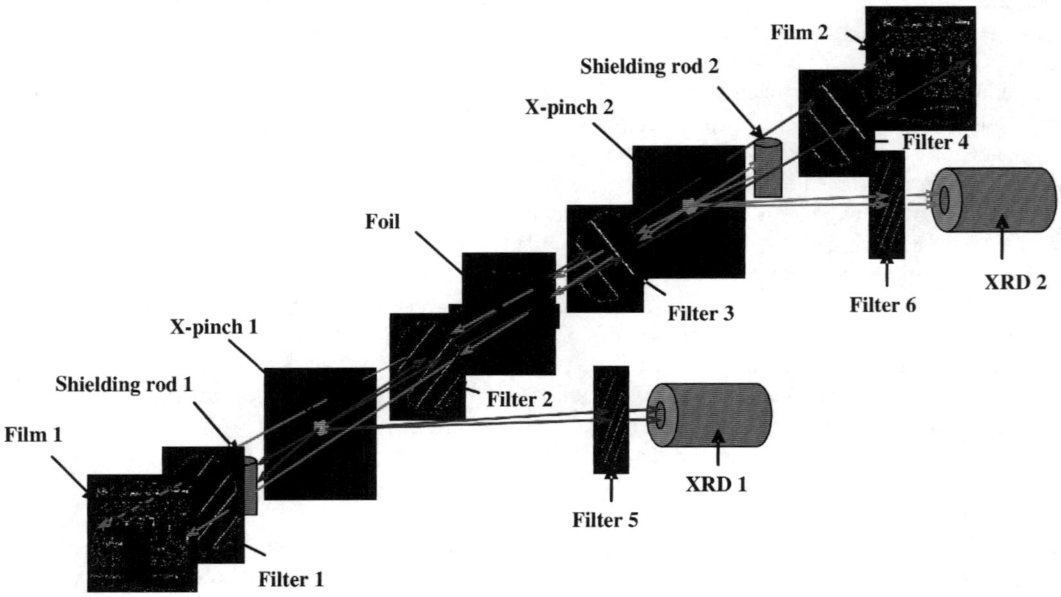

Figure 1. Two frame x-ray backlighting

interference between the x-pinches and the exploding foil additional polypropylene filters with thickness of 6 microns were mounted (filters 2 and 3 in Figure 1).

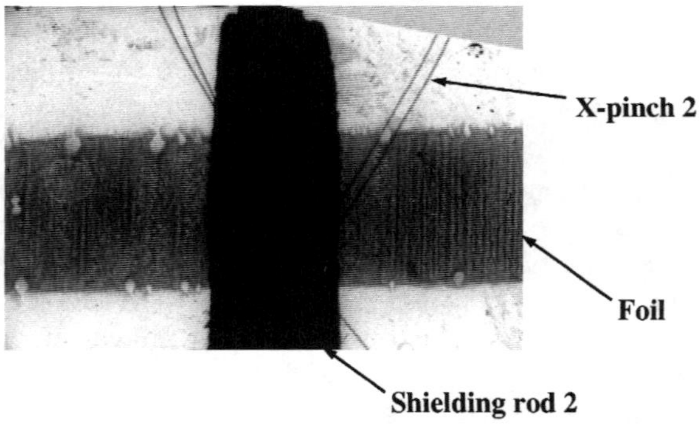

Figure 2. X-ray backlighting image of the exploded aluminum foil on the film 2. Clearly visible are the rod 2 and wires of the x-pinch 2. The bubbles formation on the edge of the foil is evident.

The metal foils under investigations had a length of 20 mm, thickness of 5 microns and width of 1 mm. The foils were soldered to the electrodes of the special holder. The last one was inserted into the dielectric vacuum chamber where the holder electrodes were connected to the WEG-2 power feeder. The charging voltage of the WEG-2 was 20 kV for Al, Ti and Ni foils, while for copper foils it was 10 kV. In order to measure energy deposition to the foil both the current through the foil and the voltage on the foil were measured. An active voltage divider and a B-dot loop were used for measurements. After the WEG-2 current pulse started the generator G-2 was triggered with required delay. The delay between the WEG-2 and G-2 current pulses was varied in the range 300-1200 ns. The G-1 generator was triggered later then G-2 generator with adjustable delay which was varied from 15 to 120 ns.

When a current pulse is applied to a x-pinch a very short soft x-ray pulse could be produced [11-13]. The soft x-rays produced by x-pinches driven by G-1 and G-2 generators were used for the foil explosion backlighting imaging. The x-rays images were recorded in the spectral range higher 0.8 keV using MICRAT-300 film and thin filter. The filter consisted of Kimfol film with thickness of 4 microns with aluminum coating with thickness of 0.4 microns.

The timing of the x-ray pulses was detected by two filtered x-ray diodes (XRD) with aluminum cathodes. The filters were composed from of Kimfol film with thickness of 4 microns with aluminum coating with thickness of 0.4 microns and polypropylene film with thickness of 6 microns. Such combination allowed to have high XRD response in the spectral region higher 0/75 keV. The current, voltage and XRDs traces were registered by TDS-3054C oscilloscope (Tektronix, Inc., Oregon, USA) with bandwidth of 500 MHz.

The energy deposition into the exploding foil was estimated using experimental traces of the current derivative $dI(t)/dt$ and voltage $U(t)$. The energy deposition could be expressed as:

$$W(t) = \int I(t) \cdot (U(t) - U_L(t)) \cdot dt \qquad (1)$$

The current $I(t)$ was inferred by integrating of the current derivative trace. The B-dot loop was calibrated using a current shunt with well defined resistance. It is known that a voltage trace $U(t)$ has an active and inductive terms. In order to eliminate the inductive term $U_L(t)$ it was measured by replacing of the foil by copper rod with diameter of 1 mm which had approximately the same inductance. The energy deposition was inferred by integrating of expression (1) till the voltage drop caused by foil shunting due to discharge in the low density corona.

The backlighting images registered at different times in course of the explosion were used for counting of the bubbles number. The bubbles appeared as a result of the foil boiling. The total bubbles number was normalized on the 1 mm^2.The bubbles temporal behavior (time dependence of the bubble diameter) was determined using two x-ray images with different time delay.

3. Experimental results

Figure 3 presents time dependencies of the current and voltage, energy deposition and number of the bubbles per mm^2 for different materials of the foil. The shunt discharge time could be evidently noted by the voltage trace dropping and the current trace knee. The energy deposition was measured to reach 5 – 5.8 kJ/g (46-54% sublimation energy) for aluminum, 1.3 – 1.8 kJ/g (21-29 % sublimation energy) for nickel; 0.9 – 1 kJ/g (10-11 % sublimation energy) for titanium and 2-2.5 kJ/g (42-53 % sublimation energy) for copper.

The figure 3 shows that boiling start does not correspond to the shunt discharge appearance (voltage drop). This delay could be interpreted as upper limit of the metastable state disintegration. This delay was measured to be 110, 350 and 315 ns for aluminum, nickel and copper foils, correspondingly. No boiling of the titanium foil was observed.

From analysis of the consequent frames it was obtained that maximum diameter of the bubbles does not exceed 12 microns that is doubled width of the foil. The duration of the bubble formation was measured to be not over 40 ns. The bubbles diameter increases faster at the beginning of the bubbles rising. Then the velocity of bubbles rising drops. In general the bubble diameter time dependence could be described by expression:

$$\frac{dD_b}{dt} = 0.19 - 0.019 \cdot D_b \ [\mu\text{m/ns}], \qquad D_b \leq 10\ \mu\text{m} \qquad (2)$$

Thus, the x-ray imaging, together with the measurements of the current flowing through a conductor and voltage on the exploded conductor had allowed inferring of the energy deposited into the conductor, delay time of the bubbles formation relative to the moment of current-cutoff and the time dependence of the vapor bubbles quantity.

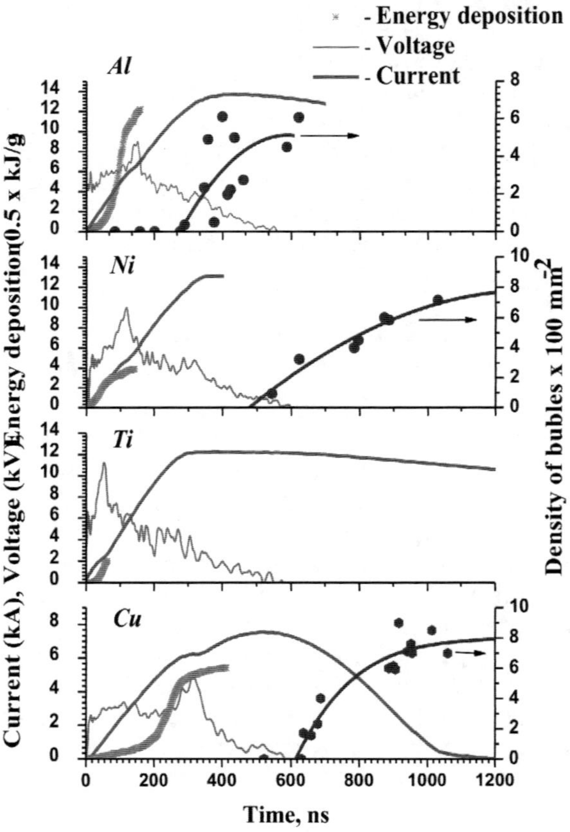

Figure 3. Time dependencies of the current and voltage, energy deposition and number of the bubbles per mm^2 for different materials of the foil

This work was supported by the Siberian Branch of the Russian Academy of Sciences, interdisciplinary integration project no. 74.

References

[1] Skripov V P 1972 *Metastable Liquid* (Moscow: Nauka)
[2] Kuznetsov V V, Oreshkin V I, Zhigalin A S, Kozulin I A, Chaikovsky S A and Rousskikh A G 2011 Metastable States and Their Disintegration at Pulse Liquid Heating and Electrical Explosion of Conductors *J. Eng. Therm* **20**, № 3, pp. 240–248.
[3] Kuksin A Yu, Norman G E and Stegailov V V 2007 Phase Diagramand Spinodal Disintegration of Metastable States of the Lennard–Jones System *High temperature* **45**, № 1, pp. 43–55.
[4] Tkachenko S I, Khishchenko K V, Levashov P V 2005 *International Journal of Thermophysics* **26**, Issue: 4, Pages: 1167-1179.
[5] Oreshkin V I, Rousskikh A G, Chaikovsky S A and Oreshkin E V 2010 Investigation of the Transport Properties ofMetals in the Biphase Region *Phys. Plasmas,* **17**, № 7, p. 072703.
[6] Mesyats G A and Proskurovskii D I 1971 Explosion Emission of Electrons from Metal Micropoints *Pis'ma ZhETF* **13**, № 1, pp. 7–10.
[7] Vorobyev V S, Malyshenko S P and Tkachenko S I 2005 Nucleation Mechanism of Explosion Disintegration ofWires with a High Energy Density *Teplofiz. Vysok. Temp.,* **43**, № 6, pp.

905–918.

[8] Sedoy V S, Mesyats G A, Oreshkin V I, Valevich V V and Chemesova L I 1999 The Current Density and the Specific Energy Input in Fast Electrical Explosion *IEEE Trans. Plasma Sci.*, **27**, № 4, pp. 845–850.

[9] Oreshkin V I, Khishchenko K V, Levashov P R, Rousskikh A G, Chaikovsky S A 2012 *High Temperature*, **50**, Issue: 5, pp. 584-595

[10] Oreshkin V I, Zhigalin A S, Rousskikh A G and Kuznetsov V V 2013 *Journal of Engineering Thermophysics*, **22**, № 4, pp. 288–297.

[11] Rousskikh A G, Oreshkin V I, Chaikovsky S A, et. al 2008 Study of the Strata Formation during the Explosion of a wire in Vacuum *Physics of Plasmas*. **15**, 102 706

[12] Ratakhin N A, Fedushchak V F, Erfort A A, Zharova N V, Zhidkova N A, Chaikovsky S A, Oreshkin V I 2007 *Russian Physics Journal*. **50**, № 2, P. 193.

[13] Mesyats G A, Shelkovenko T A, Ivanenkov G V, Agafonov A V, Savinov S Yu, Pikuz S A, Tilikin I N, Tkachenko S I, Chaikovsky S A, Ratakhin N A, Fedushchak V F, Oreshkin V I, Fedyunin A V, Rousskikh A G, Labetskaya N A, Artemov A P, Hammer D A, Sinars D B, 2010 2010 *published in Journal of Experimental and Theoretical Physics*, **138**, № 3, pp. 411–420.

[14] Artyomov A. P., Fedyunin A. V., Chaikovsky S. A., Zhigalin A. S., Oreshkin V. I., Ratakhin N. A. and Rousskikh A. G. 2013 A double-frame nanosecond soft X-ray backlighting system based on X-pinches *Instruments and Experimental Techniques* **56**(1), pp. 66-71.

Journal of Physics: Conference Series **552** (2014) 012028 doi:10.1088/1742-6596/552/1/012028

Multi-borehole electro-blast method for concrete monolith splitting off

A S Yudin*, N S Kuznetsova, V V Lopatin and N V Voitenko
Tomsk Polytechnic University, Lenina av. 30, Tomsk, 634028, Russia

E-mail: tevn@hvd.tpu.ru

Abstract The multi-borehole method of monolith splitting off by pulsed electrical discharge is presented. The experimental data of simultaneous and step by step discharge initiation for monolith splitting off with using of the polyethylene cartridge are given. The simultaneous initiation of discharge in two boreholes with 30 cm spacing, transferred energy in each borehole about 23 kJ as well as sequential initiation of discharge with transferred energy in one borehole about 47 kJ are investigated. The concrete splitting off with the dimensions of 300×300×650 mm by six boreholes with step-by-step initiation in each borehole and with total energy of about 280 kJ as well as the concrete splitting off with dimensions of 300×300×300 mm by the four boreholes with simultaneous initiation in each two boreholes and the total energy of about 90 kJ take place. It is shown that the simultaneous multi-borehole initiation of the electrical discharge leads to the cracks formation in predefined direction and to the directional solid splitting off. The experimental results show that the simultaneous discharge initiation in comparison with step-by-step one has advantage in consumed energy.

1. Introduction

In mining industry, building and construction sector for dismantling and repair of buildings, construction or expansion of tunnels, deepening of bed, removing the debris, etc. there are tasks of demolition and splitting off monolith of large solid blocks of natural and artificial origin (lumps of rock, concrete blocks and items) – “oversizes”. Traditionally the explosive, mechanical or thermal methods for such tasks are used. These methods have some shortcomings. The mechanical method is safe for workers and the environment. It should be noted, however, that in case of high-strength materials it does not always provide the desired technological effect. Thermal methods are based on using thermodrills or termite compositions. The thermodrills are mainly used for the destruction of “oversizes” and include a sufficiently complex equipment. The process of destruction with using of termite compositions proceeds rapidly, does not lead fragment dispersion, though for the complete destruction requires additional mechanical force. Explosives due to its simplicity and high performance are used in the most industrial environments. It allows destroying the objects practically of any size and strength. The disadvantages of this method include the environmental hazard of explosives, dust and debris, during the expansion that can damage surrounding objects. In consequence of it there is a requirement to ensure the security zone and the evacuation of staff while carrying out of blasting is needed. This method also requires a precise execution order because blasting strictly regulated by state regulatory authorities. In dense urban areas usage of explosives is limited, and in most cases it is prohibited. It should also be noted that the effectiveness of explosive method falls down when dimension of destructed object decreases, especially when sizes is less than 0,8 ÷ 1m.

Content from this work may be used under the terms of the Creative Commons Attribution 3.0 licence. Any further distribution of this work must maintain attribution to the author(s) and the title of the work, journal citation and DOI.
Published under licence by IOP Publishing Ltd

One of the promising direction of splitting off and destruction of solid materials is using pulse electric-discharge method, the essence of which is application of the electro-burst wave dynamics principles [1-3]. When a current impulse flows through the plasma channel of capillary electric discharge in condensed matter, due to the rapid energy release in the channel, the shock wave of pressure is generated. It forms the field of mechanical stresses, causing the material destruction.

In this paper the comparison between simultaneous and step-by-step discharge initiation for monolith splitting off with using of pulse electric-discharge method is presented.

2. Design of the equipment

The electric circuit diagram of the generator that was used in experiments is shown in figure 1. The generator includes energy storage – nonpolar capacitor battery *C* consisting of 40 capacitors of K75-101-15kV-14uF type, and pseudo-spark switch *S* of TDI1-200k/25 type. The maximum charging voltage is 15 kV and the maximum energy stored in the battery is 63 kJ. To prolong the lifespan of the capacitors the charging voltage under working conditions is reduced to 13 kV. Elements *R* and *L* are the stray parameters of the circuit that were estimated from the short-circuit experiment. R_{ch} is the discharge channel resistance that vary with the time. Its mean value was estimated from the experiments as approximately 0.03 Ohm. The length of discharge channel is 100 mm. The channel was initiated with a thin copper wire with diameter of 0.2 mm. It was done to simplify the ignition of initial discharge channel under relatively low voltage of 13 kV.

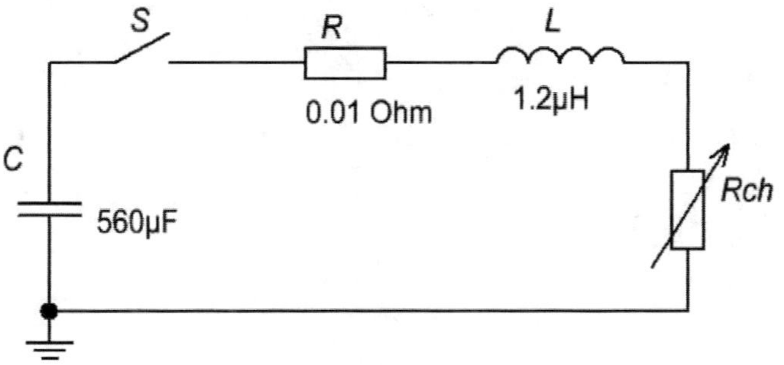

Figure 1. Electric circuit diagram of the generator.

As it was precisely described in [4], to increase the amplitude of shock waves and step up the efficiency of its transfer to the destructible material, the polyethylene cylinder with copper wire in its axis was used as a cartridge. The outer diameter of the cartridge is equal to the inner diameter of the borehole in the destructible material and is equal 26 mm.

3. Description of experiments

All experiments were carried out with concrete specimen situated in ground. It was done to decrease the amplitude of reflected shockwaves to simulate real conditions in monolith. The dimensions of specimen was (W×H×D) 1000×1000×2000 mm. Compressive strength of used concrete was 40 MPa. Tektronix TDS-2024 oscilloscope and CWT-1500 Rogowsky coil were used to record the discharge current waveform.

4. Results of experiments

Typical current waveforms for sequential and simultaneous initiation modes are shown in figure 2. Normally the two half-periods occurred during the discharge. In some cases one or three half-periods occurred and it depends on conditions around the plasma channel.

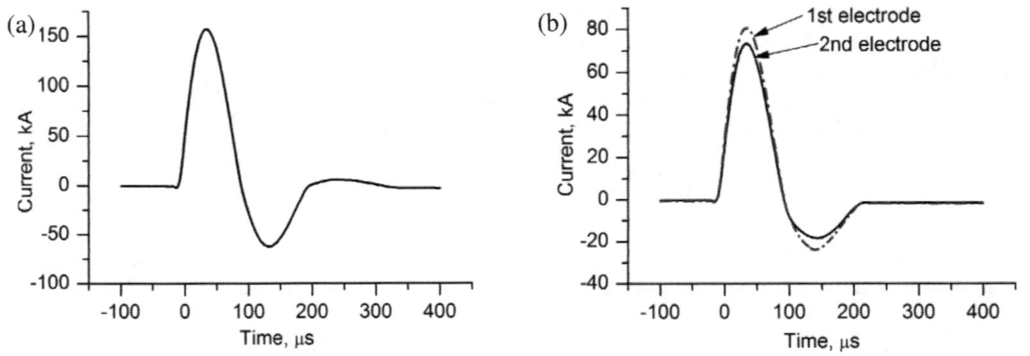

Figure 2. The current waveforms of the discharge: (a) – one electrode in sequential initiation mode; (b) – two electrodes in simultaneous initiation mode.

3.1. Simultaneous initiation in two boreholes

In the first experiment two boreholes with 35 cm spacing were bored vertically, and two boreholes – horizontally with 35 cm of depth. The initial state of concrete specimen and installed electrodes are shown in figure 3a.

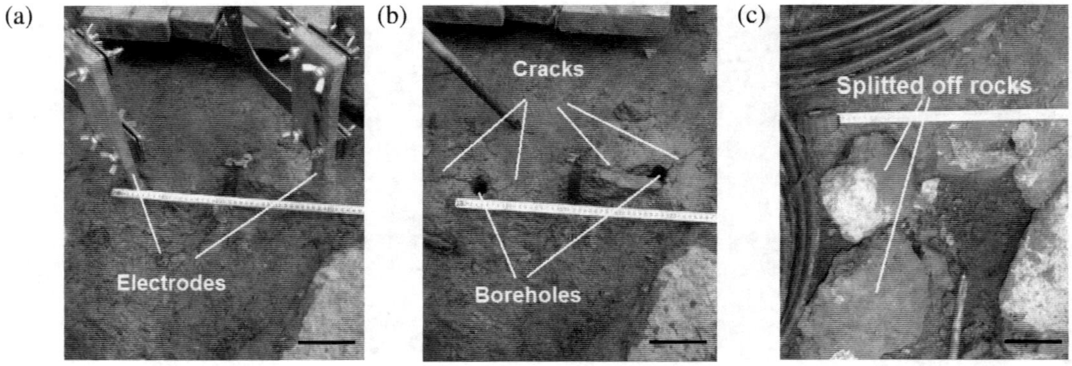

Figure 3. The result of the experiment with simultaneous initiation in two boreholes: (a) – the scheme of boreholes before discharge; (b) – cracks pattern after first discharge; (c) – fracture pattern after second discharge (Bar 100 mm).

The simultaneous initiation of discharge in two vertical boreholes was produced. As one can see in figure 3b the major cracks were formed after the discharge and they were oriented in the plane of the boreholes. Such scheme helps to produce splitting off in the predefined direction. After second simultaneous discharge in two horizontal boreholes the same cracks pattern were formed. The size of each main crack in its mouth near the borehole was approximately from 1 to 3 mm. It needs to be mentioned that final splitting off was obtained with additional manual power with the help of electric hand perforator. The result of the experiment is shown in figure. 3c. The dimensions of split part was about 300×300×300 mm. Total consumed energy in this experiment was approximately 90 kJ.

3.2. Sequential (step-by-step) initiation

Sequential initiation of the discharge was produced in accordance with the scheme shown in figure 4a. The discharge was initiated once in each borehole. The borehole depth was approximately 35 cm, the distance between the boreholes was between 30 and 35 cm, the energy in each pulse was about 47 kJ. The boreholes 1–3 was vertical and 4–6 was horizontal. The result of destruction after 5 and 6 pulses is shown in figure 4b,c.

After the discharge in each borehole the two or three main cracks were formed. The mouth width of all cracks near the borehole was about 1÷3 mm. After the second and subsequent discharges one of the main cracks was generally directed to previous borehole. The complete destruction of the concrete bounded by boreholes was induced by discharges in 5th and 6th boreholes.

There was no additional manual power in this experiment, but consumed energy is three times higher and equals to about 280 kJ.

Figure 4. The result of the experiment with step-by-step initiation in six boreholes: (a) – the scheme of boreholes, 1–6 – numbers of boreholes; (b) – fracture pattern after 5th discharge; (c) – fracture pattern after 6th discharge (Dimensions in millimeters, bar 100 mm).

5. Conclusion

The concrete splitting off with dimensions of 300×300×300 mm by the four boreholes with simultaneous initiation in each two boreholes and with total energy of about 90 kJ take place. Also the concrete splitting off with the dimensions of 300×300×650 mm by the six boreholes with step-by-step initiation in each borehole and with total energy of about 280 kJ takes place. It is shown that the simultaneous initiation of the electrical discharge leads to the cracks formation in predefined direction and to the directional solid splitting off. The experimental results show that the simultaneous discharge initiation in comparison with step-by-step one has advantage in consumed energy but needs additional manual power.

Next step in this research is to try to initiate discharge in adjacent boreholes with the time delay for wave propagation from the first borehole to the second one. This method will help to increase the pressure wave amplitude in the second borehole and will correspondingly help to form more cracks in the material.

Acknowledgement

This work was supported by Federal target program, project No 14.575.21.0059, and RFBR, research project No 14-08-01088/14.

References

[1] Kuznetsova N S , Lopatin V V, Burkin V V, Golovanevsky V A, Zhgun D V, Ivanov N A 2011 Theoretical and experimental investigation of electro discharge destruction of non-conducting materials *IEEE International Pulsed Power Conference: Digest of Technical Papers (Chicago, June 19-23, 2011)* (Chicago: University of Missouri) pp 267–271.

[2] Burkin V V, Kuznetsova, N S, Lopatin V V 2009 Dynamics of electro burst in solids: I. Power characteristics of electro burst. *J. Phys. D: Appl. Phys.* **43** 185.

[3] Bluhm H, Frey W, Giese H, Hoppé P, Schultheiß C, Sträßner R 2000 Application of Pulsed HV Discharges to Material Fragmentation and Recycling *IEEE Transactions on Dielectric and Electrical Insulation* **7** N5 625

[4] Kuznetsova N S, Burkin V V, Lopatin V V, Golovanevsky V A, Stefanov Y P, Ivanov N A, Yudin A S 2012 Blast-Hole Electrofracture of Oversize Rocks *J. Iizvestiya vuzov. Fizika.* **55** N10/3 232

Effect of electro-discharge circuit parameters on the destructive action of plasma channel in solid media

N S Kuznetsova[*], **V V Lopatin and A S Yudin**

National Research Tomsk Polytechnic University, Institute of High Technology Physics, 30 Lenin Avenue, Tomsk, 634050, Russian Federation

E-mail: tevn@hvd.tpu.ru

Abstract. The results of computer blast-hole experiments with copper wire electro-explosion in the polyethylene-concrete media has been analyzed with shock and pressure wave dynamics depending on the spatiotemporal distribution of electrical power deposition in plasma channel. Pressure and stress-wave dynamics of tensely-deformed material state has been analysed by means of physical and mathematical model, which consistently describes the operation of the discharge circuit, plasma channel expansion, generation and propagation of the shock and pressure waves in the condensed media. It has been shown the significant dependence of the stress-wave profile on the pressure pulse wave shape on the borehole wall which is determined by the rate of electrical energy release in the plasma channel and is weakly depended on the time of energy release (at given rate of its release). Analysis of stress-wave dynamics have been shown that the rapid power deposition results to the higher amplitude of compressive stresses in the wave. The lower energy deposition rate in plasma channel leads to the higher amplitude of tensile stresses both in radial and tangential direction.

1. Introduction

Over the past decade, electrically produced shocks in condensed media have been used as the means of rock/concrete fragmentation, drilling and destruction [1–6]. Most of such experiments normally involve producing an electrical discharge in a water-filled cavity drilled in the rock. One of the ways of the plasma discharge channel initiation in the condensed materials is the wire electro-explosion. Shock-wave disturbances launched from the expanding discharge plasma channel initiated by electrical wire explosion in different media are widely used in engineering and electro-discharge technologies due to possibility to manage their characteristics in a wide range by varying parameters of the discharge RLC-circuit, and the size and material of explosion wire [7-10].

A high voltage pulsed power electric discharge through a solid can create a plasma channel of high energy density of about 25-30 kJ/cm^3 [10]. Inside the channel the electric energy is transformed into internal thermodynamic energy which subsequently is used to deform and finally destruct the surrounding solid. This process is the basis for many technological applications where typically 10-100 J/cm of energy are deposited from a high voltage pulse generator in less than a few μs, leading to the plasma channel temperatures around 10^4 K and pressures of up to 10^9 Pa. The plasma channel, initially only ~10 to 50 μm wide, expands and launches a pressure wave into surrounding solid material causes the tensely-deformed material state and formation of the mechanical compressive, tensile and shear stresses. Mechanical stresses are in the range of tens to hundreds of MPa and cause the solid to be destructed. This effect is comparable to that of a chemical explosion in a preformed

Content from this work may be used under the terms of the Creative Commons Attribution 3.0 licence. Any further distribution of this work must maintain attribution to the author(s) and the title of the work, journal citation and DOI.
Published under licence by IOP Publishing Ltd

borehole. However there are few distinctive features: in an electro-explosion the relative radial expansion of the plasma channel is a factor of $10\text{-}10^3$ larger and its temperature is a factor of 10 higher. Also the quantity and energy release rate can be controlled over a much broader range by varying the time of electrical energy release from a few to hundreds of microseconds.

The destructive effect of the surrounding solid material depends on the pressure versus time waveform, $P_g(t)$, produced on the walls of the borehole. This pressure, in turn, depends on the spatiotemporal distribution of electrical power deposition (electrical energy release rate) in the plasma channel, and on shock wave propagation through the media. Thus the amplitude and profile of generated pressure wave causing the material destruction are basically determined by the power of energy release in the channel and the expanding wave energy. The main goal in raising the efficiency of electro-discharge technologies is therefore matching of energy release rate conditions during the discharge and characteristics of the generated pressure wave.

The lack of experimental information about the general laws determining the amplitude and profile of the pressure wave generated by the plasma channel expansion in the condensed media does not allow estimation of the generated pressure in the surrounding material according to the measured electrical discharge characteristics (current and voltage). Some attempts have been made in case of the discharge in liquids by the authors of monographs [8, 9], but formulation of their practical problems did not assume the recording of the shock wave profile. The analysis of the examined processes, an understanding of shock wave propagation is thus essential for improvement of electro-discharge technologies, optimizing pulse generator parameters and the discharge modes coordinated with the wave dynamics in the destructible material and conditions of its destruction. The objective of this work is to develop physical insight into shock wave generation and propagation in condensed media at electrical discharge. To solve this problem a physical and mathematical model has been presented, which consistently describes the operation of the discharge circuit, plasma channel expansion, generation and propagation of the shock and pressure waves in the condensed media. In our research, the copper wire electro-explosion in the polyethylene-concrete media has been analyzed with shock and pressure wave dynamics depending on the spatiotemporal distribution of electrical power deposition in plasma channel.

2. Numerical physics-based research method of shock and pressure wave dynamics and solid destruction

The electric circuit diagram of the pulse generator that was used in experiments and simulation is shown in figure 1.

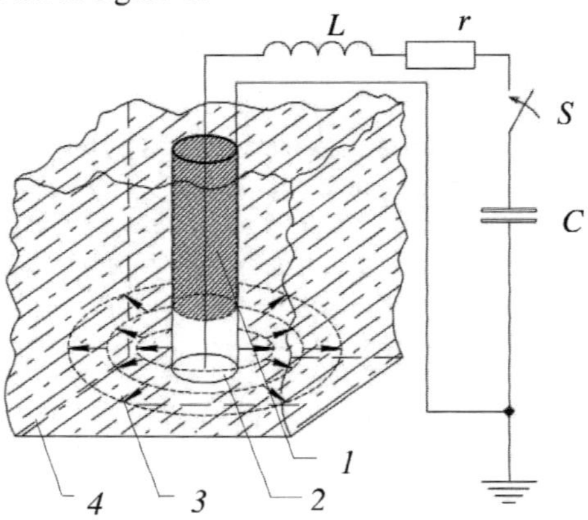

Figure 1. Scheme of plasma-blasting fracture of concrete: *1* – coaxial cable; *2* – electro-blast cartridge; *3* – shock-wave disturbances, generated by expanding plasma channel; *4* – concrete test specimen.

It was shown in [11] that for increasing the efficiency of electrical energy conversion into the pressure wave energy, the amplitude of shock wave and efficiency of its transfer to the destructible material the polyethylene cartridge with copper wire in its center line was used. The cartridge was placed in a drilled hole in the concrete sample. During the discharge the wire explosion and plasma channel formation occur. After the electrode gap bridging by plasma channel the energy accumulated into the capacitor bank of the power supply *C* is released into it. Resistance *r* and inductance *L* are the stray parameters of the circuit. Joule heating of the plasma channel results in extremely rapid channel pressure buildup and subsequent plasma expansion. A steep pressure rise takes place on the walls of the hole, generating a shock wave in the surrounding material. The propagation of shock and pressure wave disturbances forms tensely-deformed solid state and results to the mechanical compressive, tensile and shear stresses which are in the range of tens to hundreds of MPa and therefore cause its further destruction. An enhancement of mechanical fracture work (pressure wave energy) produced by the plasma channel is the main goal of energy optimization.

The physical and mathematical model of investigated system consist of equations describing the wave conversion of electrical energy, shock-wave effects and dynamics of deformed solid material.

Electrodynamic oscillating processes in the discharge circuit were characterized by the voltage balance equation:

$$L\frac{di(t)}{dt} + (R_{ch} + r)i(t) + \frac{1}{C}\int_0^t i dt = U_0 \,, \tag{1}$$

where L is the circuit inductance, μH; R_{ch} is channel resistance, Ω; *i(t)* is the current in the circuit, A; U_0 is the charge voltage of capacitor bank, V; *r* – active resistance of the generator, Ω.

Electrical plasma channel energy conversion into the plasma energy and mechanical work, performed by expanding channel:

$$\frac{1}{\gamma-1}\frac{d}{dt}(P_{ch}V_{ch}) + P_{ch}\frac{dV_{ch}}{dt} = i(t)^2 R_{ch} \,, \tag{2}$$

where γ is the effective ratio of specific heats; P_{ch} is the channel pressure, Pa; $V_{ch} = \pi r_{ch}(t)^2 l_{ch}$ – channel volume, m^3; r_{ch} and l_{ch} are the channel radius and length accordingly, m.

Dynamics of shock and pressure waves in polyethylene and concrete was described with the hydrodynamical equations in the form of equations of conservation of momentum, mass, and energy in the Lagrangean coordinates for cylindrical symmetry [13, 14, 10].

Time dependences of discharge current and voltage have been calculated according to the Kirchhoff's equations along with the empirical equations for resistances of copper wire and plasma channel. The resistance of the discharge channel before wire explosion on each time step is determined as [8]:

$$R_{ch}(t) = R_{ch0}(1 + 68.065\frac{w}{W_s} - 233.188\left(\frac{w}{W_z}\right)^2 + +1526428\left(\frac{w}{W_z}\right)^3 - 57.867\left(\frac{w}{W_z}\right)^4, \tag{3}$$

where R_{ch0} is the initial resistance, *w* is specific energy of the copper wire. The initial resistance is defined as

$$R_{ch0} = \frac{l_{ch}}{s}\rho_{cop} \,. \tag{4}$$

Here *s* is the wire cross section area and ρ_{cop} is specific resistance of copper. Specific energy of the copper wire is defined as follows:

$$w = \frac{1}{m}\int_0^t i^2 \cdot R_{ch} \cdot dt \,, \tag{5}$$

where *m* is mass of wire.

The breakdown channel resistance $R_{ch}(t)$ after wire explosion is determined through the integral of current action in form of Vaicel-Rompe [12]:

$$R_{ch}(t) = K_{ch} \cdot l_{ch}\left(\int_0^t i^2(t)dt\right)^{-\frac{1}{2}}, \tag{6}$$

where K_{ch} is a spark constant (in case of polyethylene K_{ch}=210 $V \cdot s^{0.5} \cdot m^{-1}$ [5]).

Polythene compressibility has been described by the state equation in the form of shock adiabat $D = a + bU$, where D – shock wave velocity, U – mass speed at the wave front. This equation by means of Rankine-Hugoniot equations [15] was transformed into:

$$P = \rho_0 a^2 \left(1 - \frac{\rho_0}{\rho}\right) \bigg/ \left[1 - b\left(1 - \frac{\rho_0}{\rho}\right)\right]^2, \tag{7}$$

where P – pressure in material covered with a wave, ρ_0, ρ – initial and current material density. For polyethylene $\rho_0 = 0.94$ g/cm^3, $a = 2901$m/s, $b = 1.481$ [15].

3. Simulation results

Several plasma blasting computer simulation scenarios were run and examined on concrete with diameter of the drilled hole of 26 mm and its depth of about 300 mm. The hall was drilled in the distance of 300 mm from the concrete block surface. The length and diameter of copper wire initiating the discharge were 100 mm and 0.27 mm, respectively. For instance three values of electrical energy release rate were chosen in the experiments: U_0=13 kV, L=1.22 μH, C_1=560 μF, C_2=840 μF, C_3=1120 μF.

The distractive effect of concrete is determined by the maximal pressure developed in the plasma channel, shock and pressure wave energy, and the field of tensile mechanical stresses. As a result of Joule energy-release in plasma channel its internal energy and conductivity raise, which leads to the increase in the current amplitude. Positive feedback between the heat build-up and conductivity causes an increase in the internal energy of plasma channel (figure 2, a). The electrical power deposition reaches (6–8)·10^5 W (figure 2, b).

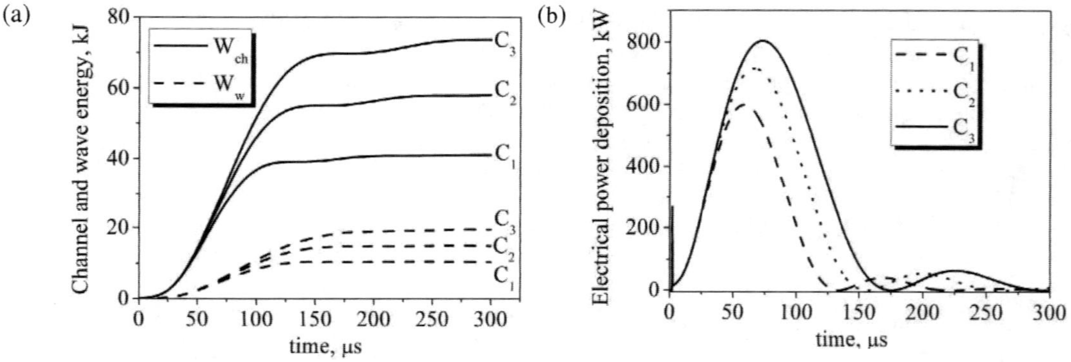

Figure 2. Channel (W_{ch}), pressure wave (W_w) energy (a) and electrical power deposition (b) versus time for different pulse generator capacitance.

Explosive electrical energy release in plasma channel leads to a sharp increase in the channel pressure up to 5.5 GPa, leading to its expansion (figure 3). The channel pressure is not substantially stabled; the increase in the channel volume causes the pressure reduction and consequently the slowdown of its expansion. The speed of pressure reduction depends on the dynamics of electrical power deposition in the channel. The time-history of the pressure pulse, produced on the borehole walls versus electrical energy release rate is shown in figure 4.

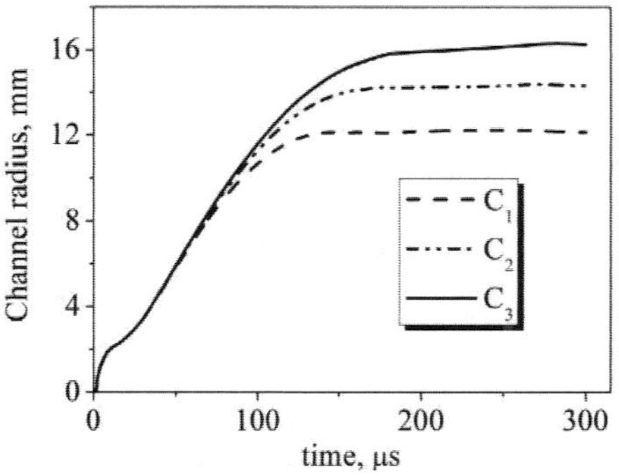

Figure 3. Plasma channel radius versus time for different rates of electrical energy release.

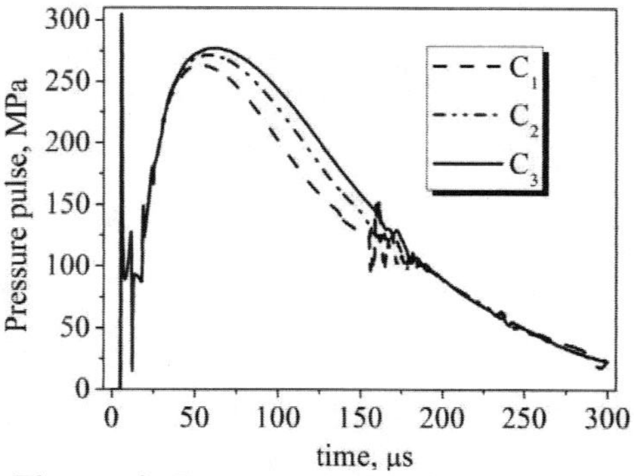

Figure 4. Pressure versus time waveforms, produced on the borehole walls for different rates of electrical energy release.

The expanding plasma channel generates shock waves. The geometrical divergence and dispersion result in shock wave damping along the way of its propagation. In the vicinity of the channel the shock wave transforms into the mechanical stress-waves propagating in combination of polyethylene–concrete materials. The stress-wave profile essentially depends on the pressure pulse wave shape on the borehole wall, which in its turn is determined by the rate of electrical energy release in the plasma channel. Figure 5 illustrates the radial and tangential stress profiles at time lying near the peak in the power level (see figure 2 (b)). It is clear from figures 4 and 5 that rapid power deposition result to the higher amplitude of compressive stresses.

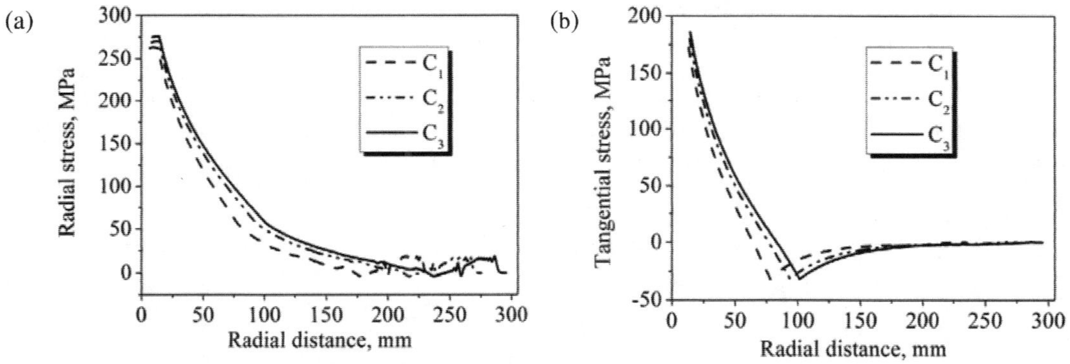

Figure 5. Distribution diagrams of radial (a) and tangential (b) stresses in the pressure wave in concrete at time of peak in the rate of electrical energy release.

The fracture pattern of solid is determined by the resulting tensely-deformed solid state, generated at interaction of direct and reflected waves. Dynamics of radial and tangential mechanical stress distribution in concrete after the wave reflection is shown in figure 6.It shows that stress distribution diagrams are of intricate profile formed both by direct and reflected waves.

The necessary condition for the crack initiation in concrete is presence of the region of tensile stresses both in radial and tangential directions. The crack was initiated when one of the stress components has exceeded the tensile strength of concrete, which is about (5–10) MPa according to the literature. Wave dynamics in figures 5 and 6 shows that the lower energy deposition rate leads to the higher amplitude of tensile stresses both in radial and tangential direction. By contrast, in the rapid deposition case the region of tensile stresses in the wave is formed later. Efficiency of generator energy transmission to the wave energy for different rates of energy release is about 22 %, 20 % and 19 % respectively.

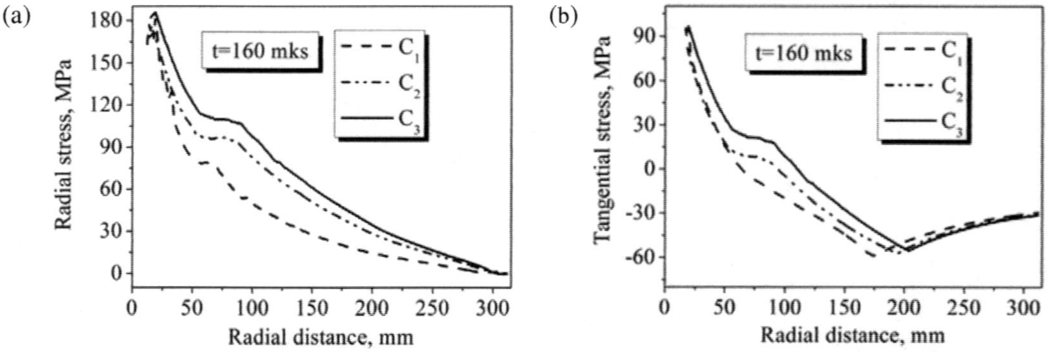

Figure 6. Distribution diagrams of radial (a) and tangential (b) stresses in concrete after the wave reflection from the concrete surface at time of 160 mks.

Thereby, the results of plasma blasting computer simulations have shown that the rapid power deposition produces an increase in peak pressure on the borehole wall. However, because it is produced by a shock, pressure on the borehole walls, falls off rapidly, whereas slow deposition produces sufficiently sustained pressure pulse.

4. Conclusion

Results of computer experiments demonstrate the necessity for complex analysis of the electrical energy balance in electrical discharge system from the pulse generator to the pressure and stress wave energy. It is shown the significant dependence of the stress-wave profile on the pressure pulse wave shape on the borehole wall which, in its turn, is determined by the rate of electrical energy release in the plasma channel and is weakly depended on the time of energy release (at given rate of its release). Analysis of stress-wave dynamics have been shown that the rapid power deposition results to the higher amplitude of compressive stresses in the wave. The lower energy deposition rate in plasma channel leads to the higher amplitude of tensile stresses both in radial and tangential direction. By contrast, in the rapid deposition case the region of tensile stresses in the wave is formed later.

Acknowledgment

This work is supported by Russian Foundation for Basic Research (grant №14-08-01088\14) and the Federal Target Programme (project № 14.575.21.0059).

References

[1] Pronko S, Schofield G, Hamelin M and Kitzinger F Megajoule pulsed power experiments for plasma blasting mining applications 1993 *Proc. 9th IEEE Int. Pulsed Power Conf.* (Albuquerque) p 15–18

[2] Nantel J and Kitzinger F Plasma blasting techniques 1990 *Pros. 3rd Int. Symp. Rock Fragmentetion by Blasting* (Brisbane) p 79

[3] Bluhm H, Frey W, Giese H, Hoppe P, Schultheis C and Sträbner R Application of pulsed power HV discharges to material fragmentation and recycling 2000 *IEEE Trans. On Dielecrics and Elecr. Insul.* **7** pp 25–636

[4] Weise G and Loffler M Experimental investigations on rock fractioning by replacing explosives with electrically generated pressure pulses 1993 *Pros. 9th IEEE Int. Pulsed power Conf.* (Albuquerque) p 19

[5] Sjomkin B, Usov A and Kurets V 1995 *The principles of electric impulse destruction of materials* (St. Petersburg: Nauka) p 276

[6] Timoshkin V, Mackersie J and MacGregor S 2004 Plasma channel miniature hole drilling technology *IEEE Trans. Plasma Sci.* **5** 2055

[7] Burtsev V, Kalinin N and Luchinsky A 1990 *Electrical explosion of wires and its application in electroohysical installations* (Moscow: Energoatomizdat) p 288
[8] Naugolnyh K and Roy N 1971 *Electric discharges in water* (Moscow: Nauka) p 238
[9] Krivitskij E *Dynamics of electro explosion in a liquid* (Kiev: Naukova Dumka) p 208
[10] Burkin V, Kuznetsova N and Lopatin V 2009 *J. Phys. D: Appl. Phys.* **43** 185204
[11] Kuznetsova N, Lopatin V, Burkin V, Golovanevskiy V, Zhgun D and Ivanov N Theoretical and experimental investigation of non-conducting materials 2011 *Pros. 17th IEEE Int. Pulsed power Conf.* (Chicago) pp 267–273
[12] Wilkins M 1964 *Calculation of elastic-plastic flow, methods in computational physics* (New York: Academic) pp 211–264
[13] Rihtmyer R and Morton K 1967 *Difference methods for initial-value problems* (New York: Interscience)
[14] Rompe R and Weizel W 1944 Zs. Physik B 122 pp 9–12
[15] Kinslow R 1970 *High-velocity impact phenomena* (New York: Academic Press) p 533

International Congress on Energy Fluxes and Radiation Effects (EFRE-2014) IOP Publishing
Journal of Physics: Conference Series **552** (2014) 012030 doi:10.1088/1742-6596/552/1/012030

Measurements of high voltage pulses with subnanosecond rise time

N M Bykov[1], I K Kurkan[1, 2] and A S Stepchenko[1]
[1] Institute of High Current Electronics, 2/3 Akademichesky Ave., Tomsk, 634055, Russia
[2] National Research Tomsk Polytechnic University, 30 Lenin Ave., 634050 Tomsk, Russia

E-mail: ikk@lfe.hcei.tsc.ru

Abstract. The paper deals with the application features of capacitive probes for subnanosecond voltage pulse measurements. The relation for the amplitude of distortions determined by the finite electrical length of a capacitive divider was obtained. This relation could serve for quantitative estimation of capacitive divider suitability. Probes of various designs were used for recording the high voltage pulses with subnanosecond voltage changes. The pulse shapes with voltage change durations of 0.2 - 20 ns from different voltage probes were compared. It was shown that the use of capacitive divider with high- and low-voltage arms filled with the same material is more appropriate. This divider provides the same voltage ratio for voltage changes of durations from tens to fractions of nanoseconds.

1. Introduction

There are some applications requiring the high-voltage pulses with duration of several tens of nanoseconds and with a voltage rise time in subnanosecond range, when a voltage pulse amplitude of about 1 MV. The problem for recording such pulses appears in different areas: in investigations of non-steady states and transient processes in gas discharge gaps for power switching [1-4], when researching and developing of high power microwave sources with phase lock and phase control [5,6], RF-pulses sources based on gyromagnetic nonlinear transmission lines [7], etc. High noise level unavoidably accompanies a measurement procedure at high voltage setups. It determines most requirements for a probe: high electric strength of construction, a high signal output level and low stray reactance.

The approach presented in paper [8] is quite acceptable for subnanosecond pulse measurements. The divider arms were considered as two coupled transmission lines; therefore the voltage divider ratio was defined by the wave impedances ratio of them. The divider makes it possible to record pulses almost without distortion within the time interval limited by the double transit time through the low-voltage arm. Therefore, its implementation for long pulse monitoring is problematic.

A capacitive divider satisfies the most requirements we mentioned above. However, conventional approach forces one to reduce the measuring elements of probe with the shortening of pulse rise time. Its dimensions should be of a few millimeters for measurements of subnanosecond rise time. Such miniature probe has shortcomings in our applications. The main disadvantage is that the output signal amplitude is small within relatively long monitoring intervals that causes a significant increase of the

Content from this work may be used under the terms of the Creative Commons Attribution 3.0 licence. Any further distribution of this work must maintain attribution to the author(s) and the title of the work, journal citation and DOI.
Published under licence by IOP Publishing Ltd

random noise fraction in the final result. Another disadvantage is relative difficulty to calculate the division ratio with acceptable accuracy.

We developed a new interpretation of capacitive divider, making it possible to design the probe which is free of limitations mentioned above. The main idea is based on the known fact. A superconductive infinitely thin layer of arbitrary dimensions placed along an equipotential surface does not disturb the electromagnetic wave propagation. It allows one to choose measuring surface dimensions of the divider arbitrarily when it is placed in a homogeneous medium. Simple analysis of capacitive divider model presented in Section 2 proves a zero level of distortions for capacitive divider with arms filled with the same material. The relation for the amplitude of distortions determined by probe finite electrical length was obtained. It demonstrates that the level of the distortions depends not only on the ratio of the divider electric length to the pulse rise time, but also on the difference between electric lengths of the divider arms. The experimental setup is described in Section 3. Experimental results presented in Section 4 attest the validity of the proposed approach. Some details of applications of different probes are discussed in Section 5.

2. Estimation of distortion amplitude for capacitive divider

Let us begin from the main equations for conventional capacitive divider. We consider an axially symmetric divider in a coaxial line filled with material of permittivity ε_h (Fig. 1).The voltage divider is a thin-walled hollow metal cylinder *2* separated from an outer conductor *1* by a layer of permittivity $\varepsilon_l \geq \varepsilon_h$. The cylinder length is equal to l, and the cylinder wall thickness and the layer thickness are so small that the divider hardly affects the pulse propagation in the line. The low-voltage arm is loaded on a measuring circuit with impedance ρ_s. The connection point of the measuring circuit is equally spaced from the divider ends.

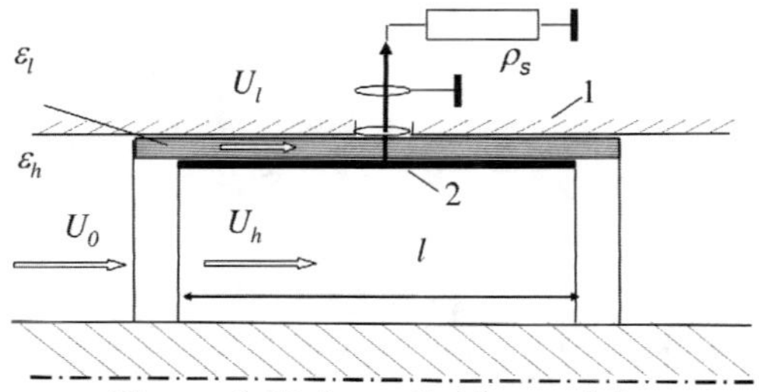

Figure 1. Voltage divider model.

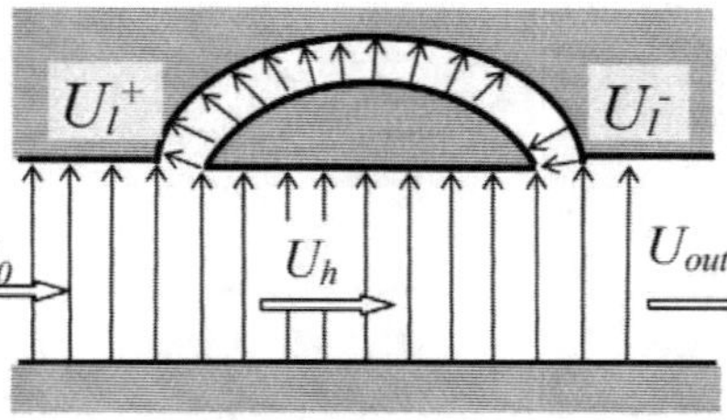

Figure 2. Dynamics of pulse propagation.

Let a near-rectangular signal with an envelope $U_0(t)$, rise time τ_{rise}, and duration τ_{pulse} be applied at the left. When the pulse transit time through the divider length is sufficiently small compared to τ_{rise}, the divider can be considered as a lumped element composed of two series-connected capacitances. The metal cylinder and the outer coaxial conductor create a low-voltage divider arm with capacitance C_l, and the inner conductor and cylinder create a high-voltage arm with capacitance C_h.

The capacitances C_h and C_l and the measuring circuit impedance ρ_s represent an elementary circuit for which the applied voltage $U_0(t)$ and the output signal voltage $U_l(t)$ are related as

$$U_0(t) = k_c\left(U_l(t) + \frac{1}{\tau_0}\int_0^t U_l(t')dt'\right), \quad k_c = \frac{C_l + C_h}{C_h}, \tag{1}$$

where k_c is the capacitive division ratio, $\tau_0 = (C_l+C_h)\rho_s$ is the divider time constant, and ρ_s is the measuring circuit impedance. Formula (1) takes into account the influence of the measuring circuit impedance and allows reconstruction of an actual pulse. In practice, the capacitance C_l is two or three orders of magnitude higher than C_h, and the integration constant τ_0 can be chosen large enough to ignore the voltage drop at the capacitance C_l. In this case, we derive the relations for an idealized capacitive divider as

$$U_0(t) \cong k_c \cdot U_l(t), \quad k_c \cong C_l / C_h, \quad C_h << C_l, \quad \tau_{pulse} << \tau_0 \tag{2}$$

According to the accepted approach, the divider electrical length τ_l (the signal transition time trough the low-voltage arm) should be much smaller than the rise time τ_{rise}. The waveform of voltage $U_0(t)$ obtained from (1) or (2), describes the pulse shape with some methodical error, which is determined by the finite electrical length of the divider.

Let us estimate this error. Suppose the conditions (2) are satisfied and there are no losses. Let's test the divider by pulses having infinitely short rise time, considering the divider to consist of two (low- and high-voltage) lines with wave impedances ρ_l and ρ_h and electrical lengths τ_l and τ_h, respectively.

For analysis, it is convenient to consider the divider geometry (Fig. 2), where the difference between the electrical lengths of the divider arms is represented by an increase in the geometrical length of the low-voltage arm. The waves excited in the low-voltage arm at divider ends have opposite polarities, as it follows from the voltage matching conditions at the interfaces: $U_0 = U_h + U_l^+$, $U_h = U_{out} + U_l^-$. At the left end of the divider, the incoming wave U_0 is divided into two waves U_l^+ and U_h of the same polarity. At the right end of the divider the wave of the high-voltage arm U_h is divided into the wave U_l^- and the output wave U_{out}. The output wave is lower than the wave U_h; therefore the wave exited in the low-voltage arm U_l^- has to be of opposite polarity to U_h.

In the accepted simplifications, the waves U_l^- and U_l^+ are equal in amplitude and will be fully reflected at the divider ends. The total signal U_l at the load of the measuring circuit for a long pulse with a step-like rise front is shown in Fig. 3.

The difference between the electrical lengths of the divider arms (τ_l - τ_h) is designated as $\Delta\tau$. The result of testing the divider by a pulse of duration much shorter than the electrical length τ_l is a sequence of pulses of opposite polarity (Fig. 4). The first pulse is a signal corresponding to the actual pulse. The rest of the pulses results from multiple reflections of the waves U_l^- and U_l^+ from the divider ends. The pulse with a steep rise time excites parasitic oscillations in the divider with a period equal to τ_l.

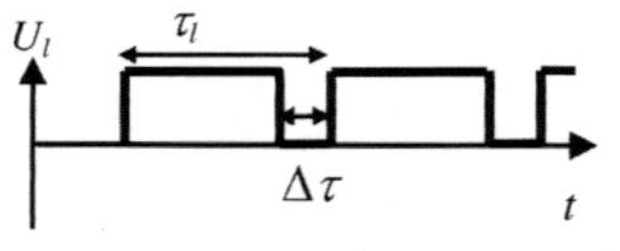

Figure 3. Tests of the divider by a long pulse with steep rise time.

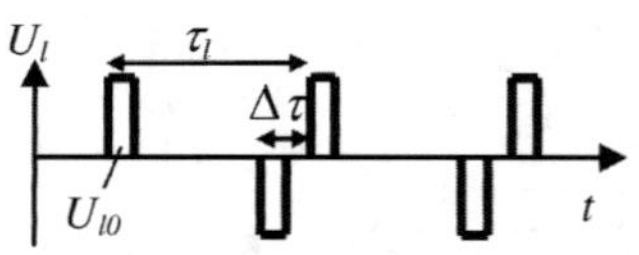

Figure 4. Tests of the divider by a short pulse with steep rise time

Note that the results presented in Fig. 3 and 4 can be obtained if we connect two generators with high output impedance and opposite polarities to the divider ends shifted by the $\Delta\tau$. In this case, we can estimate the parasitic oscillation amplitude A_m for an arbitrary pulse in terms of the difference $U_l(t)$ - $U_l(t - \Delta\tau)$. For the linear steepness the following relations are obtained

$$A_m = (k_\rho)^{-1} U_0 \frac{\Delta\tau}{\tau_{rise}} \cong (k_c)^{-1} U_0 \frac{\Delta\tau}{\tau_h} \frac{\tau_l}{\tau_{rise}}, \quad \Delta\tau = \frac{l}{c}\left(\sqrt{\varepsilon_l} - \sqrt{\varepsilon_h}\right), \tag{3}$$

where $(k_\rho)^{-1} \cong (\rho_h/\rho_l)^{-1}$ is the impedance division ratio, and c is the vacuum speed of light. When the rise time τ_{rise} is smaller than $\Delta\tau$, the amplitude A_m reaches the maximum value.

If the difference $\Delta\tau$ is equal to zero, parasitic oscillations are not exited by pulses of arbitrary duration.

3. Experimental setup

The purpose of experimental verification was to confirm the foregoing considerations. The experimental setup comprised a SINUS high-voltage pulse generator [9], a transmission line, a sharpening spark gap, and a resistive load. The pulse generator was switched onto the coaxial transmission line which was connected through the sharpening spark gap to the matched resistive load.

The transmission line was filled with transformer oil and had the wave impedance of 30 Ohm at the outer conductor radius of 4.6 cm. The pulse generator formed a "smooth" quasi-rectangular pulse with τ_{pulse} of 18 ns, τ_{rise} of 3 ns, and U_0 of 150 kV at the matched load. The sharpening spark gap closing time was changed within 0.15 - 0.6 ns. Voltage steep drop of subnanosecond duration τ_{st} appeared in the transmission line on the smooth pulse when the sharpening spark gap closed. Thus, the formed pulse had voltage changes of duration from 0.15 to 20 ns.

We used two main capacitive dividers: the reference divider C_f and the test divider $C_{g\text{-}f}$. The capacitive dividers were made of copper foil 0.1 mm thick and had the equal areas. The divider length l along the coaxial line axis was 5.5 cm. The high-voltage arms of the dividers C_f and $C_{g\text{-}f}$ were filled with fluoroplastic in the form of cylinder of the same 5.5-cm length. The low-voltage arm of the test divider $C_{g\text{-}f}$ was made of glass-fiber laminate 0.2 mm thick and the reference divider C_f was made of fluoroplastic film 0.08 mm thick. The capacitances of the dividers measured at a frequency of 1 kHz were C_f= 1.47 nF and $C_{g\text{-}f}$ = 1.63 nF and corresponded to the permittivity of ~ 2.1 for fluoroplastic and ~ 5.8 for glass-fiber laminate. Thus, the arms of the reference divider C_f were filled with materials similar in properties, and the arms of the test divider $C_{g\text{-}f}$ were filled with materials differing in dielectric properties. The capacitive division ratios for both dividers were close to each other and were about 300.

In the transmission line, there were two additional probes: a differential voltage probe C_E and a differential current probe C_H which were used for verification. The probes were located in drilled holes of diameter l = 7 mm near the surface of the outer conductor of the coaxial line. The designs of the differential probes are shown schematically in Fig. 5. The measuring element of the voltage probe was a metal disk *1* of diameter 6 mm and thickness 0.4 mm which was coaxially connected to the end of measuring cable.

Figure 5. Schematics of differential probes: **a)** - voltage C_E, **b)** - current C_H, *1* - measurement element.

Note that the use of the measuring element with a diameter close to the hole diameter was chosen to minimize the risk of transmission line breakdown to the measuring circuit and to increase the output signal. The measuring element of the current probe was the conductor *1* with diameter and length of 2 mm, which was connected to the end of measuring cable and to the outer conductor of the coaxial line. All main and control probes were located in one measuring unit.

A TDS7404 digital oscilloscope with a band of 4 GHz and sampling interval of 50 ps was used. All probes were calibrated by a relatively smooth quasi-rectangular pulse (τ_{pulse} of 18 ns, τ_{rise} of 3 ns, U_0 of 150 kV) after averaging over 300 pulses. The waveforms taken from different probes coincided with a high accuracy.

4. Experimental results

Figure 7 shows waveforms of the voltage *U(t)* measured using the reference divider U_f and the test divider $U_{g\text{-}f}$ with the shift of 0.35 ns relative to each other. The waveforms correspond to a single pulse. There is the drop with duration of $\tau_{st} \approx 0.15$ ns on the both waveforms. It can be seen that in the test divider $C_{g\text{-}f}$ the drop excites parasitic oscillations. The oscillation amplitude is close to the voltage drop amplitude. Increasing τ_{st} decreased the oscillation amplitude. When τ_{st} was longer than 0.7 ns, no regular oscillations were detected. The oscillation period equal to ≈ 0.35 ns corresponds to the electrical length of the low-voltage arm filled with material of permittivity $\varepsilon_l \approx 4$. In the reference divider C_f, irregular oscillations of small amplitude were observed. Similar oscillations were observed on the waveforms from differential probes. It leads us to the conclusion that the oscillatory process does actually occur.

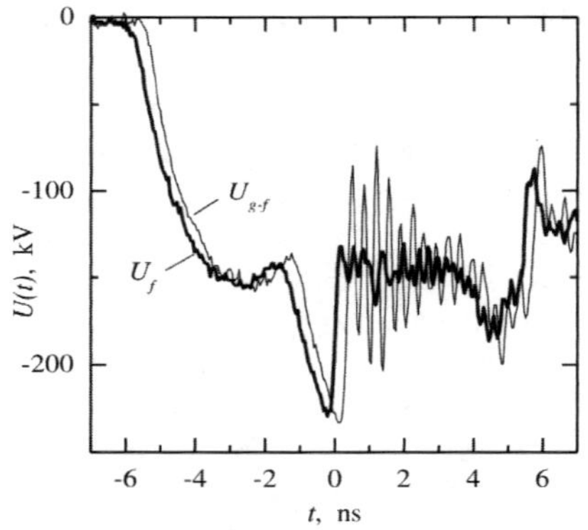

Figure 7. Waveforms from U_f - reference divider and $U_{g\text{-}f}$ - test divider. Single voltage pulses with drop τ_{st} of 0.15 ns

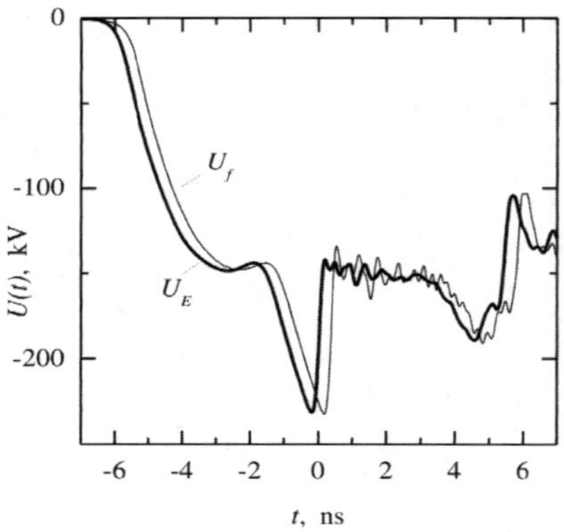

Figure 8. Waveforms from U_E - differential probe and U_f - reference divider averaged out of 300 events.

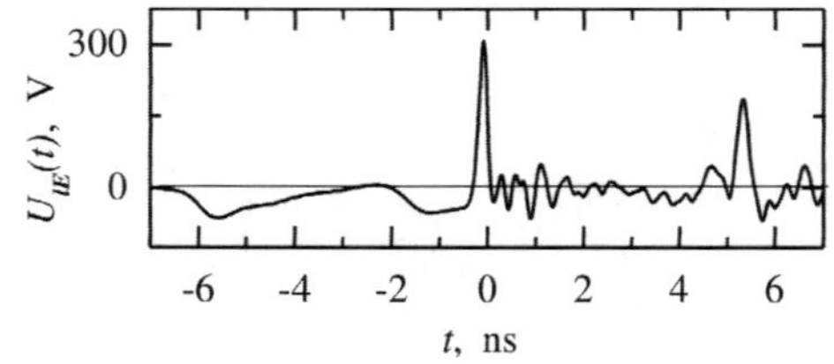

Figure 9. Waveform of U_{IE} from the differential voltage probe averaged over 300 events

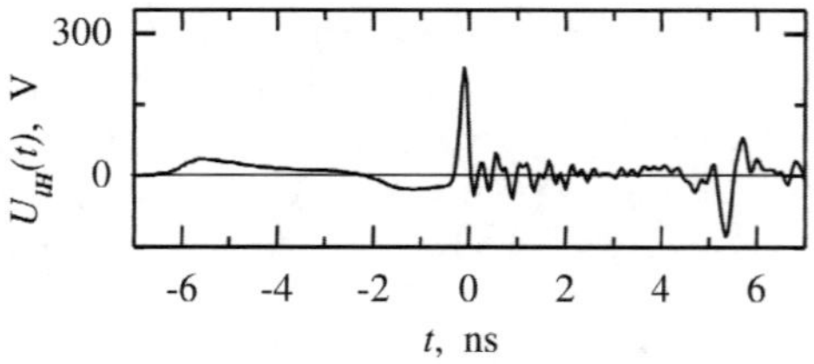

Figure 10. Waveform of U_{IH} from the differential current probe averaged over 300 events

Figure 8 shows the result of reconstructed waveforms from the reference divider U_f and from differential voltage probe U_E for a monitoring interval of 14 ns. Reconstruction of the waveform from the differential voltage probe was performed according to the formula (1), taking into account only the integral term. The whole formula was applied for the reference divider waveforms. Both waveforms are averaged over 300 events. The waveforms of U_f and U_E are shifted by 0.35 ns relative to each other to provide better presentation.

The average waveform U_{IE} from the differential voltage probe without reconstruction is shown in Fig. 9. A similar waveform of U_{IH} from the current probe is shown in Fig. 10. Both waveforms were averaged over 300 pulses. It allowed us to reject a random noise in the data from probes. Reconstruction of current pulse shape from the current probe data has to be performed according to formula similar to (1), taking into account both the integral and linear terms. Since the reactance of the measuring element was estimated as 10 Ohm at a frequency of 1 GHz, and it is comparable to measuring circuit impedance of 50 Ohm.

It was found that the waveforms of U_f and U_E describe the pulse shape *U(t)* almost identically in a monitoring interval of 20 ns. Therefore, the voltage division ratios of the differential probe C_E and capacitive divider C_f are almost the same in the cases of voltage changes with duration of 0.25-20 ns. The waveforms from the probe C_E contained considerable random noise that caused the strong distortion of the reconstructed waveform from pulse to pulse. At the same time, the noise weight in data from the divider C_f was so small that the reproducibility of a pulse shape was high.

5. Discussion of results and conclusion

As follows from (1), the noise level increases proportionally to τ_{pulse}/τ_o for data from the differential probe and could reach an unacceptable value by the end of the monitoring interval. At the same time, for idealized capacitive divider (2) the noise level remains insignificant. Another shortcoming of differential probe is the relative difficulty to calculate the division ratio with acceptable accuracy. This fact becomes especially important when the experimental calibration of the probe is difficult.

Despite the shortcomings of the differential probes, their use for monitoring the pulses with subnanosecond voltage changes is justified where arrangement of other type of probes is problematic. Excluding the problem of random noise, the applicability ranges of the differential probe C_E and reference capacitive divider C_f in our design should be considered as identical. This applicability range is determined by the diameters of the measuring cable.

The formula (1) provides the powerful tool for quantitative estimation of capacitive divider suitability. Once the amplitude of distortions is predefined, the shortest rise time for the given divider is known. Another valuable fact is that the impedance division ratio is equal to the capacitive division ratio for a divider which arms are filled with the same material. If the divider is calibrated by any available testing pulse source, then the obtained division ratio is valid for pulses with an arbitrary rise time.

The developed approach makes it possible to expand the divider length to prevent a substantial voltage drop, without any losses to suitability for recording the subnanosecond voltage changes.

In the study, we obtained the relation for quantitative estimation of distortion amplitude in capacitive voltage divider. It is shown that the distortion amplitude is proportional to the electrical length difference between the low- and high-voltage arms of a divider. If the difference is close to zero, the divider electrical length could be chosen arbitrarily.

Measurements of relatively long pulses with subnanosecond voltage changes by a capacitive divider with arms filled with the same material are more appropriate. This provides the same voltage ratio for voltage changes of durations from tens to fractions of nanoseconds.

Acknowledgments
The authors thank Yu.D. Korolev, V.F. Landl, A.I. Klimov, and A.V. Gunin.

References

[1] Efremov A M, Kovalchuk B M and Korolev Y D 2012 Momentary interruption of current passing through zero in subnanosecond high-pressure gas-discharge switches *Tech. Phys.* **57** 478-86

[2] Korolev Y D and Bykov N M 2012 High-voltage spark gap in a regime of subnanosecond switching *IEEE Trans. on Plasma Sci.* **40** 2443-48

[3] Korolev Y D, Bykov N M and Ivanov S N 2008 Subnanosecond processes in the stage of breakdown formation in gas at a high pressure *Plasma Phys. Rep.* **34** 1022-27

[4] Korolev Y D, Frants O B, Geyman V G, Kasyanov V S and Landl N V 2012 Transient processes during formation of a steady-state glow discharge in air *IEEE Trans. Plasma Sci.* **40** 2951-60

[5] Afanas'ev K V, Bykov N M, Gubanov V P, El'chaninov A A, Klimov A I, Korovin S D, Rostov V V and Stepchenko A S 2006 A high-power periodic nanosecond pulse source of coherent 8-cm electromagnetic radiation *Tech. Phys. Lett.* **32** (11) 925-27

[6] Rostov V V, El'chaninov A A, Klimov A I, Konev V Yu, Romanchenko I V, Sharypov K A, Shunailov S A, Ul'maskulov M R and Yalandin M I 2013 Phase control in parallel channels of shock-excited microwave nanosecond oscillators *IEEE Trans. on Plasma Sci.* **41** 2735-41

[7] Romanchenko I V, Rostov V V, Gubanov V P, Stepchenko A S, Gunin A V and Kurkan I K 2012 Repetitive sub-gigawatt rf source based on gyromagnetic nonlinear transmission line *Rev. of Sci. Instr.* **83** 074705

[8] Efremov A M and Kovalchuk B M 2004 A subnanosecond voltage divider on coupled lines *Instrum.Exp. Tech.* **47** 61-62

[9] Korovin S D and Rostov V V 1996 High-current nanosecond pulse-periodic electron accelerators utilizing a Tesla transformer *Russian Phys. J.* **39** (12) 1177-85

Journal of Physics: Conference Series **552** (2014) 012031 doi:10.1088/1742-6596/552/1/012031

Foil-less plasma-filled diode for HPM generator

A A Eltchaninov[1], B M Kovalchuk[1], I K Kurkan[1, 2] and A A Zherlitsyn[1]
[1] Institute of High Current Electronics, 2/3 Akademichesky Ave., Tomsk, 634055, Russia
[2] National Research Tomsk Polytechnic University, 30 Lenin Ave., 634050 Tomsk, Russia

E-mail: ikk@lfe.hcei.tsc.ru

Abstract. Plasma-filled diode regarded as perspective source of electron beam feeding HPM generator of GW power level, comparing to conventional explosive emission vacuum diode. Electron beam generation occurs in plasma double layer, where plasma boundary plays as an anode. It allows cancelling the usage of anode foils or grids in HPM generators with the virtual cathode, which could limit its life time to few shots. The presence of ions in the e-beam drift space could raise the limiting current for a drift space, but it could affect to microwave generation also. Sectioned plasma-filled diode with beam current of about 100 kA, electron beam energy of about 0.5 MV and beam current density of 1-10 kA/cm^2 was realized. Cylindrical transport channel with the diameter of 200 mm and the length of about 30 cm was attached to the diode. Beam current measurements in a drift space were performed. Computer simulations of electron beam transport with the presence of ions were carried out with the 2.5D axisymmetric version of PiC-code KARAT. Obtained results would help optimizing electrodynamic system of HPM generator subjected to the presence of ions.

1. Introduction

An increasing of HPM generators output power requires increasing in the feeding electron beam power and its current density. So, the plasma-filled diode looks as a good candidate to use in the HPM generators. The presence of plasma in the accelerating diode gap removes the charge current limitation. This allows an increasing of an electron beam power by one or two order comparing to the vacuum diode with the same accelerating voltage. Low impedance pulsed power generator, such as linear transformer driver (LTD) [1], could be utilized efficiently to feed these diodes [2]. The usage of a plasma-filled diode gives another benefit which is an absence of anode meshes or foils in HPM generators with the virtual cathode (vircators). These foils limiting the gigawatt level vircators operation time to a few shots [3, 4].

There are few tasks to be investigated and solved then using plasma-filled diode in HPM generators. The main one is beam transport conditions and formation of virtual cathode (VC) in presence of ions, as the electron beam tends to contract by self-magnetic fields. Plasma-filled diode with voltage of about 0.5 MV, beam current of about 100 kV and its density of 1-10 kA/cm^2 was realized for further microwave applications. It was attached directly to cylindrical drift channel with no anode mesh in between. Current evolution in drift space was measured in experiment. Numerical modelling of electron beam transportation with the presence of ions by the 2.5D symmetric version of Particle-in-Cell (PiC) code KARAT [5] shown good agreement with experimental results giving a base for further development and optimization of electrodynamic system of HPM generator.

Content from this work may be used under the terms of the Creative Commons Attribution 3.0 licence. Any further distribution of this work must maintain attribution to the author(s) and the title of the work, journal citation and DOI.
Published under licence by IOP Publishing Ltd

2. Experiment

Plasma-filled diode was fed by LTD with six sections connected in series. Equivalent capacitance was 53 nF, inductance was 260 nH, stored energy was 7.8 kJ with sections charging voltage of 90 kV. Experiment setup arrangement represented on Figure 1. Diode was split to six separate sections to decrease beam current self focusing, as magnetic field of section defined by its current instead of full diode current. The diameter of anode hollow (1) for each section was 38 mm. Every section has the cylindrical cathode (2) of 12 mm diameter made of stainless steel. Cathodes were placed on diameter of 80 mm. Drift channel (3) with diameter of 200 mm and length of 290 mm was attached to the sectioned diode.

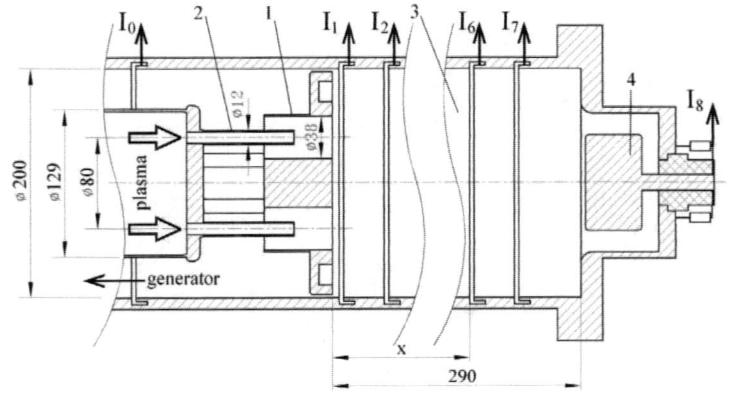

Figure 1. Experimental arrangement: 1 – anode, 2 – cathode, 3 – drift channel, 4 – electron beam collector

Six capillary plasma guns used for plasma generation in diode. Dielectric material of plasma guns was polyethylene. The diameter of capillar was 1.3 mm and its length was 15 mm. Guns were placed into the cathodes (2) and injected plasma towards to the drift channel (3). Guns were fed by separated pulse generator with capacitance of 40 nF and charging voltage of 80 kV. The current for a single gun was of 1.5 kA, first half-wave duration was of 800 ns. LTD pulse to the diode was delayed for time t_d from plasma gun generator to change plasma density in the diode.

The full diode current I_0 was measured by the inductive sensor. The output LTD voltage was measured by the active resistive voltage divider. Diode voltage was calculated taking into account inductive voltage drop on the section between the LTD output voltage sensor and the cathode. Beam currents I_1-I_8 in the drift channel (3) at different distances x from the anode (1) were measured by inductive sensors spaced by the 40 mm step. Collector with diameter of 84 mm was loaded by resistive shunt for collector current measurements.

The time delay t_d between the main LTD pulse and the pulse of the plasma generator defines the plasma-filled diode operation and conditions of beam transportation in the drift channel (3). Figure 2 represents waveforms of diode current and voltage as well as beam current in the drift channel at different lengths x from the anode for t_d=0.7 and 1.0 μs. High voltage stage of electron beam generation in the diode foreran by low impedance stage with impedance much lower than the LTD's one, then the voltage at the diode gap is small and the current close to the short circuit one. At the low impedance stage all current goes through the anode (1) with no current through the drift channel (3). After that diode impedance increases forming the voltage pulse of 450 kV and electron beam is generated with a current going through the drift channel.

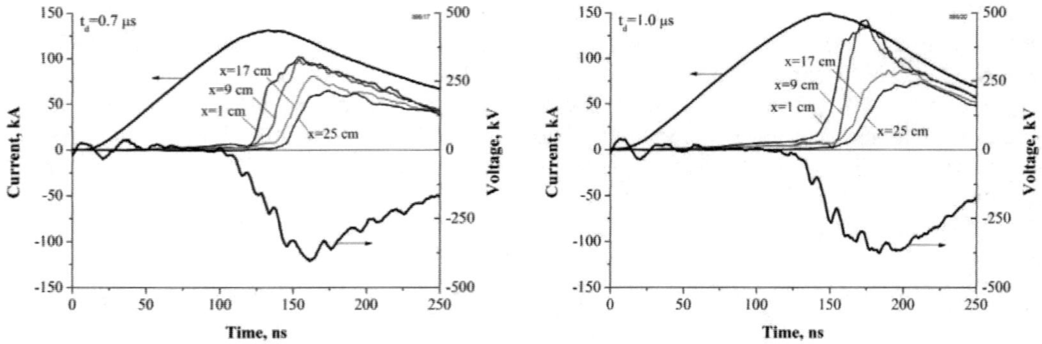

Figure 2. Plasma-filled diode voltage and current waveforms for t_d=0.7 and 1.0 μs. Beam currents in the drift channel at distances x=1,9,17,25 cm from the anode

For t_d=0.7 μs current at low impedance stage was of 130 kA. Electron beam current with amplitude of 100 kA and rise front duration of about 20 ns went to x=9 cm without losses. Near the right end of the drift channel beam current amplitude dropped almost twice. With increasing t_d to 1 μs current at low impedance stage was increased to 150 kA. Beam current at distance x=9 cm increased to about 130 kA with the same loss rate for a bigger distances.

3. Numerical modelling

Numerical experiment by PiC method recognised as powerful tool in research of plasma and dense particle beams. Modelling of drift channel was carried out by symmetric 2.5D version of code KARAT. The goal of this modelling was to obtain qualitative demonstration of processes in the drift channel of experimental setup.

The electron beam was injected through the left wall with the shape similar to the experimental one (Fig.3). Plasma was modelled by the PIC plasma consisting of electron and proton macroparticles. The electron beam particles were injected with the energy corresponded to the experimental voltage pulse. The current of injected beam were equal to the total LTD current. The PIC plasma particles were injected with the energy of 3 eV. The pre-ionized area was used to simulate a delay between the plasma generator pulse and the main LTD pulse.

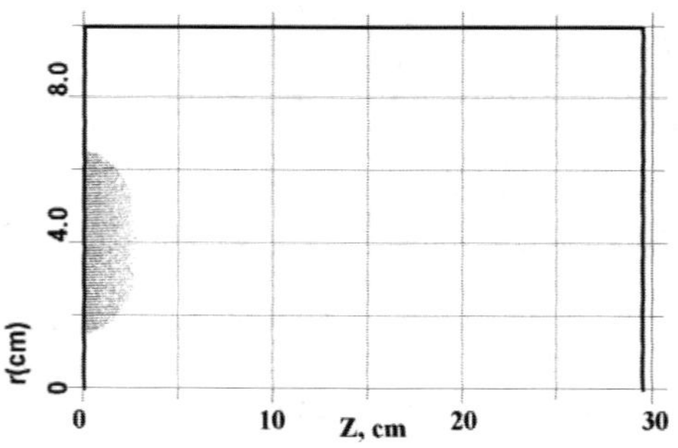

Figure 3. Layout for numerical modelling by 2.5D PiC code KARAT using electron beam and PIC electron-proton plasma injection

The transported current for this geometry without PIC plasma was limited to about 15 kA. We were rising the PIC plasma density from one numerical experiment to another. The transporting beam current with the amplitudes similar to the experimental one were obtained at PIC plasma density of

$5x10^{13}$ cm^{-3}. The waveforms of passed current at different crossections of the drift channel are shown at Figure 4.

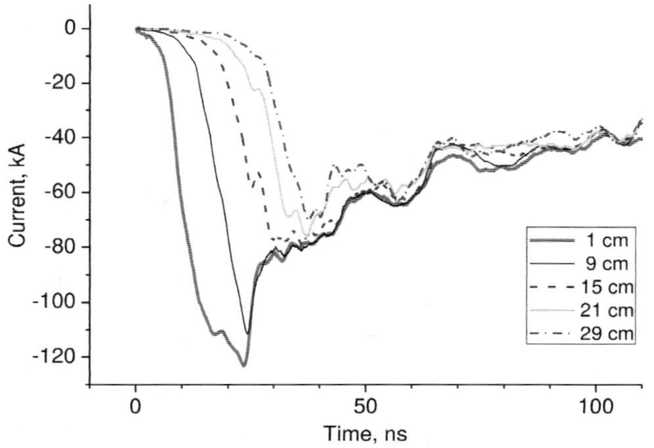

Figure 4. Result of numerical modelling by 2.5D PiC code KARAT with PIC electron-proton plasma ($n=5x10^{13}$ cm^{-3}). Transporting current at different distances

The electron transportation channel was formed in the drift space of 30 cm length in less than 40 ns, as the low-energetic protons were accelerated by the virtual cathode formed by the electron beam close to the injection area. Figure 5 demonstrates distributions of electrons and protons in real and phase spaces at 50 ns from the beginning of voltage pulse, describing the "stable" stage.

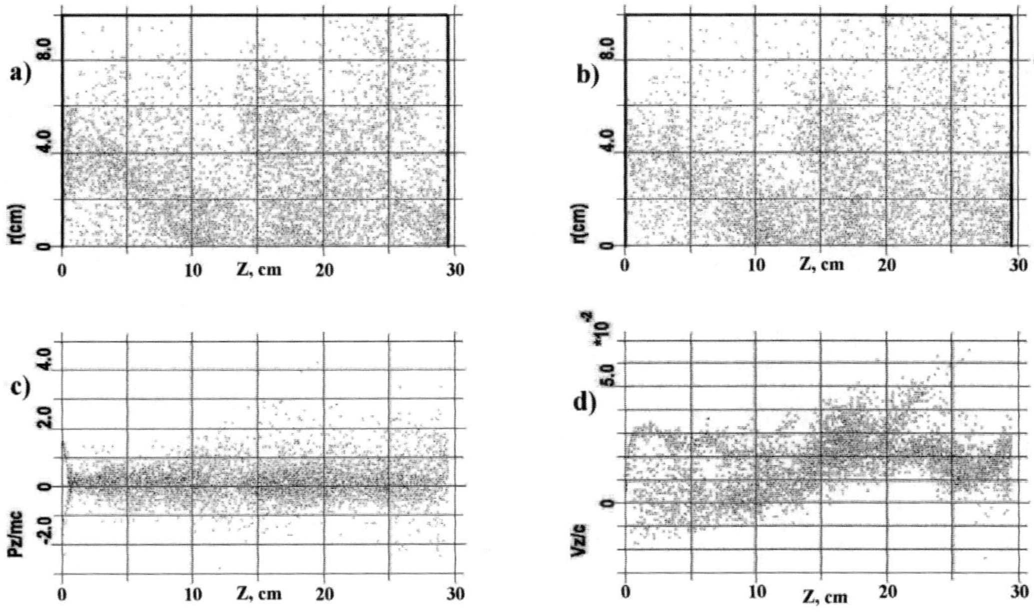

Figure 5. The distributions of electrons (on the left – **a** and **c**) and protons (on the right – **b** and **d**) in real (top) and phase (bottom) spaces at 50 ns from the beginning of voltage pulse

The electrons and protons has similar space distribution (the upper images in Figure 5). The electron and proton beams that are hollow initially pinching at the 10 cm distance. Then they are spreading but the most part of particles reaching the collector is within the radius of 40 mm. Distributions of longitudinal momentum of electrons and velocity of protons are represented in the lower images of Figure 5. The electron beam injected in the drift channel forms the virtual cathode just near the injection area and there is the two counter stream state of electron beam behind the VC. The longitudinal component of the velocity of electrons is well below the one corresponding to the starting energy. The protons injected with low energy are efficiently accelerated by the VC to the energies similar to the starting energy of the electrons. The proton beam surprisingly forms virtual anode about the centre of the drift channel with two counter streams before it and the downstream behind. There are some protons in the downstream beam with the energies above the starting energy of electrons.

4. Conclusion

Sectioned plasma-filled diode with beam current of about 100 kA, electron beam energy of about 0.5 MV and beam current density of 1-10 kA/cm^2 was realized within this work. This diode is proposed to feed a HPM generator with the virtual cathode, allowing to obsolete foils in its design.

The results of the numerical modelling of electron beam transportation with the presence of ions by the 2.5D symmetric version of PiC code KARAT are in good agreement with experiment, giving us powerful tool for further design and optimization of a vircator. It demonstrated the complicated dynamic of partially compensated beam. The presence of virtual cathode allows the protons to fill in the drift channel in a short time.

A possible electromagnetic structure of vircator with the plasma-filled diode should not use long transportation channel as it might be a reason of great losses of microwave energy. A possible changes in the electromagnetic structure due to the presence of dense plasma should be also taken into account.

Acknowledgments

The authors are grateful to Prof. Vladimir P. Tarakanov, the author of KARAT code.
The work was supported by RFBR grant 14-08-01123-a.

References

[1] Kovalchuk B M, Kharlov A V, Kumpyak E V and Zherlitsyn A A 2013 Pulse generators based on air-insulated linear-transformer-driver stages *Phys. Rev. Spec. Top.-Accel. Beams* **16** 050401

[2] Kovalchuk B M, Zherlitsyn A A and Pedin N N 2010 Plasma-filled diode in the electron accelerator on base of a pulsed linear transformer *Laser Part. Beams* **28** 547-52

[3] Korovin S D, Kurkan I K, Loginov S V, Pegel I V, Polevin S D, Volkov S N and Zherlitsyn A A 2003 Decimeter-band frequency-tunable sources of high-power microwave pulses *Laser Part. Beams* **21** 175-85

[4] Shlapakovski A S, Kweller T, Hadas Y, Krasik Y E, Polevin S D and Kurkan I K 2009 Effects of Different Cathode Materials on Submicrosecond Double-Gap Vircator Operation *IEEE Trans. Plasma Sci.* **37**, 1233-41

[5] Tarakanov V P 1992 *User's manual for code Karat,* (Springfield, VA: Berkley Research Associate. Inc)

A study into light scattering and absorption by aluminum nanoparticles in PETN

B Aduev, D Nurmukhametov*, A Zvekov, A Nikitin

Institute of Chemistry of Coal and Material Science, Sovetsky av., 18, Kemerovo, Russia, 650000

E-mail: lesinko-iuxm@yandex.ru

Abstract. The paper is devoted to experimental and theoretical research into nanoparticles' optic properties in pentaerythritol tetranitrate (petn) matrix. A photometric sphere was applied for the transmittance and sum of transmittance and reflectance measurement of petn pressed pellets containing aluminum nanoparticles at the light wavelength 643 nm. The theory of light propagation in terms of spherical harmonics solution of radiative transfer equation in the slab geometry with Fresnel boundary conditions was developed. The properties of aluminum nanoparticles were evaluated in terms of Mie theory. The absorbed energy distribution inside the sample was calculated. It was shown that the Beer's type law is applicable approximately. The apparent light absorption cross section determined, which takes into account both scattering and absorption, is bigger than the geometrical one. The aluminum refractive index value, estimated during comparison of theory with the experimental data, agrees well with the handbook's data.

1. Introduction

A lot of studies have been devoted to research into optical properties of metal nanoparticles [1-6] because they could be utilized in many fields including optical detonators [7-10] solar cells [1,2] design, hyperthermia cancer therapy [5] and nanophotonics devices development [3]. As the light interaction with nanoparticles concerned on both the effects of scattering and absorption, the problem is a challenging one. The effect of multiple scattering was pointed out in [11]. Though there are a lot of papers discussing the optical properties of nanometals, most of the results were obtained using simple spectral approaches where there were no differentiation of scattering and absorption. In the present paper we developed a new approach of the optical results obtained with an integrating sphere processing that made possible to estimate the scattering and absorption parameters of metal nanoparticles.

2. Experimental section

The pressed pellets of pentaerythritol tetranitrate (petn) contained aluminum nanoparticles were used as the experimental samples because they are perspective explosive materials for optic detonators. The grains' average size of the petn powder was 1 – 2 μm [7,8]. The aluminum nanoparticles were synthesized in the Institution of Physics of Metals with the gas-phase method. The size distribution of the nanoparticles showed the maximum at 100 – 120 nm. The aluminum oxide (Al_2O_3) content

Content from this work may be used under the terms of the Creative Commons Attribution 3.0 licence. Any further distribution of this work must maintain attribution to the author(s) and the title of the work, journal citation and DOI.
Published under licence by IOP Publishing Ltd

estimated with electron microscope JEOL JSM63901A equipped with a JEOL JED2400 feature for the element analysis was 24 wt%.

The samples were prepared as follows. The weighted amounts of aluminum and petn were mixed, than the substance was ultrasonificated in hexane, which leads to uniform distribution of the nanoparticles inside the sample. After that the hexane was evaporated and the mixture dried. The weighted amount of the mixture was pressed into the pellets with the diameter 3 ± 0.01 mm and variable thicknesses. The pressure increased up to 1.8 GPa during 30 min when a pellet was being pressed. This procedure allowed us to obtain samples with density value of the monocrystal (1.73±0.03 g/cm^3). The pellets did not have any visible defects which were able to contribute to the scattering [7].

An integrating photometric sphere was used to determine the reflectance and transmittance. In the first blank experiment the intensity inside the sphere I_0 was measured without the sample. Then the sample was placed on the entrance hole and inside the sphere in order to measure transmitted intensity I_T and sum of the transmitted and reflected intensity I_{T+R} correspondingly. These values divided by I_0 gave T and T+R correspondingly. The method was tested with the standard glass light filters [12] and its consistency was verified.

3. Theoretical section

The scattering medium is characterized by scattering phase function $x(\hat{\mathbf{s}}',\hat{\mathbf{s}})$ and linear coefficients of extinction k, absorption k_{abs} and k_{sca} scattering [13]. The albedo of every light-medium interaction is $\Lambda = k_{sca}/k$. The radiative transfer equation, describing the light propagation in scattering and absorbing medium, is:

$$(\hat{\mathbf{s}},\nabla)I(r,\hat{\mathbf{s}}) - kI(r,\hat{\mathbf{s}}) = \frac{k\Lambda}{4\pi}\int I(r,\hat{\mathbf{s}}')x(\hat{\mathbf{s}}',\hat{\mathbf{s}})d\hat{\mathbf{s}}' + q(r,\hat{\mathbf{s}}) \quad (1)$$

where $I(r,\hat{\mathbf{s}})$ is specific intensity of light in the direction $\hat{\mathbf{s}}$ at the coordinate r, $q(r,\hat{\mathbf{s}})$ is the sources function. This equation could be simplified in the one dimensional case to

$$\mu\frac{dI(\tau,\mu)}{d\tau} - I(\tau,\mu) = \frac{\Lambda}{2}\int_{-1}^{1} I(\tau,\mu')x(\mu',\mu)d\mu' + q(\tau,\mu) \quad (2)$$

where μ is cosine of the spherical angle, $\tau = kz$ is dimensionless coordinate.

The radiation transfer equation was solved with the spherical harmonics approach requiring splitting of the radiance into scattered I_s and incident I_0 parts [14]:

$$I = I_0 + I_s \quad (3)$$

The incident part obeys the Beer's law. One is able to write the following formula for I_0 taking into account the reflectance from the bottom boundary of the sample (slab):

$$I_0 = J\cdot\left(\exp(-\tau) + R_f\exp(\tau - 2L)\right) \quad (4)$$

where $J = J_0(1 - R_f)$, R_f is the Fresnel reflectivity from the upper sample's surface, J_0 is the intensity of the incident irradiation outside the sample. The equation system describing spatial dependence of the spherical harmonics expansion coefficients of the intensity (C_m) reads as [14]:

$$\frac{1}{2m+1}\cdot\left[(m+1)\frac{dC_{m+1}}{d\tau} + m\frac{dC_{m-1}}{d\tau}\right] + \left(1 - \frac{\Lambda x_m}{2}\right)C_m = \frac{J\Lambda x_l}{2}\left[\exp(-\tau) + R_f\exp(\tau - 2L)\right] \quad (5)$$

where x_l are expansion coefficients of the scattering phase function into spherical harmonics series. The boundary conditions for the equation (5) are the Fresnel ones [15]:

$$I_s(0,\mu) = R(\mu) I_s(0,-\mu), \qquad 0 \le \mu \le 1 \tag{6}$$

$$I_s(L,-\mu) = R(\mu) I_s(L,\mu), \qquad -1 \le \mu \le 0 \tag{7}$$

where $R(\mu)$ is angular dependence of the Fresnel reflectivity coefficient, $\mu = \cos\theta$ is cosine of the spherical angle. Formulae (9) and (7) give the conditions for the upper and bottom sample's surfaces. One obtains the boundary conditions for the harmonics' contributions applying the $R(\mu)$ expansion into Legendre polynomials:

$$\sum_{m=0}^{N} (N_{lm} - R'_{lm}) C_m(0) = 0 \tag{8}$$

Where the matrix elements read as:

$$R'_{lm} = (-1)^m \frac{2m+1}{2} \int_0^1 P_m(\mu) R(\mu) P_l(\mu) d\mu \tag{9}$$

$$N_{lm} = \frac{2m+1}{2} \int_0^1 P_m(\mu) P_l(\mu) d\mu \tag{10}$$

The solution of (5) is represented as:

$$C_m(\tau) = \sum_{l=0}^{N} a_{ml} \tilde{C}_l \exp(\gamma_l \tau) + C_p^1 \exp(-\tau) + C_p^2 \exp(\tau) \tag{11}$$

Where coefficients C_p^1 and C_p^2 are:

$$C_p^1 = -J\Lambda \cdot \sum_{m=0}^{N} \left[\delta_{pm} + A_{pm}\right]^{-1} B_m \tag{12}$$

$$C_p^2 = J\Lambda R_f \exp(-2L) \cdot \sum_{m=0}^{N} \left[\delta_{pm} - A_{pm}\right]^{-1} B_m \tag{13}$$

where the upper index -1 denotes the matrix inversion.

The coefficients $\tilde{C}_l$ were determined from the boundary conditions (8) forming an overdetermined system, so the least squares method was used. The following equation written in matrix form was derived for the coefficients $\tilde{C}_l$:

$$\tilde{C} = \left[Z^T Z\right]^{-1} Z^T b \tag{14}$$

where $Z = \begin{pmatrix} (N - R')a \\ (\tilde{N} - \tilde{R}')a \exp(\gamma L) \end{pmatrix}$ and $b = \begin{pmatrix} (N - R')(C_p^1 + C_p^2) \\ (\tilde{N} - \tilde{R}')(C_p^1 \exp(-L) + C_p^2 \exp(L)) \end{pmatrix}$ are auxiliary matrixes.

For the comparison with the experiment, the values of absorption coefficient (*A*), reflectivity (*R*) and transmittance (*T*) were evaluated:

$$A = (1-\Lambda) \cdot \left(\frac{1}{J_0} \int_0^L C_0 d\tau + (1 - R_f) \cdot \left[1 - \exp(-L) + R_f \exp(-L) \cdot (1 - \exp(-L))\right] \right) \tag{15}$$

$$T = \left(\frac{1}{J_0} \sum_{m=0}^{N} C_m(\tau)\xi_m + \left(1 - R_f\right)^2 \exp(-L) \right) \quad (16)$$

$$R = \left(\frac{1}{J_0} \sum_{m=0}^{N} (-1)^m C_m(0)\xi_m + R_f + R_f\left(1 - R_f\right)\cdot \exp(-2L) \right) \quad (17)$$

where the vector ξ_m is

$$\xi_m = \frac{2m+1}{2} \int_0^1 P_m(\mu)\cdot[1 - R(\mu)]\cdot \mu d\mu \quad (18)$$

The theoretical results were compared with the experiment in assumption that the Mie theory of light scattering is applicable. The details of the absorption and scattering coefficients as well as the scattering phase function evaluation could be found in [8,10,13].

4. Results and discussion

Two series of measurements were carried out. In the first one the samples of width 0.13 ± 0.01 mm containing different weight percentage of nanoparticles were used. In the second one the sample width was varied at constant weight percentage of aluminum 0.017%. The results a shown on the figures 1 and 2 correspondingly. The light wavelength was 643 nm.

The reflectivity value for the pure samples was 0.10, which is close to the monocrystal's one. The theoretical margin in the case of Fresnel refraction at n=1.54 (petn) is 0.09. Thus the petn matrix is a transparent medium.

The samples having some aluminum content showed the increasing in the reflectivity up to 15%. This value does not depend on the nanoparticles' weight percentage and the sample's width at the level of the experimental accuracy. The increasing in the reflectivity is the consequence of the light scattering inside the sample.

The theoretical dependencies of the reflectivity and sum of the reflectivity and transparency are also presented on the figures 1 and 2. The aluminum refractivity index was varied during the calculations. It is evident from the data presented on the figures 1 and 2 that the theory matches experiment at the refractive index value 1.28-5.90i.

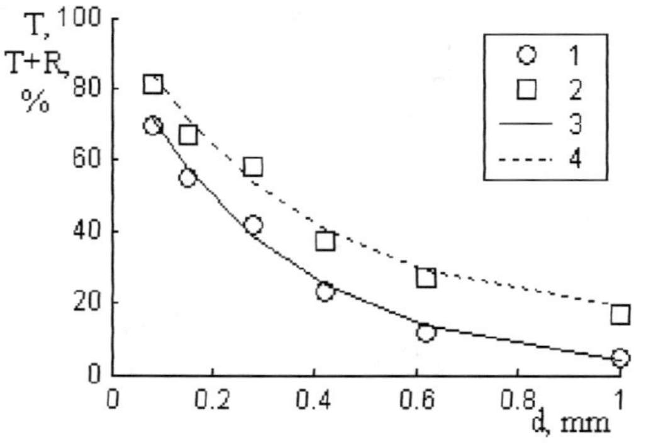

Figure 1. The dependence of transmittance (1,3) and sum of the transmittance and reflectance (2,4) on the sample's width (the weight percentage of aluminum was 0.017%). 1,2 – experiment, 3,4 – calculation.

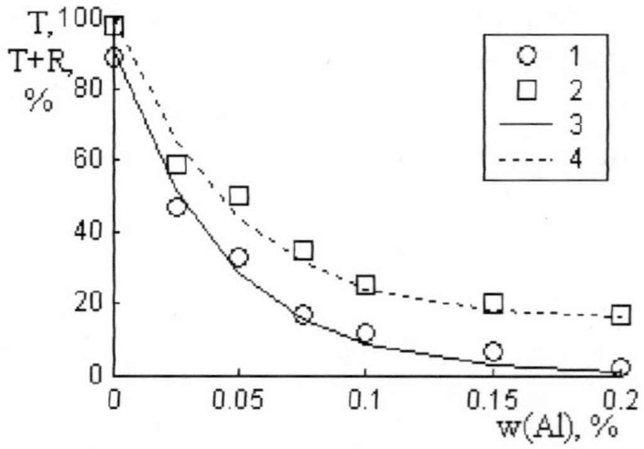

Figure 2. The dependence of transmittance (1,3) and sum of the transmittance and reflectance (2,4) on the aluminum weight percentage (the sample's width was 0.13 mm). 1,2 – experiment, 3,4 – calculation.

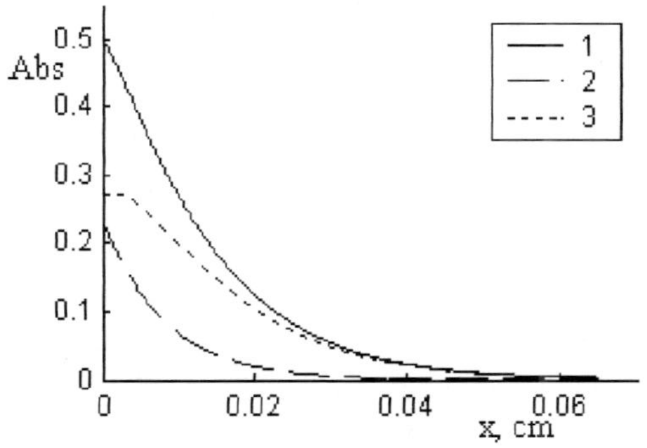

Figure 3. The calculated absorbed energy distribution inside the sample in relative units. 1 – the absorbed energy distribution, 2 – contribution of the incident light, 3 – scattered light contribution.

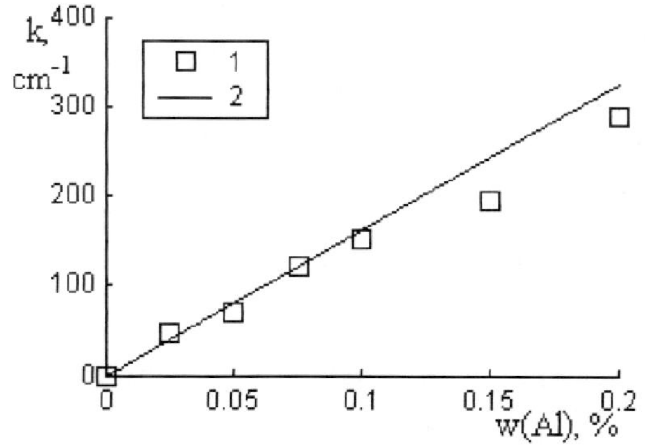

Figure 4. The apparent absorption coefficient dependence on the aluminum weight percentage. 1 – experiment, 2 – calculation.

The handbook [16] gives the values of (1.14 ± 0.16) - (6.54 ± 0.57)*i* for the aluminum refractive index at the wavelength 650 nm which agrees well with the margin obtained in the present paper. Thus the bulk metal properties, such as refractive index, could be
extrapolated into the range of small particles' sizes. The average values of the aluminum nanoparticles absorption and scattering effectivenesses at the light wavelength 643 nm are 0.6614 and 2.1561 correspondingly. At the histogram's maximum (50 nm) these values are 0.7022 and 1.8506 correspondingly. The difference is concerned on the finite size distribution width.

The calculated absorption energy distribution inside the sample (1) in relative units is shown on the figure 3 in the case of weight percentage of aluminum nanoparticles equal to 0.05%. The contributions of the incident (2) and scattered (3) light are also presented. It is worth mentioning that they are comparable near the sample's surface. As the distance from surface increases, the contribution of the scattered part becomes more significant. The maximum on the scattered light contribution in the absorbed energy is a typical one. It arises because the scattered light partly penetrates the sample's boundary. The absorbed energy distribution inside the sample is nearly exponential, so the law of the Beer's type is applicable. The apparent absorption coefficient is determined by both scattering and absorption processes in the medium. The substantial deviation from the Beer's type law during the simulations was observed in the case of the very thin samples only, where the refraction from the bottom sample's surface is pronounced.

The figure 4 presents the experimental (1) and calculated (2) apparent absorption coefficients dependent on the aluminum weight percentage. The experimental margins were estimated as $d^{-1}\ln[(1-R)/T]$. The theoretical ones were calculated as the energy distribution curve slope in the semilogarithmic coordinates. The experimental and calculated results agree well (figure 4). Thus the Beer's type dependence of the apparent absorption coefficient on the aluminum concentration arises in terms of the both experiment and theory (figure 4).

The assumption that the light scattering is able to provide substantial increasing in the energy absorbed by the nanoparticles in the transparent medium was suggested in [11]. The apparent nanoparticle light absorption cross section could be estimated as $k/C = 1.52\cdot10^{-10}$. The obtained value incorporates the scattering effects. The geometrical cross section of the 50 nm nanoparticle corrected by the refraction from the upper sample's surface is $(1-R)\cdot k/\pi r_{eff}^2 C =$ 1.64 ($R \approx 0.15$). Thus, the multiple scattering is able to increase the amount of energy absorbed by a nanoparticle, so it becomes bigger than falls on its surface. Though the effect is not as substantial as it was predicted earlier (12 – 24 times) [11], if one assumes the small values of the absorption effectiveness there is 2.48-fold increasing in the absorbed energy.

5. Conclusion

A photometric sphere was applied for the transmittance and sum of transmittance and reflectance measurement of petn pressed pellets containing aluminum nanoparticles at the light wavelength 643 nm. The theory for these values calculation in terms of spherical harmonics solution of radiative transfer equation in the slab geometry with Fresnel boundary conditions was developed. The Mie theory was applied for nanoparticles properties evaluation. The absorbed energy distribution inside the sample was calculated. It was shown that the Beer's type law is applicable approximately. The apparent light absorption cross section determined, which takes into account both scattering and absorption, is bigger than the geometrical one. The obtained aluminum refractive index agrees well with the handbook's data.

Acknowledgements

The authors thank prof. Viktor Yakovlev for the hired integrating sphere. The work was partially supported by Russian Foundation for Basic Research (13-03-98032_r_sibir_a).

References

[1] Wang H, Su Ch, Wu Ch, Tsai H, Li Ch, Li W 2013 Preparation of composite light-scattering layer for dye sensitized solar cells *Thin Solid Films* **529** 15-18

[2] Moulin E, Sukmanowski J, Luo P, Carius R, Royer F, Stiebig H 2008 Improved light absorption in thin-film silicon solar cells by integration of silver nanoparticles *J. Non-Cryst. Solids* **354** 2488-91

[3] Geints Yu E, Zemlyanov A A, Panina E 2010 Nanophotonics of Isolated Spherical Particles *Rus. Phys. Journal* **53** 410-20

[4] Svetlichnyi V A, Lapin I N 2013 Structure and properties of nanoparticles fabricated by laser ablation of Zn metal targets in water and ethanol *Rus. Phys. Journal* **56** 581-87

[5] Dombrovsky L A, Timchenko V, Jackson M, Yeoh G H 2011 A combined transient thermal model for laser hyperthermia of tumors with embedded gold nanoshells *Int. J. Heat and Mass Trans.* **54** 5459–69

[6] Rutberg F G, Gusarov V V, Kolikov V A, et al. 2012 Analysis of physicochemical properties of nanoparticles obtained by pulsed electric discharges in water *Tech. Phys.* **57** 1641-45

[7] Aduev B P, Belokurov G M, Nurmukhametov D R et al. 2012 Photosensitive material based on PETN mixtures with aluminum nanoparticles *Comb. Expl. and Shock Waves* **48** 361-66

[8] Aduev B P, Nurmukhametov D R, Furega R I et al. 2013 Explosive decomposition of PETN with nanoaluminum additives under the influence of pulsed laser radiation at different wavelengths *Rus. J. Phys. Chem.* B **7** 453-6

[9] Aduev B P, Nurmukhametov D R, Tsipilev V P, et al. 2013 Effect of ultrafine Al-C particle additives on the PETN sensitivity to radiation exposure *Comb. Expl. and Shock Waves* **49** 215-8

[10] Kriger V G, Kalenskii A V, Zvekov A A, et al. 2012 Effect of laser radiation absorption efficiency on the heating temperature of inclusions in transparent media *Comb. Expl. and Shock Waves* **48** 705-8

[11] Aleksandrov E I, Voznyuk A G, Tsipilev V P 1989 Effect of Absorbing Impurities on Explosive Initiation by Laser-Light *Comb. Expl. and Shock Waves* **25** 1-6

[12] Zinchenko A D, Pogrebov A I, Tarzhanov V I, et al. 1992 Optical Characteristics Of Some Powdered High Explosives *Comb. Expl. and Shock Waves* **28** 524-30

[13] Van de Hulst H C 1981 *Light Scattering by Small Particles* (NY: Dover Books on Physics)

[14] Chandrasekhar S 1950 *In: Radiative transfer* (London: Oxford University Press)

[15] Garcia R D M 2013 Radiative transfer with polarization in a multi-layer medium subject to Fresnel boundary and interface conditions *J. Quant. Spectrosc. Radiat. Transfer* **115** 28–45

[16] Zolotarev V M, Morozov V N, Smirnova E V 1984 *Opticheskie Postoyannye Prirodnych i Technicheskich Sred*. (Leningrad: Himiya) (in Russian).

Journal of Physics: Conference Series **552** (2014) 012033 doi:10.1088/1742-6596/552/1/012033

Theoretical approach to modelling the low-barrier chemical reactions initiated by pulsed electron beam

N E Aktaev*
National Research Tomsk Polytechnic University, Tomsk, Russia

E-mail: nurkenaktaev@gmail.com

Abstract. A possibility to analyze low-barrier chemical reactions induced by electron beam is proposed within the framework of generalization of classical Kramers approach. A relationship for calculation of chemical reaction rate is received for potential barriers comparable with energy of system. It has been shown that results of our approach agree with results of dynamical modeling within 2% not only in the area of applicability of classical (Kramers) approach but also more widely.

1. Introduction

At the present time fair quantity experimental [1-3] and theoretical [4-6] studies is devoted to investigation of physical properties of nonequilibrium plasma. Interest in this kinds of studies is the possibility to employ of results in perspective applied directions: transportation of electron beams in gas of variable pressure, plasma-chemical gas cleaning, synthesis of nanoparticles and others. One of the possible variant of nonequilibrium plasma formation is the influence of pulsed electron beam with the power flow of 10^6 – 10^9 W/cm^2 on neutral gas medium. Consequently of the influence fast processes appear besides slow. So in molecular nitrogen under the influence of short pulsed (10 – 100 nsec) electron beam channels of chemical reactions are initiated. In this case a dominating channel is a relaxation of vibrational mode. It should be noted that the presence of impurity can accelerate significantly this relaxation. Thereby the system (molecular nitrogen) in a state of excitation (quasistationary state) relatively fast relaxes to the more stable (quasistationary or stationary) state with lower energy.

In terms of physical kinetics quasistationary states of system are separated by potential barrier B_r. For the first time a problem of potential barrier overcoming was decided in 1940 by H.A. Kramers in his original work [7], however general-theoretical studies on the subject attract wide interest to this day [8-11]. The main result of [7] is the formula (Kramers formula) for the potential barrier overcoming rate (reaction rate R_r). However boundary applicability's of the formula do not permit to describe a wide range of tasks in plasma chemistry. One of the limitation is impossibility of calculation of fast process reaction rate on the assumption that the barrier height B_r comparable with system energy proportional to the temperature T. This condition is observed when the rate of energy input exceeds appreciably a cooling rate.

If a study of the fast process is involved then Kramers approach should be generalized for $B_r \approx T$. In present paper the universal relation for reaction rate is obtained. The relation allows to describe the overcoming process both $B_r >> T$ and $B_r \approx T$. Universality of the obtained relation is demonstrated by means of comparison with results of dynamical modeling.

Content from this work may be used under the terms of the Creative Commons Attribution 3.0 licence. Any further distribution of this work must maintain attribution to the author(s) and the title of the work, journal citation and DOI.
Published under licence by IOP Publishing Ltd

2. Generalization of Kramers approach

In the Born-Oppenheimer approximation a potential energy (potential) of chemical reaction could be considered as a function depending on generalized coordinate q. Typical shape of the potential is shown in Figure 1. It is assumed that at the initial time the system is located in a quasi-stationary state q_{qs}, which corresponds to a local minimum on the potential curve U_{qs}. Overcoming of the potential barrier B_r located at the saddle point q_{sd} and the transition to a position with lower potential energy than U_{qs} corresponds to the end (often intermediate) of chemical reaction. Corresponding coordinate on the potential curve is indicated as q_d. Dynamic of nonequilibrium system that overcomes the potential barrier is described by the master equation, which is a balance equation for the probability of each state of the system

$$\frac{\partial\rho(\xi,t)}{\partial t}=\int\left[\Pi(\xi\,|\,\xi')\rho(\xi',t)-\Pi(\xi'\,|\,\xi)\rho(\xi,t)\right]d\xi', \tag{1}$$

where ρ is the probability density, t stand for time, $\Pi(\xi\,|\,\xi')$ and $\Pi(\xi'\,|\,\xi)$ are the transition probabilities of system from ξ'-state to ξ-state and from ξ-state to ξ'-state respectively. But frequently it does happen to move to Fokker-Plank equation by the agency of the Kramers-Moyal forward expansion

$$\frac{\partial\rho(p,q,t)}{\partial t}=-\frac{p}{m}\frac{\partial\rho(p,q,t)}{\partial q}+\frac{\partial}{\partial p}\left\{\rho(p,q,t)\left[\frac{dU}{dq}+\beta p\right]\right\}+D_p\frac{\partial^2\rho(p,q,t)}{\partial p^2}, \tag{2}$$

where p is the conjugate momentum, $D_p=\eta T$ is the diffusion coefficient in the momentum space, the friction parameter η is related to the damping coefficient β and to the inertia parameter m as $\eta=m\beta$. We are interested in the task of establishing a quasistationary current through the barrier B_r. In terms of quasistationary state it can be assumed that the probability density ρ ceases to be a time function. Consequently the expression (2) for above conditions takes the form

$$0=-\frac{p}{m}\frac{\partial\rho(p,q)}{\partial q}+\frac{\partial}{\partial p}\left[\rho(p,q)\left(\frac{\eta}{m}p-m\omega_{sd}^2\tilde{q}\right)\right]+D_P\frac{\partial^2\rho(p,q)}{\partial p^2}, \tag{3}$$

where $\tilde{q}=q-q_{sd}$, ω_{qs} and ω_{sd} are the absolute values of the angular frequencies of the collective motion around the maximum (the saddle point) and the minimum (the quasistationary point) of the potential energy. In this case it can be supposed that quasi-equilibrium distribution is performed $\rho(p,q)=\zeta(p,q)\cdot\Omega(p,q)$, here $\Omega(p,q)$ is the Maxwell-Boltzmann distribution and $\zeta(p,q)$ characterizes a deviation from the equilibrium distribution. Solving equation (3) as (4) in the neighbourhood of the saddle point we obtain

$$\rho_{sd}(p,q_{sd})=\exp\left[-\frac{p^2}{2mT}\right]\int^{p}\exp\left[-\frac{\hat{a}-\eta}{2mD_p}h^2\right]dh, \tag{4}$$

and near quasistationary state we have

$$\rho_{qs}(p,q)=\left[1+erf(j)\right]\Omega(p,q_{qs})\sqrt{\frac{\pi m D_p}{2(\hat{a}-\eta)}}\exp\left(\frac{B_r}{T}\right), \tag{5}$$

where $\hat{a}=\frac{\eta}{2}\pm\left[\frac{\eta^2}{4}+(m\omega_{sd})^2\right]^{1/2}$ and $j=\omega_{sd}(q_{qs}-q_{sd})\sqrt{\frac{m}{2T}}$. Expressions (5) and (6) are required to find the reaction rate. The rate according to the transition state method [12] has the form

$$R_r=m^{-1}\int_{-\infty}^{+\infty}p\rho_{sd}(p,q_{sd})dp\left[\int_{-\infty}^{+\infty}dp\int_{-\infty}^{q_{sd}}\rho_{qs}(p,q)dq\right]^{-1}. \tag{6}$$

Substituting in (6) the expression for the probability density (4) and (5) we have

$$R_r=\omega_{sd}^{-1}\sqrt{\frac{2T}{\pi m}}\left(\sqrt{\omega_{sd}^2+\frac{\beta^2}{4}}-\frac{\beta}{2}\right)\exp\left(-\frac{B_r}{T}\right)\left\{\left[1+erf(j)\right]\times\int_{-\infty}^{q_{sd}}\exp\left[-\frac{m\omega_{qs}^2(q-q_{qs})^2}{2T}\right]dq\right\}^{-1}. \tag{7}$$

Expression (7) is a modified formula with respect to the classical Kramers formula

$$R_K=\frac{\omega_{qs}}{2\pi\omega_{sd}}\left(\sqrt{\omega_{sd}^2+\frac{\beta^2}{4}}-\frac{\beta}{2}\right)\exp\left(-\frac{B_r}{T}\right). \tag{8}$$

Modification of the expression (7) in relation to (8) is the ability to obtain the adequate numerical values of the reaction rate in situations: i) the barrier height of the potential is comparable and sometimes less than the temperature (energy) of the system; the adequacy of the results provides the integral in (8) which defines the diffuseness of the initial state of the system; ii) saddle point q_{sd} is close enough to quasistationary state q_{qs}. Such nearness and the small width of the potential barrier are observed in the analyze of reactions with a quantum effects. The error function $erf(x)=2\pi^{-1/2}\int_0^x\exp(-t^2)dt$ takes into account the distance between these points. In detail the limits of applicability (8) and their influence on the numerical value of the rate have been analyzed in e.g. [13-15].

3. Dynamical modelling

Comparison of the rates calculated according to expressions (7) and (8) should be carried out with the results obtained in the framework of numerical simulation. In the role of the dynamical equations we chose the Langevin stochastic differential equations

$$d\begin{bmatrix}p\\q\end{bmatrix}=\begin{bmatrix}-dU/dq & -\beta\\0 & m^{-1}\end{bmatrix}\times\begin{bmatrix}dt\\pdt\end{bmatrix}+\begin{bmatrix}\sqrt{2D_p}dw_t\\0\end{bmatrix}, \tag{9}$$

where w_t is the Wiener process whose increment dw_t possesses the normal distribution with the variance dt. This process obeys the conditions

$$\langle w_n(t)\rangle = 0; \qquad \langle w_n(t') - w_n(t'')\rangle\langle w_m(t') - w_m(t'')\rangle = \delta_{nm}\delta|t' - t''|, \tag{10 a, b}$$

where δ_{nm} is the Kronecker delta, $\delta(t)$ is the delta-function. Within the framework of the Langevin formalism, the time-dependent decay rate $R_D(t)$ can be calculated by counting the number of trajectories $N_D(t)$ that reach the point q_d before the time moment t

$$R_D(t) = \left[N_{tot} - N_D(t)\right]^{-1}\frac{dN_D}{dt}, \tag{12}$$

where N_{tot} is the total trajectories number modeled. Typical time evolution $R_D(t)$ can be found in [13-16]. A distinctive feature of this evolution is the presence of longtime quasistationary limit. The value of this limit we will compare with the results calculated according to expressions (7) and (8). The comparisons will be made using the fractional difference $\xi = \left(\left[R_r \text{ or } R_K\right]R_{dqs}^{-1} - 1\right)\cdot 100\%$. Simulation parameters are taken from [15].

Figure 2 illustrates two fractional differences. The lower curve corresponds to the results of [15]. It can be seen that a good agreement (within 2%) is observed in the case where the barrier height is at least 1.5 times higher than the temperature of the system. In the area $T^{-1}B_r < 1{,}5$ there is a significant disagreement of rates calculated in the framework of the classical expression (8) with the result of dynamical simulations R_{dqs}. The upper curve represents the relative difference between R_{dqs} and R_r. Analyzing this curve it may be concluded that two 2% agreement between these rates is observed not only in the field $T^{-1}B_r > 1{,}5$ but also in the area which corresponds to the small barriers as compared with the reaction system temperature.

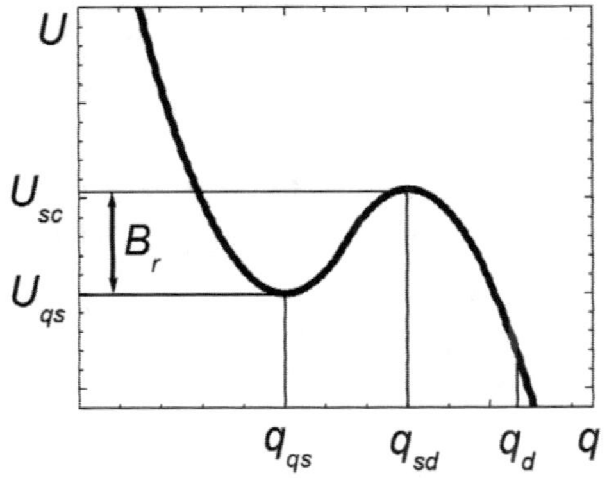

Figure 1. Deformation dependence of potential energy

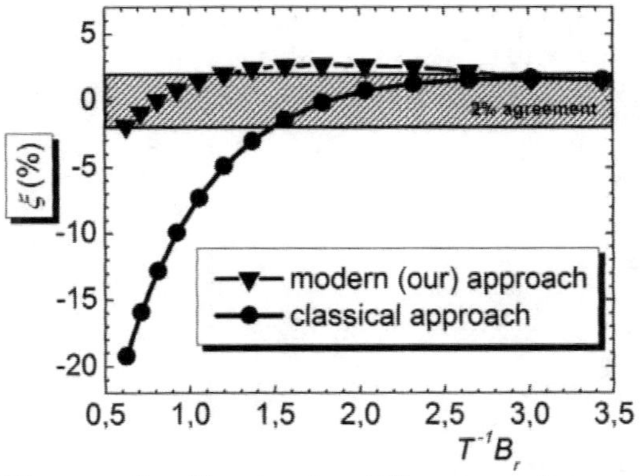

Figure 2. Fractional differences between dynamical and analytical rates versus $T^{-1}B_r$. Statistical errors do not exceed symbol size.

4. Conclusion

By solving the Fokker-Planck equation the relation of the chemical reaction rate is obtained. Universality of the resulting expression with respect to the classical Kramers formula is the ability to obtain adequate numerical values under the condition that the potential barrier height is comparable to the energy of the system. This situation is typical for the fast processes when the rate of energy input is much higher than the rate of cooling. Accuracy of the relation is analyzed by comparing with the results of dynamical modeling. It is shown that for the area $B_r >> T$ the results obtained using the expressions (7) and (8) within statistical error do not differ. However for $B_r \approx T$ accuracy of our approach is 2 %, whereas classical Kramers approach gives an error up to 20%.

Acknowledgement

This work was supported by the mega grant “Materials for extreme conditions” (VIU-IFVT-85-2014) and funded within the framework of the Programme on National Research Tomsk Polytechnic University Competitiveness Enhancement in the Group of Top Level World Research and Academic Institutions.

References

[1] Mesyats G A 2006 *Plasma Phys. Control. Fusion* **47** A109.
[2] Remnev G E, Uemura K, Kozyrev A V, Lopatin V V 2014 *Journal of Physics: Conference series*, accepted.
[3] Zhuo H B, Chen Z L, Sheng Z M et al 2014 *Physical Review Letters* **112** 215003.
[4] Lotov K V, Sosedkin A P, Petrenko A V 2014 *Physical Review Letters* **112** 194801.
[5] Avinash K, Kaw P K 2014 *Physical Review Letters* **112** 185002.
[6] Lin Yang, Xiaohua Tan, Xiang Wan et al 2014 *Journal of Applied Physics* **115** 163106.
[7] Kramers H A 1940 *Physica* **7** 284.
[8] Arrayas M, Kaufman I Kh, Luchinsky D G et al 2000 *Physical Review Letters* **84** 2556.
[9] Sarracino A 2013 *Physical Review E* **88** 052124.
[10] Pollak E, Lanconescu R 2014 *Journal of Chemical Physics* **140** 154108.
[11] Aktaev N E, Gontchar I I 2010 *Bulletin of the Russian Academy of Sciences: Physics* **74** 513.
[12] Hanggi P, Talkner P, Borkovec M 1990 *Reviews of Modern Physics* **62** 251.
[13] Gontchar I I, Chushnyakova M V, Aktaev N E et al 2010 *Physical Review C* **82** 064606.
[14] Aktaev N E, Gontchar I I 2011 *Bulletin of the Russian Academy of Sciences: Physics* **75** 994.
[15] Pavlova E G, Aktaev N E, Gontchar I I 2012 *Physica A* **391** 6084.
[16] Gontchar I I, Kuzyakin R A, Pavlova E G et al 2012 *Journal of Physics: Conference Series* **381** 012089.

Journal of Physics: Conference Series **552** (2014) 012034 doi:10.1088/1742-6596/552/1/012034

Portable device for the detection of nitro-explosives based on optical properties of sensor's material

A A Baranova, K O Khokhlov*
Ural Federal University named after the first President of Russia Boris Yeltsin, Yekaterinburg, Russia

E-mail: k.o.khokhlov@urfu.ru

Abstract. The aim of this study was to design a device for explosives detection. The study design is based on excited steady-state luminescence quenching registration. Sensor's material luminescence intensity reduction occurs due to an interaction of explosives vapours contained in the air. The decrease rate of the luminescence intensity indicates the concentration of vapours. To study the luminescent properties of the sensor element, its luminescence spectra excited by photons with energies in the range 280 - 425 nm were measured. The excitation photoluminescence spectra for luminescence bands of the sensor element were also measured. Excitation source was light emitting diode (375 nm) and luminescent signal receiver was a photodiode (430 - 650 nm) in device designed. The device is operated under control of a program. The algorithm provides multiple operating modes (configuration, calibration, measurement etc.). Thus this device is referred to the class of devices with increased sensitivity to the explosives vapors. The advantages of device are autonomic power, small weight and sizes, simplicity of device operation for measurements.

1. Introduction

The luminescent properties of various compounds are widely utilized in chemical, biological, and medical engineering due to high sensitivity and unique opportunities to study molecules' excited states, photochemical reactions, dynamics of fast molecular processes, as well as structure and properties of complex chemical and biological objects.

Fluorescence quenching is a process that reduces the induced light flow intensity. Several processes can lead to this phenomenon, including reactions in excited state, energy transport, the formation of complexes at quencher molecules collisions with fluorescent substance molecules (molecules-sensors). Prerequisite of this effect demonstration is contact between them. Currently, fluorescence quenching is used in various practical applications, including explosives detection. The development of new devices that are capable detecting traces of explosive materials (HE) may help to eradicate terrorism

Therefore, the goal of this study was to evaluate luminescent characteristics of sensor materials to describe parameters that may be considered during a development process of a mobile device for detection of explosive materials traces.

Content from this work may be used under the terms of the Creative Commons Attribution 3.0 licence. Any further distribution of this work must maintain attribution to the author(s) and the title of the work, journal citation and DOI.
Published under licence by IOP Publishing Ltd

2. Experimental part

A method for the synthesis of new monomolecular fluorescent compounds that consist of poly-aromatic heterocyclic systems with flat and branched architecture fragments was utilized in this study. The details of this technology are described [2-4]. Chemical compounds that were obtained as a result of the utilization of the above described technology are sensitive to vapours of trinitrotoluene (TNT) and its derivatives as well as to hard-detected nitro HE-like hexogen.

Powder of polyaromatic π-conjugated system was dissolved in an organic solvent. The obtained colloidal solution was electrospinned with the help of viscose matrix. Thereafter the obtained samples were dried and used in the experimental procedure. Photoluminescence (PL) and PL excitation spectra of studied material, containing the molecules-sensors are presented in Figure 1 и 2.

It can be seen that the full range of luminescence is in the limit of 400 - 650 nm, the efficient luminescence of the material is observed in the range of 425 - 550 nm, and the most intensive luminescence bands are with maximum range of 447 to 474 nm.

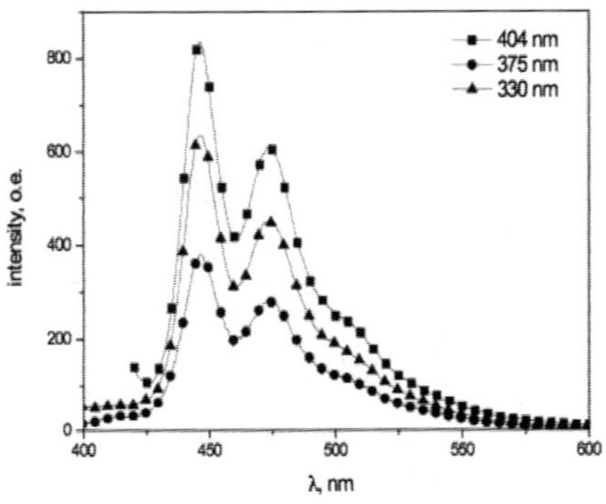

Figure 1. PL spectra of material at excitation by photons in wavelength range of 330-404nm. Luminescence excitation wavelengths are indicated in figure legend.

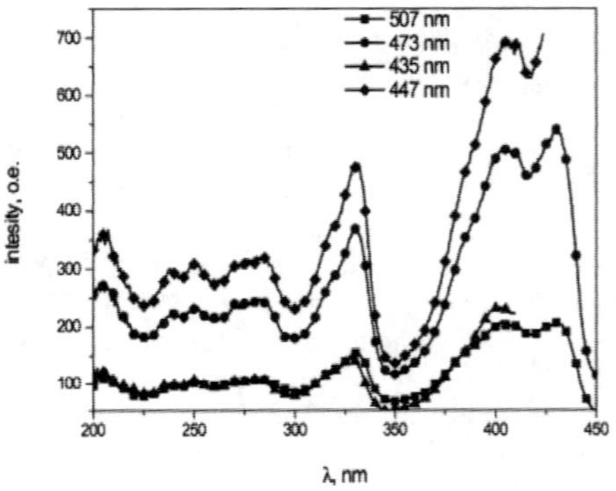

Figure 2. Excitation spectra of the main luminescence bands of sensor.

There were observed luminescence bands with maxima at 447, 473 and 507 nm under excitation by photons with energy of 330 nm and 375-450 nm.

3. Theoretical part

The nitroaromatic compounds luminescence detection method is based on the change of fluorescence quantum yield or excited state lifetime of the donor molecule of sensor due to the total or partial electron transport in contact with nitroaromatic compounds.

During interaction with flourescent molecule-sensor (chemosensor) the analyzed molecule reduces the intensity of light. This phenomenon is obeyed by the law, known as Stern-Volmer equation, which is a mathematical description of fluorescence quenching phenomenon [5]:

$$\frac{I0}{I} = (1 + KA\,[Q])(1 + kq\,\tau 0[Q]),$$

where:

I0 - luminescence intensity in theabsence of analyzed substance;
I - luminescence intensity in the presence of analyzed substance;
[Q] - substance concentration (I depends on it);

KA - binding constant - represents the contribution of so-called static quenching, which occurs through the formation of donor-acceptor complex between the sensor in the ground state and nitroaromatic compound;
τ0 - lifetime of the excited state of the molecule-sensor photoluminescence at photoexcitation for dynamic quenching process;
kq - interaction rate between the molecule of analyzed substance and molecule-indicator.

In most cases, the quenching constants for Stern-Volmer I0/I are determined by the contribution of prevailing in a particular case type of quenching.

In the case of both static and dynamic quenching the fluorescence decrease is due to electron transport from the molecule-sensor in the ground or excited state to molecule of nitroaromatic compound in the ground state.

A simplified diagram of the fluorescence quenching by nitro compounds is shown in Figure 3.

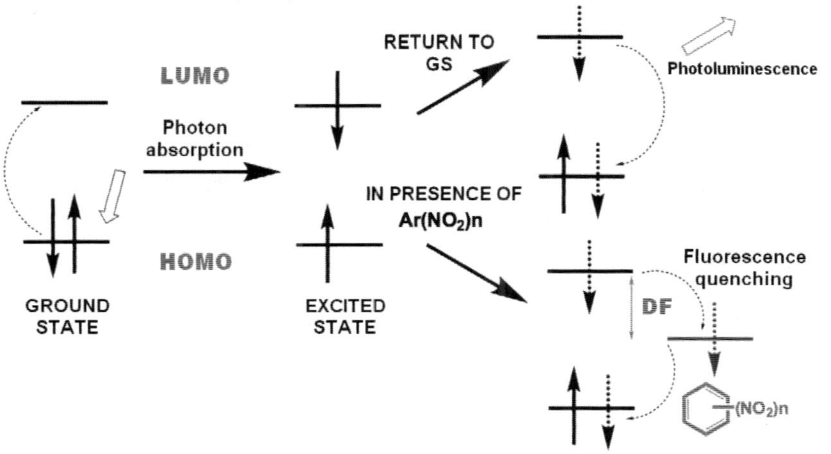

Figure 3. The mechanism of fluorescence quenching by contact of chemical sensor and nitro compounds. LUMO - the lowest unoccupied molecular orbital, the HOMO - the highest occupied molecular orbital, DF - driving force. The arrows indicate the radiative and non-radiative transitions of electrons.

At photoexcitation of molecule-sensor the electron makes a transition to the molecular orbital of higher energy. In the absence of quencher, the reverse transition is accompanied by photoluminescence. Upon contact with the nitro compound the excited electron of molecule-sensor returns to the ground state not participating in the radiative transition, which is observed experimentally as fluorescence quenching. The energy difference between HOMO and LUMO is driving force (DF) of fluorescence quenching process (Figure 3).

In most cases, the static fluorescence quenching is prevalent and therefore it is often used in nitroaromatic compounds luminescent detection. The quenching presence of static nature primarily is judged by high values (10^3 M^{-1} or higher) of complexation constants (quenching constants) between nitroaromatic compounds and molecules of photoluminescent sensors (typically electron-rich (hetero) aromatic compounds). The lifetime of excited state of photoluminescence chemosensors with increasing concentration of nitroaromatic compounds remains unchanged.

4. Practical part

During development of device for explosives traces detection the experimental results and the estimated model of quenching process shown in Figure 3 are taken into account. It is necessary to use photons with energy at 330 nm, or in the integral mode in the wavelength range of 375-450 nm to obtain maximum sensitivity of material to HE and to excite the efficient luminescence.

Based on the fluorescent material characteristics, a device that meets the following requirements has been designed: air pumping with constant rate through the sensor element comprising luminescent material (chemosensors);

- impact of light source with wavelength corresponding to excitation of luminescence on the surface and in the surface layer of sensor element;
- luminescence intensity measurement by photosensitive element; it concludes vapor concentration of nitro HE on the dynamics of concentration change in time.

As the source of luminescence excitation in device there were selected a light emitting diode with indicated energy range and luminescent signal photodetector - photodiode with sensitivity in the range of 430 - 1100 nm. To prevent exciting photons to photodetector, it was placed in front of photodetector the cutting optical filter with transmission wavelength above 425 nm.

Table 1. The device has the following characteristics.

Characteristics	Value
measurement range of luminous flux:	0.0015 - 0.1 lm;
measurement error of luminous flux:	not less than 0.1 %;
luminescence excitation wavelength:	375 nm;
luminescence detection range:	from 430 to 650 nm.

Block diagram of device is shown in Figure 4.

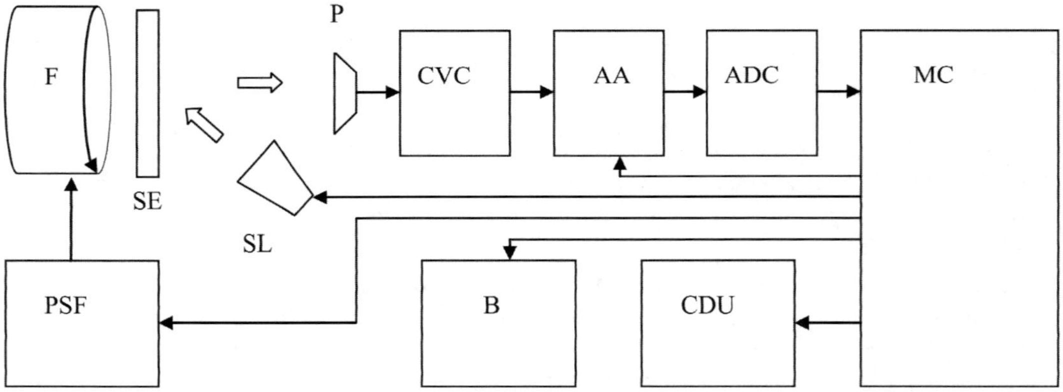

Figure 4. Block diagram of device for explosives and narcotics detection.

The device comprises the following basic structural units: fan (F); sensor element (SE); source of exciting the luminescence light (SL); photocell for luminescence intensity registration (P); current-to-voltage convertor (CVC); amplifier with adjustable amplification factor (AA); analog-to-digital converter (ADC); microcontroller (MC); buzzer (B); control and display unit (CDU); fan power supply (PS).

CVC converts a current of P that is proportional to the intensity of light into a signal of voltage that is the most suitable for further converting. AA is used for further signal amplification and to bring it to the range of optimal value of electric signal applied to ADC.

The device functioning is controlled by microcontroller unit. Device switching on and off, mode selection, display and reading information is produced by CDU. Multilevel menu is displayed on device indicator. The upper level corresponds to device selection mode, the bottom level - to mode value. All information about measurement time, results and their status (test, search, fighting etc.) is stored in non-volatile memory of device. You can connect device to PC via USB interface, built-in MC. In this mode, it is transferred the device control to PC and provided access to device database. B is sound alarm to detect HE and serves as main indicator of this event. B can be either built-in or plug-in to the standard connector.

PS provides a constant fan speed in case of significant supply voltage change during battery discharge in operation process. The device is powered by rechargeable battery. Its charge is made from specialized charger.

5. Conclusion

Experimental results and estimated model of sensor photoluminescence quenching processes allowed us to develop and make the layout version of device for explosive vapors detection. The testing with HE-DNT simulator and real devices will be carried out.

6. References

[1] Caron T, Guillemot M, Montmeat P, Veignal F, Perraut F, Prene P, Serein-Spirau F 2010 *Ulrta trace detection of explosives in air: Delelopment of a portable fluorescent detector* (Talanta 81) pp 543-548

[2] Anzenbacher Jr P, Mosca L, Palacios M.A., Zyryanov G V 2012 *Chem. Eur. J.* **18** **(40)** pp 12712-718

[3] Mosca L, Koutnнk P, Lynch V M, Zyryanov G V, Esipenko N A, Anzenbacher P 2012 *Cryst. Growth* **12** (12) pp 6104-09

[4] Kopchuk D S, Egorov I N, Tseitler T A, Khasanov A F, Kovalev I S, Zyryanov G V, Rusinov V L, Chupakhin O N 2013 *Chem. Heterocycl. Compd.* **49(3)** pp 503-505

[5] Harvey D 2000 *Modern Analytical Chemistry* (Boston) p 798

Journal of Physics: Conference Series **552** (2014) 012035 doi:10.1088/1742-6596/552/1/012035

On Channeling Potential for Relativistic Electrons in Crossed Laser Field

EN Frolov[1,2], AV Dik[1], SB Dabagov[1,3,4] and KP Artyomov[5]

[1] Lebedev Physical Institute of RAS, Leninsky pr., 53, Moscow, 119333, Russia
[2] Tomsk Polytechnic University, Lenina st., 30, Tomsk, 634050, Russia
[3] INFN Laboratori Nazionali di Frascati, Via E. Fermi, 40, Frascati (RM), 00044, Italy
[4] NRNU MEPhI, Kashirskoe sh., 31, Moscow, 115409, Russia
[5] Institute of High Current Electronics SB RAS, Akademichesky av., 2/3, Tomsk, 634055, Russia

E-mail: frol.onn@gmail.com

Abstract. In this work the expression for the ponderomotive potential responsible for electron channeling in the field of crossed laser beams is analyzed. The influence of lasers crossing angle and electron velocity on the ponderomotive force is described in two plane waves approximation revealing "*nonrelativistic inversion*" phenomenon. Some estimations for the beam density of electrons channeled in such system are given.

1. Introduction

Channeling is a phenomenon which takes place when a relativistic projectile propagates in a periodic structure of some sort (e.g. crystals) and its interaction with the structure (crystal lattice nodes) could be effectively averaged forming well defined potential channels (either planar or axial). Particles with transverse energy less then the channel barriers become trapped in a channel potential, i.e. channeled, and perform oscillations transverse to the channel direction. While one can find more information on crystal channeling elsewhere [1, 2] we only note here that the term "*channeling*" is widely used nowadays describing a variety of phenomena such as neutral particles beams channeling [3, 4], particles channeling in crystals, capillaries and nanotubes [5, 6, 7] and electrons channeling in plasma channels [8]. Moreover, motion of charged particles in crossed electromagnetic waves [9, 10, 11, 12, 13, 14] can be described withing channeling phenomenology, which is the main topic of this paper.

In general, both laser-electron and laser-plasma interactions are very important research areas being nowadays under active development. Hence, clear understanding of electron dynamics in an external laser field is no doubt crucial for many applications. Some of them rely on electron beam motion in the field of a standing electromagnetic wave or crossed laser beams. Being first to consider the problem, Kapiza and Dirak have statistically derived the possibility of electron deflection by standing electromagnetic field [15]. Later with the development of laser technologies this problem was considered in terms of channeling in series of works by Bertolotti *et al.* [9, 10]. In those works the detailed dynamics of an electron channeled in the field of standing electromagnetic wave and crossed laser beams were discussed. And also the proposal for a new type radiation source based on particle channeling oscillations in such systems was made. These ideas could be found in other separate studies (see [11, 16, 17] and Refs. therein).

Content from this work may be used under the terms of the Creative Commons Attribution 3.0 licence. Any further distribution of this work must maintain attribution to the author(s) and the title of the work, journal citation and DOI.
Published under licence by IOP Publishing Ltd

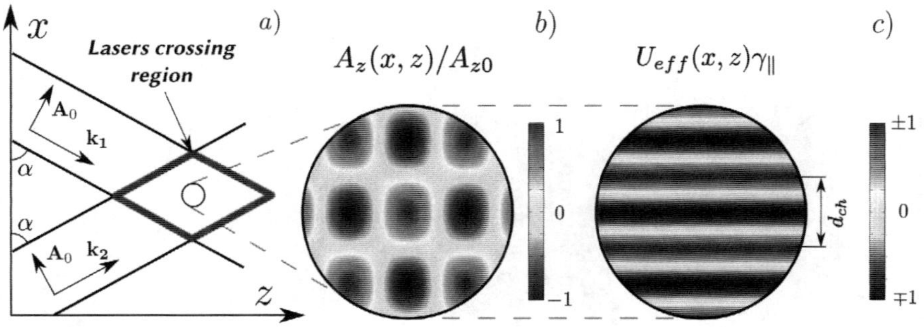

Figure 1. (a) Schematic geometry of two crossed p-polarized laser beams. (b) The longitudinal component of summarized vector potential (arb. un.) in the central area of lasers crossing region, in two plane waves approximation. Here $\alpha \in (0, \pi/2)$. The peaks of A_z migrate in the direction of Oz-axis. (c) Normalized effective potential distribution for the same area as in Fig.1.b. The potential absolute value depends on lasers parameters, angle α and velocity of an electron propagating in such a field. Sign ambiguity characterizes possible "*inversion*" described below.

And finally, in the last decade this topic gains additional attention with the latest advance of achievable laser intensities, growing interest to laser wakefield acceleration techniques and demand of new beam diagnostic tools.

Kaplan *et al.* [18, 19] reconsider a well known ponderomotive force introducing its "*inversion*" for particular lasers polarization and projectiles energy. Following their work, Smorenburg *et al.* [20] has derived generalized formula for the ponderomotive potential in the region of two counterpropagating laser beams of arbitrary polarization. According to the article, ponderomotive force sign inversion for the p-polarization case is possible for particular electron energy.

Lately Balcou, Andriyash *et al.* (see [21, 22, 23] and Refs. therein) in a series of articles have again made a proposal for a radiation source utilizing channeling of electrons in crossed laser beams basing it on both numerical and analytical results of charged particles dynamics and radiation in such system. They also consider the phenomenon to be a promising tool for electron beam diagnostics allowing emittance measurements [22].

2. Electron dynamics in the field of two laser beams

The case of electron dynamics in standing electromagnetic wave created by two counterpropagating laser beams of arbitrary polarization is thoroughly covered in [18, 19, 20]. But directing lasers so that they become crossed at some angle (see Fig. 1.a) causes the potential to change dramatically and differ from the expressions presented in mentioned papers. The case of crossed plane electromagnetic p-polarized waves is covered here.

2.1. General information

Let us consider two p-polarized plane electromagnetic waves of equal frequency ω and amplitude A_0, $\mathbf{k_1} = w/c\{-\cos\alpha;\ 0;\ \sin\alpha\}$ and $\mathbf{k_2} = w/c\{\cos\alpha;\ 0;\ \sin\alpha\}$. Therefore, their superposition gives $A_x = A_{x0}\sin(\omega t - zk\sin\alpha)\cos(xk\cos\alpha)$ for transverse vector potential and $A_z = A_{z0}\cos(\omega t - zk\sin\alpha)\sin(xk\cos\alpha)$ for longitudinal vector potential components as shown in Fig. 1.b, where $k = \omega/c$. Let us call planes with $xk\cos\alpha = \pi(n+1/2)$ the "*antinodes*" and planes with $xk\cos\alpha = \pi n$ the "*nodes*", meaning those are nodes and antinodes of A_z. Obviously, for $\alpha = \pi/2$ the system is similar with the one described in [18, 19, 20]. Electrons with sufficiently small transverse momentum and longitudinal momentum $|p_{\|}| < m_e c$ (where m_e is the electron rest mass and c is the speed of light) will oscillate

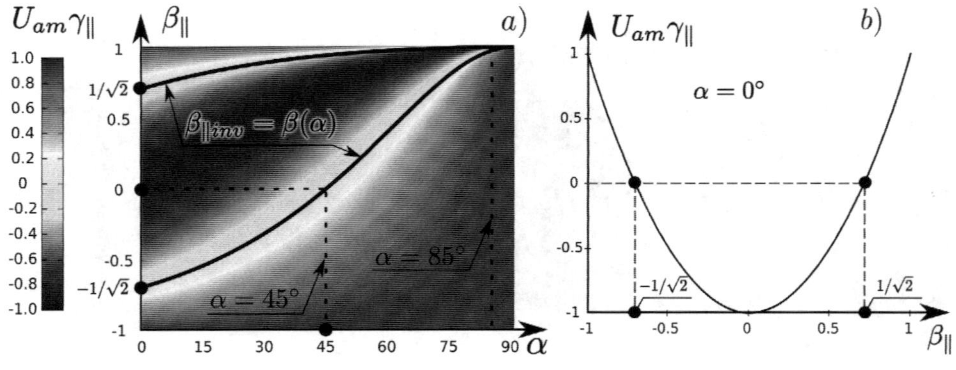

Figure 2. (a) The amplitude of $U_{am}\gamma_{\|}$ normalized by $(2m_e\omega^2)/(e^2A_0^2k^2)$ as a function of the angle α is shown in color. The black curves show the electron speed $\beta_{\|inv}$ for which no ponderomotive force affects its motion. (b) The same but for a fixed $\alpha = 0$, which corresponds to counterpropagating electromagnetic waves.

near the nodes due to ponderomotive force. On the other hand, for electrons with the same transverse momentum and longitudinal momentum $|p_{\|}| > m_e c$ the nodes will be scattering, while the antinodes — attracting. Finally, electrons with $|p_{\|}| = p_{\|inv} \equiv m_e c$ will "*feel*" no ponderomotive potential and perform no averaged oscillations. Thus, according to [18], "this potential inversion is considered to pin down a line between relativistic and nonrelativistic motions".

2.2. *Crossed electromagnetic waves*

Notable, the description given in [18, 19, 20] is not applicable to the case of arbitrary α. As shown in [24], the expression for the averaged effective potential in laboratory frame is $U_{eff} = U_{am}\cos(2kx\cos\alpha)$ with its amplitude defined by

$$U_{am} = \frac{e^2A_0^2k^2[(1+\cos^2\alpha)\beta_{\|}^2 - \cos(2\alpha) - 2\beta_{\|}\sin\alpha]}{2\gamma_{\|}m_e\omega^2(1-\beta_{\|}\sin\alpha)^2}, \qquad (1)$$

where e is the electron charge, $\beta_{\|}$ is the electron velocity projection on Oz-axis normalized by c, $\gamma_{\|} = 1/\sqrt{1-\beta_{\|}^2}$. The averaged effective potential has a spatial period of $d_{ch} = \pi/(k\cos\alpha)$ equal to the distance between two nodes planes of A_z depicted in Fig. 1.c.

Finding roots of the quadratic equation on $\beta_{\|}$ in the numerator of Eq. (1) one can see that for the case of counterpropagating electromagnetic waves ($\alpha = 0$) they are $\beta_{\|inv} = \pm 1/\sqrt{2}$, which gives $p_{\|} = m_e c$. But, in general, the value of longitudinal electron velocity, for which $U_{am} = 0$ and no channeling oscillations occur, is a function of α

$$\beta_{\|inv}(\alpha) = \frac{\sin\alpha \pm \cos\alpha\sqrt{[1+\cos(2\alpha)]}}{1+\cos^2\alpha} \qquad (2)$$

This dependence for the *inversion speed* is shown in Fig. 2.a (black curves) together with the normalized effective potential amplitude $U_{am}\gamma_{\|}$ (in color). In Fig. 2.b one can find the values of $\beta_{\|inv}$ for the case considered in the papers mentioned above ($\alpha = 0$) and they are similar to those reported by other authors. However, they change significantly if *crossed* lasers are considered, e.g. an electron with zero longitudinal velocity ("*nonrelativistic inversion*") is affected by *no* ponderomotive force in the field of normally aligned lasers (in Fig. 2.a $U_{am}\gamma_{\|} = 0$ for $\alpha = 45°$ and $\beta = 0$). The dependence of normalized

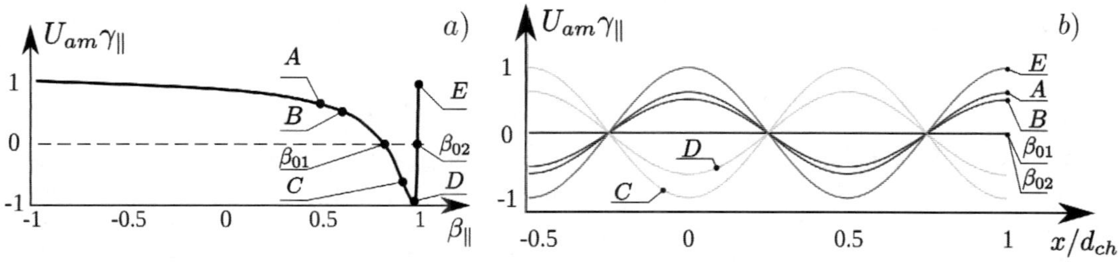

Figure 3. (a) The averaged potential amplitude normalized by $(e^2A_0^2k^2)/(2m_e\omega^2)$ versus the particle longitudinal velocity $\beta_\parallel$. The potential calculated for $\alpha = 85°$ is shown. β_{01} and β_{02} are the electron longitudinal velocities, for which $U_{am}(\beta_{01}) = U_{am}(\beta_{02}) = 0$. Negative $\beta_\parallel$ corresponds to the electron motion in opposite to the $\mathbf{k}_s = \mathbf{k_1} + \mathbf{k_2}$ direction, while for positive $\beta_\parallel$ the electron moves being co-directional to $\mathbf{k}_s$. (b) The normalized potential distribution as a function of transverse coordinate normalized by d_{ch} for electrons with different longitudinal velocities. Curves A, B, C, D, E correspond, respectively, to the particle with $\beta_A = 0.5$, $\beta_B = 0.6$, $\beta_C = 0.92$, $\beta_D = 0.967$, $\beta_E = 0.99986$.

channeling potential amplitude as a function of electron longitudinal velocity $U_{am}\gamma_\parallel(\beta_\parallel)$ is shown in Fig. 3.a for $\alpha = 85°$. One can notice that the potential amplitude there strongly depends not only on the value of longitudinal velocity but also on its direction. The spatial distribution of the potential across Ox-axis is shown in Fig. 3.b where its inversion could be easily seen for different $\beta_\parallel$ values. Indeed, as aforementioned, some electrons (with velocities β_A, β_B and β_E) are scattered on the nodes of A_z and oscillate around the antinodes of A_z, whereas for β_C and β_D the potential is inverted. And β_{01} and β_{02} are the inversion velocities here.

We also note here that all the reported results were checked with numerical simulations of electrons dynamics in crossed electromagnetic waves. Electron trajectories in the region of waves overlapping were calculated with the help of forth-order Runge-Kutta method validating the analytically derived expressions.

2.3. Beam confinement

As regards to the question of electron beam dynamics in the field of crossed electromagnetic waves, we provide some estimations for the electron density value, for which channeling of the majority of electrons is still possible in the considered system. For simplicity, the cylindrical beam (uniform distribution) of diameter d stretched along Oz-axis is considered. An electron on the beam surface will be affected by the space charge force normal to Oz $F_\perp = 2\pi e^2 n_0(1-\beta_\parallel^2)d$. For particles of the beam to remain trapped in the channel the ponderomotive force in a channel $F_{ch} \approx U_{am}/d_{ch}$ needs to be greater then the space charge force

$$\frac{U_{am}}{d_{ch}} \geq 2\pi e^2 n_0 \frac{d_{ch}}{\gamma_\parallel^2} \rightarrow n_0 \leq \frac{\gamma_\parallel^2 U_{am}}{2\pi e^2 d_{ch}^2}, \tag{3}$$

where n_0 is the electron density of a channeled beam. Expressing U_{am} with external electromagnetic field intensity, we get

$$n_0 \leq \frac{2\gamma_\parallel I}{m_e c\omega^2 d_{ch}^2} \tag{4}$$

Thus, for the intensities $I = 10^{14} - 10^{22}\ Wcm^{-2}$ the limiting density is $n_0 \sim 10^{19} - 10^{27}\ cm^{-3}$ ($\gamma_\parallel \sim 10^3$) that could make electron channeling in the field of crossed laser beams a promising technique for future applications (such as mentioned in [13, 21, 22, 24, 25]) when the feasible laser intensities will overcome $10^{20}\ Wcm^{-2}$. Also we have to highlight that laser beam divergence and an electron beam divergence will both toughen the channeling condition since the provided expression is applicable to the plane waves approximation.

3. Conclusion

Though the considered problem has been a subject of detailed research for a long time, some new peculiarities of ponderomotive potential for the field of crossed electromagnetic waves are presented. This becomes more and more important today because of many applications of laser-electron interaction. The most notable of them are laser wakefield acceleration, new radiation sources and diagnostic tools (e.g. one can find a description of beam diagnostics tool known as Shintake Monitor in [26]) based on electron dynamics and radiation in a strong external field. An interesting example is recently reported in the paper [25] on the "*anomalous radiation trapping*" of electrons in crossed laser beams with emission of high-energy photons. Ponderomotive potential inversion should be taken into account in all of those cases. And *nonrelativistic potential inversion* requires magnetic component of a Lorenz force (usually neglected) to be taken into account even for very slow electrons. Since the potential changes drastically depending on laser crossing angle it can open new possibilities (or possibly restrictions) for some applications (such as reported in [25]), which will be the subject of our future research.

Acknowledgments

This work was supported by the Ministry of Education and Science of RF in the frames of Competitiveness Growth Program of National Research Nuclear University MEPhI, Agreement 02.A03.21.0005 and the Dept. for Education and Science of LPI RAS.

References

[1] Lindhard J 1965 *K. Dan. Vidensk. Selsk. Mat.-Fys. Medd.* **34** 1–64
[2] Gemmell D S 1974 *Rev. Mod. Phys.* **46**(1) 129–227
[3] Bukreeva I, Popov A, Pelliccia D *et al.* 2006 *Phys. Rev. Lett.* **97**(18) 184801
[4] Dabagov S B 2003 *Phys.-Usp.* **46** 1053
[5] Zhevago N K and Glebov V I 2003 *Phys. Lett. A* **310** 301 – 310
[6] Klimov V V and Letokhov V S 1996 *Phys. Lett. A* **222** 424 – 428
[7] Karabarbounis A, Sarros S and Trikalinos C 2013 *Nucl. Instr. Meth. Phys. Res., Sect. B* **316** 160 – 170
[8] Dik A V, Ligidov A Z and Dabagov S B 2013 *Nucl. Instr. Meth. Phys. Res., Sect. B* **309** 210 – 213
[9] Bertolotti M, Sibilia C and Fuli L 1986 *Trends in Quantum Electronics* ed Prokhorov A M and Ursu I (Springer Berlin Heidelberg) pp 155–159
[10] Fuli L, Bertolotti M, Sibilia C *et al.* 1986 *Laser Part. Beams* **5** 557–564
[11] Andreev A V and Akhmanov S A 1991 *JETP Lett.* **53** 18–21
[12] Artru X, Ispirian K A and Ispiryan M K 2007 *ArXiv e-prints* (*Preprint* 0707.0148)
[13] Frolov E N, Dik A V and Dabagov S B 2013 *Nucl. Instr. Meth. Phys. Res., Sect. B* **309** 157 – 161
[14] Dabagov S B, Dik A V and Frolov E N 2014 *ArXiv e-prints* (*Preprint* 1404.3307)
[15] Kapitza P L and Dirac P A M 1933 *Math. Proc. Cambridge* **29**(02) 297–300
[16] Andreev A V and Akhmanov S A 1991 *Zh. Eksp. Teor. Fiz* **99** 1668–1678
[17] Fedorov M V, Oganesyan K B and Prokhorov A M 1988 *Applied Physics Letters* **53** 353–354
[18] Kaplan A E and Pokrovsky A L 2005 *Phys. Rev. Lett.* **95**(5) 053601
[19] Pokrovsky A L and Kaplan A E 2005 *Phys. Rev. A* **72**(4) 043401
[20] Smorenburg P W, Kanters J H M, Lassise A, Brussaard G J H, Kamp L P J and Luiten O J 2011 *Phys. Rev. A* **83**(6) 063810
[21] Balcou P 2010 *The European Physical Journal D* **59** 525–537
[22] Andriyash I A, d'Humières E, Tikhonchuk V T and Balcou P 2013 *Phys. Rev. ST Accel. Beams* **16**(10) 100703
[23] Andriyash I A, d'Humières E, Tikhonchuk V T and Balcou P 2013 *J. Phys. Conf. Ser.* **414** 012008
[24] Frolov E N, Dik A V and Dabagov S B 2014 *J. Phys. Conf. Ser.* **517** 012002
[25] Gonoskov A, Bashinov A, Gonoskov I, Harvey C, Ilderton A, Kim A, Marklund M, Mourou G and Sergeev A 2014 *Phys. Rev. Lett.* **113**(1) 014801
[26] Shintake T 1992 *Nucl. Instr. Meth. Phys. Res., Sect. A* **311** 453 – 464

Radiation optical effects in commercial SiO_2:Ge fibers

A V Ishchenko[1,*], B V Shulgin[1], A S Vokhmintsev[1], I A Weinstein[1], L N Shalimov[2], N G Manko[2], A N Shtykov[2], G V Shestakov[2], E K Chistyakova[1,2], A A Shonokhova[1,2]

[1]Ural Federal University named after the first President of Russia B.N. Yeltsin, Yekaterinburg, Mira 19, Russia
[2]FSUE Scientific and Production Association of automatics named after academician N.A. Semikhatov, Mamin-Sibiryak str. 145, Yekaterinburg, Russia

E-mail: a-v-i@mail.ru

Abstract. Present work is dedicated to investigation of radiation optical properties of SiO_2:Ge glass fibers produced by Fujikura and Corning Glass. Distribution of Ge dopant characteristics in core and cladding of quartz fibers have been determined. Ge dopant rings shaped luminescence in optical fibers has been observed. New method of determination the Ge dopant distribution in core and cladding cross sectional profile of silica glass fibers is presented. The thermoluminescence and light storage processes in fibers affected by x-ray radiation have been investigated. The photoluminescence investigations shows that there are two bands of luminescence with 400 nm and 550 nm maxima, which appear in different temperature stimulation conditions. Obtained results are discussed.

1. Introduction

For over than 40 years doped quartz optical fibers were used in optical communication systems and different optical fiber devices. On the basis of these devices the fiber-optic gyroscopes and distributed fiber-optic sensors of external physical effects for continuous monitoring of controlled parameters are produced [1–14]. Fiber technologies are increasingly used for registration of ionizing radiation [15]. Largest research centers in Russia, Europe, USA, China, Japan, as well as manufacturers of fiber optical devices search for solution to the fiber optics problems. Quartz optical fibers doped with Ge, P, F, N, Cr, Bi and other transition metals are being studied and developed. While germanium doped quartz glass is most widely used, however, some properties of these glass fibers have been researched insufficiently.

This paper is focused on photo-, radio-, cathodo- and thermoluminescence properties of germanium doped Fujikura and Corning silicate glass fibers. Our studies have been performed by taking into account the distribution of germanium dopant profile in the core center and cladding in glass fibers basing on the original method [16].

2. Samples and research methodology

Germanium dopant distribution profile of the core center and cladding of quartz fiber s research was carried out for a number of samples of germanium glass fibers Fujikura LWP and Corning SMF-28e+. The fiber samples were cut into a length of about 5 cm formed in the fiber bundles of the certain type and placed in a cylindrical form with epoxy resin. After the hardening process, one of the cylinder

Content from this work may be used under the terms of the Creative Commons Attribution 3.0 licence. Any further distribution of this work must maintain attribution to the author(s) and the title of the work, journal citation and DOI.
Published under licence by IOP Publishing Ltd

surfaces of epoxy resin was grinded and polished until the surface was of the required quality. Then conductive carbon layer of thickness no more than 5 nm was deposited on the polished side by vacuum plasma spray on the device Quorum 150T.

Profile measurements and map of the dopant ion distribution in the glass fibers were acquired using an electron microscope Zeiss Sigma VP equipped with energy dispersive analysis attachment INCA Oxford Instruments. The accelerating voltage of the electron beam was equal 20 kV at beam current 20 nA. Image registration was carried out using secondary electron detector SE2, backscattered electron detector CZ BSD and cathode luminescent detector CL (spectral range of 200–850 nm).

Fujikura LWP and Corning SMF-28e+ fiber glass samples were cleaned of polymer coating and treated with acetone to investigate luminescence properties. Then the fiber was grinded in agate mortar to obtain homogeneous fine powder which was deposited on a metal substrate with ethanol.

Photoluminescence (PL) and photoluminescence excitation spectra (PLE) of the glass fibers formed as fine fibers were measured using Perkin Elmer LS55 spectrophotometer equipped with a Hamamatsu photomultiplier R928. A pulsed xenon lamp was used as the excitation source. The signal registration was carried out in fluorescence mode.

Thermoluminescence (TL) was recorded using the same Perkin Elmer LS55 spectrophotometer upgraded with built-in high-top box for measurements at linear heating at the range of 300–770 K. The original quartz fiber powder samples were annealed beforehand at 770 K, and then irradiated with X-ray Oxford Instruments Eclipse IV device with dose of 15 Gy at 300 K. Registration of TL was carried out at a constant heating rate of 2 K/s. Three-dimensional spectral distribution of TL was registered in the range of 290–650 nm with a scan rate of 800 nm/min at a heating rate of 0.1 K/s. The sample temperature was varied for 4 K during the measurement of one spectrum.

3. Experiment Results

3.1. Cathodoluminescence and germanium dopant distribution profile

Cross-sectional images of the fiber surface of Fujikura LWP and Corning SMF-28e+, obtained by CZ BSD detector, are presented in figure 1. The figure shows quartz cladding of diameter of 125 μm and a polymeric protective coating of diameter of 224 μm. The region of the fiber core center of 9 μm diameter is clearly visible due to the contrast in atomic number of the chemical elements produced by registered CZ BSD backscattered electron detector. Cross-sectional surface is dotted with small recesses and cavities formed during samples grinding and polishing. EDS-detector ion concentration measurement of glass fiber core and cladding of Fujikura LWP and Corning SMF-28e+ fibers proves that their composition is almost identical to each other. The ion concentrations are shown in table 1 in mass and atomic fractions. Slight impurity content of chlorine ions has been detected in Fujikura LWP fiber sample which might have occurred as the remainder of the batch used in the preparation process.

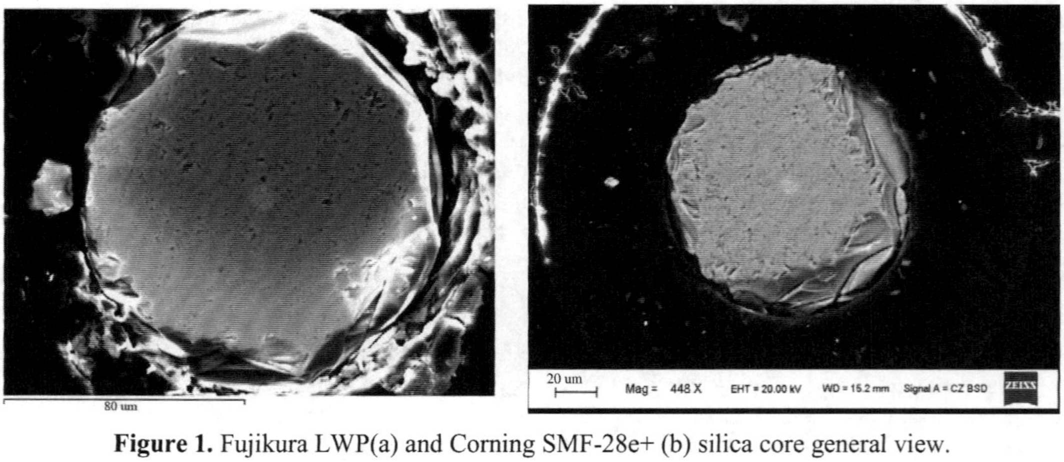

Figure 1. Fujikura LWP(a) and Corning SMF-28e+ (b) silica core general view.

Journal of Physics: Conference Series **552** (2014) 012036 doi:10.1088/1742-6596/552/1/012036

Table 1. FujikuraLWP silica fiber core ion concentration

Element	Core		Cladding	
	Mass fraction, %	At. fraction, %	Mass fraction, %	At. fraction, %
		Fujikura LWP		
O	53.80	67.91	53.87	67.21
Si	43.54	31.31	46.13	32.79
Cl	0.14	0.08	–	–
Ge	2.52	0.70	–	–
		Corning SMF-28e+		
O	51.89	66.20	53.09	66.52
Si	45.50	33.07	46.91	33.48
Ge	2.61	0.73	–	–

Figure 2 represents the Fujikura LWP and Corning SMF-28e+ fiber samples polished sections image registrations resulted in cathodoluminescence microscopy channel under the influence of the electron beam. Complex glow distribution profile formed as concentric circles near the center of the quartz core is observed in images. The profile includes three areas differing in luminescence intensity: central weakly luminescent area in diameter of 8–10 μm, strongly luminescent halo with the thickness of ~5 μm, and Fujikura LWP fiber sample ring with the thickness of ~5 μm in diameter and ~30 μm with low intensity glow. It is worth to note that the luminescence distribution profiles of the different fiber types are the same in the central area and different at the edge.

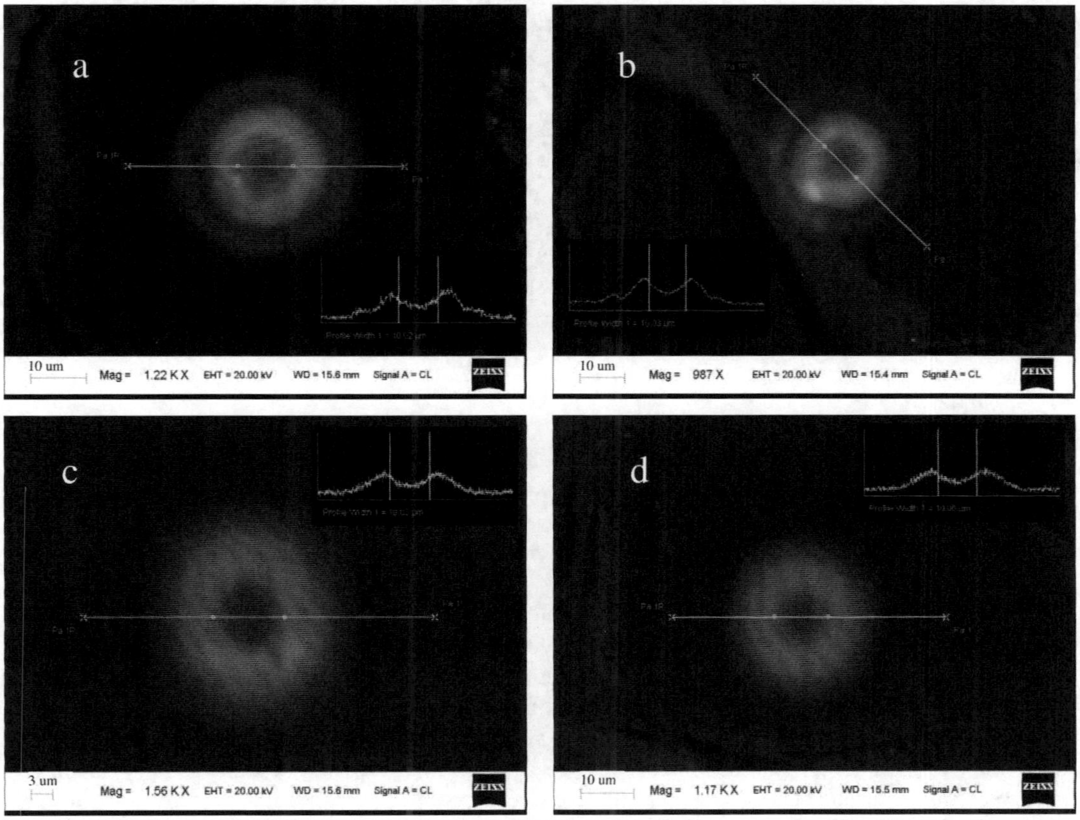

Figure 2. Fujikura LWP (a, b) and CorningSMF-28e+ (c, d) fiber sample CL-channel microscope images. Luminescence intensity distribution profiles measured along a green line are shown.

EDS ion distribution in the observed luminescence area was held using the results of measurements carried out in the CL-channel microscope. Thus, the ion concentration distribution profiles along a line passing through the center of the luminescence region have been measured and also element concentration distribution maps have been constructed. Figure 3 shows that the central region corresponding to the quartz core center provides the maximum amount of germanium dopant, to the center of the core edge its concentration decreases sharply. Highest germanium concentration region (about 0.7 at.%) has low luminescence intensity (central area). Maximum luminescence intensity is achieved in the core-cladding edge area of the quartz core in the ring width of 5 μm; the germanium concentration is reduced in this area to the limit values of EDS-spectrometer sensitivity (figure 3, c, f).

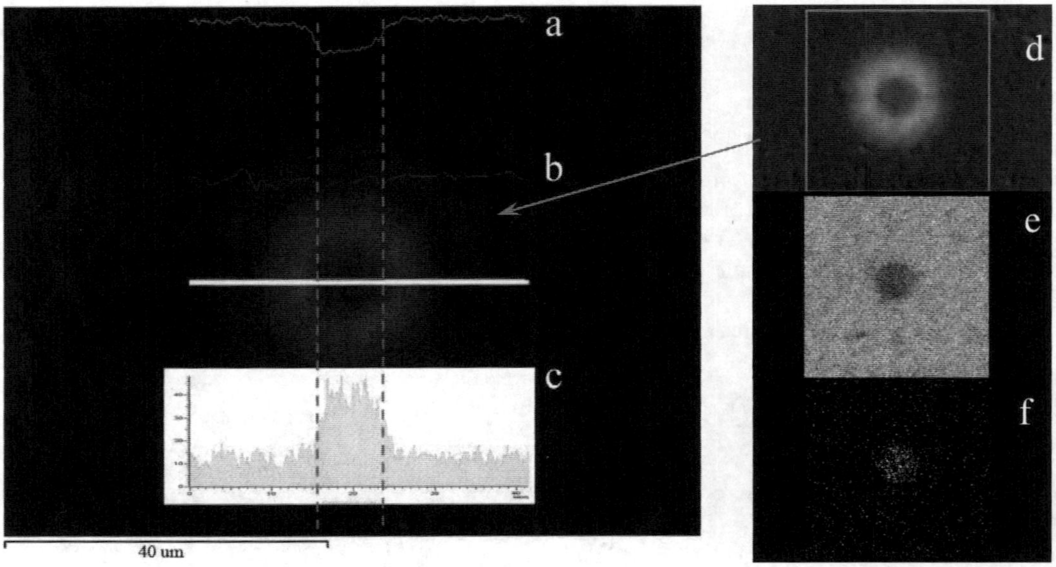

Figure 3. Typical distribution profiles (a, b, c) and distribution images as an element concentration distribution maps (d, e, f) of fibers under investigation measured with EDS-method: a, e – Si; b – O; c, f – Ge; d – mapping area is marked. For a direction of yellow line which element concentration distribution profiles were measured.

3.2. Photoluminescence

PL and PLE spectra of grinded optical fiber Fujikura LWP and Corning SMF-28e+ are the same. The typical luminescence spectrum of the Fujikura LWP fiber has a complex profile, which is a superposition of two bands (Figure 4). The PL spectrum component decomposition into the Gaussian shape bands is plotted in Figure 4c. Two luminescence bands are uniquely determined with maxima at 450 and 400 nm (which correspond to photon energy of 2.77 and 3.12 eV respectively).

As it follows from the literature [8–14], it is established that the amorphous silicon oxide α-SiO_2, doped germanium, intense luminescence band with a maximum at 400 nm (3.12 eV) can be attributed to the radiative transition in the center of luminescence consisting of two-fold coordinated germanium ion [= Ge:] in coordination of rutile (β band) [14]. Excitation of the band mentioned above is carried out in well-known $B_{2\beta}$ band with a maximum at 240 nm (5.15 eV) which was also detected in the EPL spectrum (Fig. 4a). It should be noted, however, that so-called α_E luminescence band of the center [= Ge:] with a maximum at 288 nm (4.3 eV) as described in [14] was not found.

Low-intensity luminescence band at 450 nm (2.77 eV) can be attributed to radiative transitions in the center of luminescence, which is a two-fold coordinated silicon ion [= Si:]. A detected second type of luminescent centers is suits undoped amorphous quartz [13]. However, due to the low intensity of light, these centers of luminescence in undoped core cladding of optical fiber in the electron microscope CL-channel are failed to register.

It follows from the above that the glow in the core-cladding border-area inside the center of the core of quartz fiber is determined by the presence of luminescence centers of two-fold coordinated germanium ions type [= Ge:]. However, in the center of the fiber core where the concentration of germanium is the maximum, due to the concentration quenching the intensity of luminescence is low, whereas in the core-cladding border area and in the areas of the cladding surrounding core the luminescence intensity reaches its maximum. Luminescence intensity distribution analysis in such optical fiber is complex and is a subject for the further investigation.

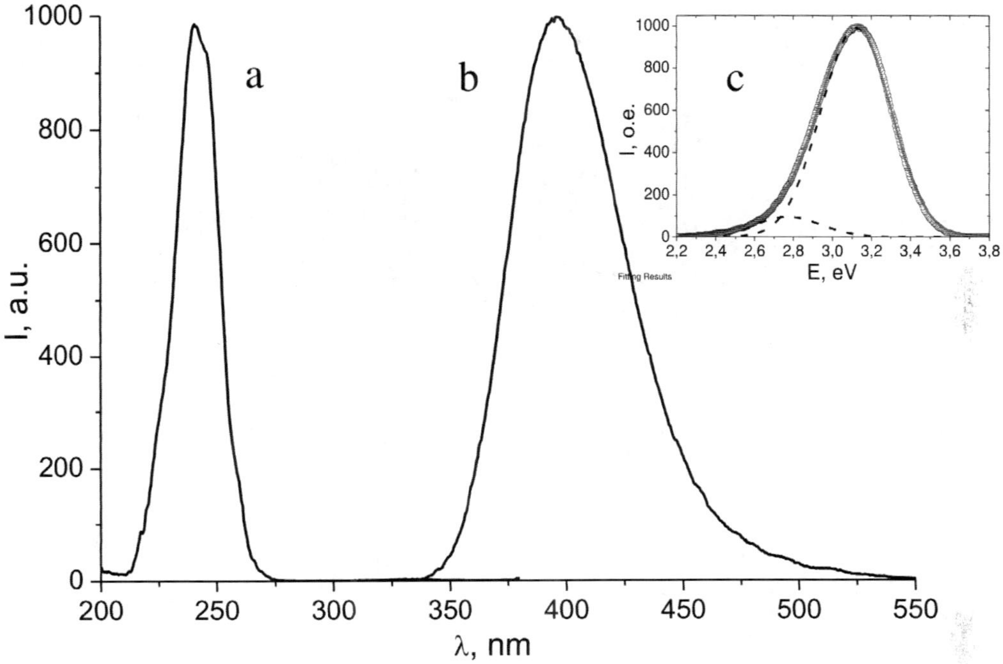

Figure 4. PL excitation spectrum (a) for400 nm and PL spectrum for excitation wavelength of 247 nm (b). PL spectrum component distribution result of the Fujikura LWP fiber powder sample (c).

3.3. Thermally stimulated luminescence

Properties investigation of light storage in silica optical fibers was conducted by measuring the TL curves in conjunction with the measurement of TL spectra. In the result of measurements a three-dimensional distribution of TL curves with spectral resolution in the range of 290–650 nm was constructed (figure 5). Under thermoluminescence emission two glow bands are observed with maxima at 400 and 550 nm, appearing at the temperature ranges of 320–600 and 300–400 K, respectively. It should be mentioned that the band with a maximum at 550 nm was not observed while measuring photoluminescence spectra.

To determinate accurately the spectral composition of TL the TL-distribution section of the wavelength distribution (figure 6) at different temperatures of stimulation were built. As the result of TL-spectra approximation, the following parameters of emission bands were determined. At stimulation temperature of 440 K one luminescence band of Gaussian shape with a maximum 3.12 eV (398 nm) and half-width of 0.54 eV were found. At stimulation temperature of 330 K two luminescence bands with maxima at 2.26 (549 nm) and 3.09 eV (401 nm) and half-widths of 0.25 and 0.55 eV, respectively were detected.

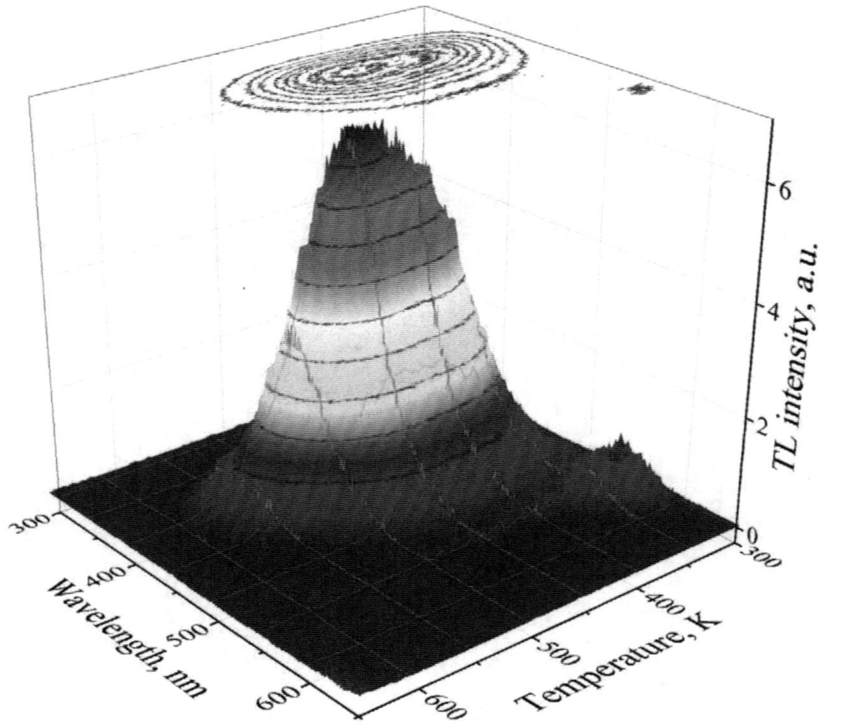

Figure 5. Three-dimensional TL distribution with spectral resolution of the Fujikura LWP optical fiber powder sample.

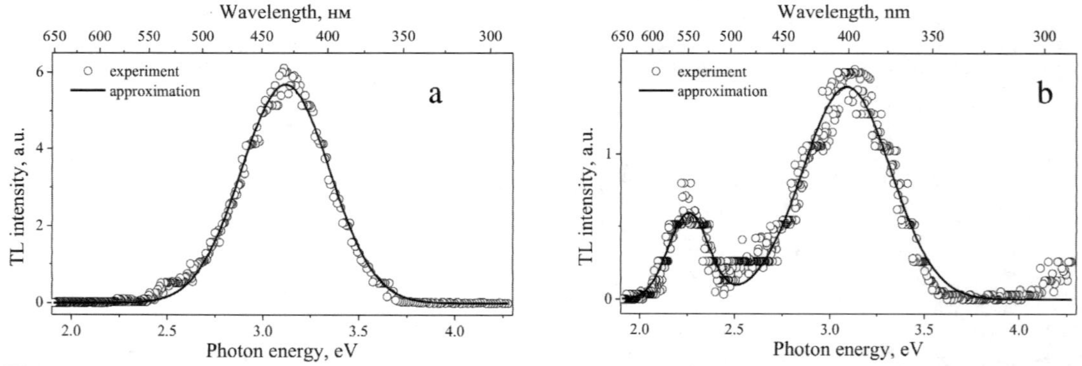

Figure 6. Thermally stimulated luminescence spectra approximation results 440 (a) and 330 K (b)

Green luminescence band with the maximum at 550 nm (2.2–2.3 eV) was previously described in [10, 13, 14]. According to [14] said band is found in newly-implanted by the germanium ions samples of the amorphous silica. After annealing, the luminescence intensity of this band is significantly reduced and practically does not appear in the luminescence spectra. In our experiments the band within the area at 550 nm appears after irradiation of the samples by X-rays. From the above it can be concluded that green luminescence band at 550 nm is to be associated with additive defects in the matrix without germanium ions. This conclusion is also confirmed in [13], where the luminescence band at 550 nm appears under mechanical indentation and laser irradiation on a sample of non-doped amorphous silica. Studies of the nature of the green band are currently underway.

TL curves obtained in our experiments have a complex profile and consist of more than two components. In contrast to the results obtained in [15], attempts of approximation TL curve by the general order theoretical curves did not lead to adequate results, which we believe is due to the presence of fading at low temperatures areas of curves. Restoring the true shape of the TL curves and the parameters determination of the trapping centers requires further experimental studies.

4. Conclusion

The parameters of cathodo-, photo-and thermoluminescence of samples of Fujikura LWP and Corning SMF-28e+ germanium doped silicate glass fibers are described taking into account the profile of germanium dopant distribution in the core and cladding of glass fiber. It is found that the maximum luminescence intensity of observed luminescence [= Si:] and [= Ge:] centers in optical fibers is shown at the area near the fiber core in the cladding. In the core the intensity of luminescence is lower than at cladding area. The experiments shows that the Ge-doped optical fiber has the intensive thermoluminescence and as can be seen from the results of measurements of the three-dimensional spectral distribution of TL in the range 290–650 nm there are two emission bands at 400 (3.1) and 550 nm (2.2–2.3 eV) involved in the energy storage processes. These bands can be attributed at two-fold coordinated [= Ge:] and [= Si:] luminescence centers in optical quartz glass fibers, respectively. It was found that the band at 550 nm ([= Si:] centers) is shown only after the external ionizing radiation damage in the optical fiber. Fiberglass of Fujikura and Corning are potentially suitable for using there as a TL radiation fiber sensors. In this paper we has been described the new investigation method of the distribution dopant germanium profile on the core-cladding border-area inside the center of the glass fiber which potentially suitable for final inspection and adjustment process in the manufacture of optical fiber.

References

[1] Berg R A, Lefevre H C, and Shaw H J 1984 An overview of Fiber Optic Gyroscopes *J. of Light-wave Tech.* **LT-2**91

[2] Dianov E M, Prohorov A M 1986 *Uspekhi fizicheskikh nauk* **148 B.2** 289

[3] Agraval G 1996 *Nonlinear optics* (Moscow: Mir)

[4] Dianov E M, Mashinsky V M 2005 *J. of Light-wave Tech.* **23** 3500

[5] Dolgov I I, Dolgov P I, Dolgov E A, Stepanov E A et al. 2013 *Photon Express* **6** 154

[6] Belovolov M I 2013 *Photon Express* **6**40

[7] Brambilla G 2010 *Journal Optics* **12(4)** 043001

[8] Lyytikainen K et al. 2004 *Optical Express* **12** 972.

[9] Bradly D A et al. 2012 *Applied Rad. and Isotopes* **7**12

[10] Zatsepin A F, Fitting H-J, Kortov V S, Pustovarov V A, Smidt B, Buntov E F 2009 *J. of Non-Cryst. Sol.***355**61

[11] Fitting H-J, Barfels T, Trukhin A N, Schmidt B, Gulans A, Czarnowski A 2002 *J. of Non-Cryst. Sol.* **303** 218

[12] Alessi A, Agnello S, Messina F, Gelardi F M 2010 *J. of Lum.***130** 1866

[13] Fournier J, Néauport J, Grua P, Juber V, Fargin E, Talaga D and Jouannigot S 2010 *Physics Procedia* **8** 39

[14] Salh R, Czarnowski A and Fitting H-J 2005 *Phys. stat. sol. (c)* **2** 580

[15] Benabdesselam M, Mady F, Girard S 2013 *J. of Non-Cryst. Sol.* **360** 9

[16] Ishchenko A V, Chistyakova E K, Shalimov L N, Manko N G, Shestakov G V, Weinstein I A, Ivanov V Y, Cherepanov A N, Shulgin B V 2013*RU Patent application number 2013142549*

Modern microcenter heat explosion model

A V Kalenskii[1, 2], V G Kriger[1,*], I Yu Zykov[1] and M V Anan'eva[1]
[1] Solid state department, Kemerovo State University, Kemerovo 650043, Krasnaya 6, Russia

E-mail: kriger@kemsu.ru

Abstract. Modernization and investigation of the microcenter heat explosion model of the energetic materials initiated by the pulse radiation was made in this paper. Absorptivity of aluminium nanoparticles in PETN-matrix was calculated and was taken into account. Dependences of the absorptivity on the particles' sizes and wave length of irradiation was also taken into account. It was shown that the particle's radius, which corresponds to the absorption maximum, and the peak value both depend on the irradiation wave length. For the first harmonic of the ND:YAG laser the absorption maximum corresponds to the nanoparticle's radius 100 nm, for the second harmonic to the radius it is 44 nm. The peak value increases from 0.2942 to 0.7064. Dependences of the critical initiation energy densities on the metal inclusions' radii were calculated for the energetic materials. It was concluded that the RDX – aluminium composite is the perspective material to use in optic detonator especially for the second harmonic of the ND:YAG laser.

1. Introduction

The microcenter model of the energetic materials explosion initiation is based on the assumption that in the sample's volume there are inclusions with the radii about 100 nm effectively absorbing laser irradiation. It is supposed that the main result of absorption is the inclusions heating which cause the growth of the temperature of its neighborhood. The temperature growth leads to a reaction zone formation. The cross section of the light absorption by the inclusion is thought to be equal its geometrical cross-section [1]. It means that the absorptivity $Q_{abs}=1$. First the model was proposed to explain the regularities of the azide heavy metals explosive decomposition process [2-3]. At the same time in works [4-6] was proposed mechanism of the branched chain reaction, which allowed to describe quantitatively the regularities of the silver azide explosive decomposition initiated by the action of the laser pulse [6-8].

In works [9-10] it was shown that the additions of aluminium nanoparticles allow to decrease the critical initiation energy density of PETN in two orders of magnitude. Due to this fact it was proved the role of the small inclusions absorbing the irradiation in the samples of the secondary explosive materials. It was shown the possibility of use of the secondary explosive materials, containing nanosized additives, as a percussion cap for the optical initiation systems. To find the material and the sizes of the additives giving the minimal critical initiation energies, it is necessary to simulate the initiation process according the materials properties, which were supposed previously negligible. The aim of this work is to work out and investigate the modern variation of the microcenter heat explosion model taking into account the absorptivities of the metal nanoparticles.

Content from this work may be used under the terms of the Creative Commons Attribution 3.0 licence. Any further distribution of this work must maintain attribution to the author(s) and the title of the work, journal citation and DOI.
Published under licence by IOP Publishing Ltd

2. Microcenter model in case of pulse laser initiation

In case of pressed pellet of PETN, containing nanosized aluminium particles, the light got into the sample undergoes multiple reflections at the grain boundaries. As the reflections have the chaotic character the illumination are averaged over all the directions, this allows to use (as it was done in works [1, 11]) spherical symmetry and suppose the inclusion is also a sphere. The main processes taking into account in the presented model is the heating of nanoparticle by the irradiation, heat removing into the energetic material, and chemical heat-generating reaction of the explosive's decomposition. The processes are described by the equations set:

$$\begin{aligned}
&\frac{\partial T}{\partial t} = \alpha \cdot \left(\frac{\partial^2 T}{\partial x^2} + \frac{2}{x} \cdot \frac{\partial T}{\partial x} \right) + k_0 \frac{nQ}{c} \cdot \exp\left(-\frac{E}{k_B T} \right), x > R \\
&\frac{\partial n}{\partial t} = -k_0 n \cdot \exp\left(-\frac{E}{k_B T} \right), x > R \\
&\frac{\partial T}{\partial t} = \alpha_M \cdot \left(\frac{\partial^2 T}{\partial x^2} + \frac{2}{x} \cdot \frac{\partial T}{\partial x} \right), x < R \\
&\frac{\partial n}{\partial t} = 0,\ n = 0,\ x < R
\end{aligned} \qquad (1)$$

the boundary condition is $x = R$:

$$J - c_M \alpha_M \cdot \left. \frac{\partial T}{\partial x} \right|_{x \to R-0} + c\alpha \cdot \left. \frac{\partial T}{\partial x} \right|_{x \to R+0} = 0 \qquad (2)$$

where T is temperature, E – decomposition activation energy, k_0 - preexponential factor, Q – thermal effect of the reaction, α and α_M are the coefficients of thermal conductivity of the matrix and inclusion materials, R in the inclusion radius, c и c_M is the volumetric heat capacity of the matrix and inclusion material, n is the share of undecomposed explosive, and J(t) is the absorbed density of the laser pulse radiation power. For the simulation the following parameters were used: c = 2.22 J/(cm^{3}K), c_M = 2.7 J/(cm^3·K), E = 165 kJ/(mol·K), k_0 = 1.2·1016 s^{-1}, α = 1.1·10^{-3} cm^{2}s^{-1}, α_M = 0.97 cm^{2}s^{-1}, Q = 9.64 kJ/cm^3 [11]. The dependence of the laser pulse radiation power on time is close to the normal distribution function [12]. Taking as a reference time position of maximum intensity of the pulse, we obtain for value $J(t)$ the expression [13]:

$$J(t) = \sqrt{\pi} \cdot Q_{abs} R^2 k_i H_0 \cdot \exp\left(-k_i^2 t^2 \right) \qquad (3)$$

where k_i = 8.325·107 s^{-1} is the parameter determining the pulse duration, H_0 is the averaged density of energy during a single pulse; Q_{abs} is the absorption efficiency equal to the ratio of intensities of radiation absorbed and incident on the inclusion, meanwhile its value depends both on the inclusion radius and radiation wavelength. Multipliers of Eq. (3) normalize the integral of J(t) over time by H_0.

Numerical solution of model (1)–(3) was made using variable-pitch grid. A step in the neighborhood of the inclusions with the radii R ≥ 30 nm was no more than 1/20 of the thickness of the inert substance heated during the pulse ($\sqrt{2\alpha/k_i}$), then the cell size increased exponentially, so that the total thickness of the surrounding material was no less than 8R. For the pulse duration used in the simulation, the heating length was equal to ≈50 nm and the size of cells near the inclusion was about 2.5 nm. Step of grid inside the inclusion exceeded the step outside in $\sqrt{\alpha_M/\alpha}$ times. The cell on the bound inclusion-matrix contained both the matrix material and the inclusion material with the thickness equal to the half step of the grid. For the mentioned characteristics the size of the cell inside the inclusion was about 10nm that is near to the thickness of the light absorption layer. This layer does not exceed 10 nm for the majority of metals. For the inclusions with the radii $20 \geq R > 10$ nm one cell was used inside the inclusion and next there was a cell containing the boundary inclusion-matrix. For the inclusions with the radii $R \leq 10$ nm there was one cell included all the inclusion material, boundary inclusion-matrix and the matrix material. This method allows to make a reasonable consideration of light absorption by means of boundary condition (2) [14].

The ordinary differential equation set got after the dividing of the space into cells was solved by the Runge — Kutta method of 1–5 orders with a variable time step. A relative error at an integration step does not exceed 10^{-9}, whereas the integral relative error estimated by the precision of performance of the law of conservation of energy did not exceed $2.5 \cdot 10^{-5}$.

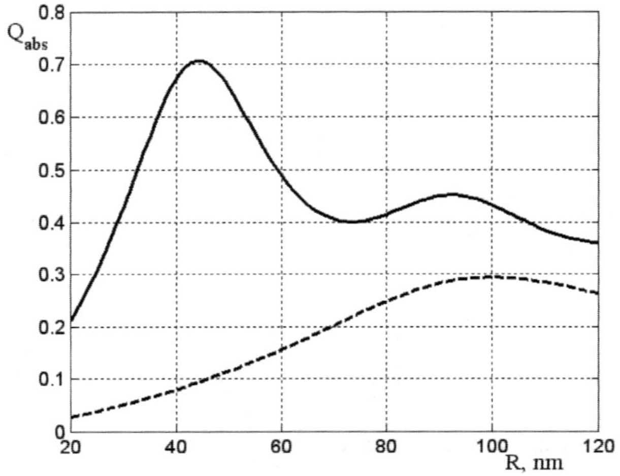

Figure 1. Calculated dependences of the absorptivities (Q_{abs}) on the aluminium nanoparticles' radii for the first (1064 nm) and the second (532 nm) harmonics of the of the ND:YAG laser.

3. Influence of the metal nanoparticles' absorptivity on the critical parameters of initiation

Theories of absorption and scattering of the electromagnetic radiation by the small particles are known for enough long time. In case of the spherical particle the Mie's theory was proposed. The method of calculation was described in work [15-16]. Figure 1 shows the calculated dependences of the absorptivities (Q_{abs}) on the aluminium nanoparticles' radii in the PETN-matrix for the first (1064 nm) and the second (532 nm) harmonics of the of the ND:YAG laser. The refractive index of aluminium at the wavelengths of 1064 nm and 532 nm was m_i=0.978 - 8.030i and 0.5559 - 4.8553i, respectively [17]. The curves have maximums, whose positions are determined mainly by the radiation wavelength. In the area of the radii smaller than the radius of the maximum the curve falls to zero. In the area of the radii greater than the radius of the maximum the arrival at the oscillation plateau occurs.

The influence of the wavelength within the Mie theory is due to the fact that the arguments of the special functions for which expansion is carried out depend on the wavelength and the complex refractive index. If the refractive index m_i does not depend on the wavelength, the dependences $Q_{abs}(R/\lambda)$ would coincide. The real and imaginary parts of m_i can change by several times if the wavelength varies, which complicates the calculations. In the case of aluminium, in the transition from the first to the second harmonic of the ND:YAG laser, the real part of the refractive index decreases by 1.76 times and the imaginary part decreases by 1.65 times [17]. Because of both factors, the absorption efficiency maximum is shifted toward the small diameters of the inclusions (see Figure. 1). For the first harmonic the maximum is observed when inclusion's radius is 100 nm, for the second harmonic when the inclusion's radius is 44 nm. Furthermore, the maximum amplitude increases from 0.2942 at a wavelength of 1064 nm up to 0.7064 at a wavelength of 532 nm.

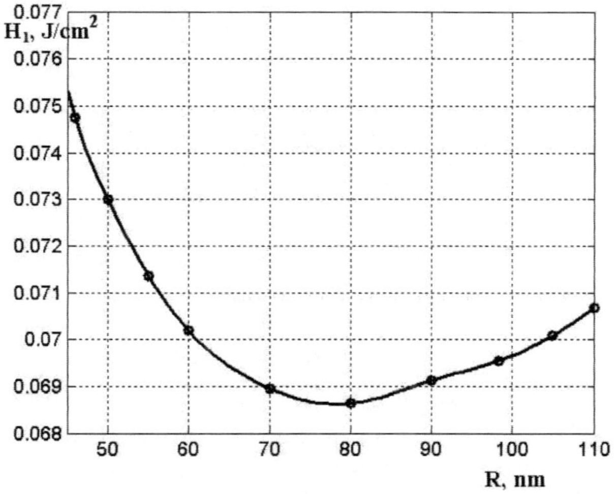

Figure 2. Calculated dependence of the critical initiation energy density on the radius of aluminum inclusions in the PETN matrix without taking into account the absorption particularities.

Figure 2 shows the calculated dependence of the critical initiation energy density on the aluminium nanoparticle's radius in PETN without taking into account the absorption particularities ($Q_{abs} = 1$). The value of the critical initiation energy density does not depend on the wavelength in the area of the impurity absorption of the matrix, this statement includes the values of the first and the second harmonic of the ND:YAG laser. The curve's minimum is located in the radii range 70-80 nm. The minimum appears due to the heating particularities of the spherical nanoparticle. The maximum temperature change δT of the inclusion with the radius R during the heating by the laser pulse with the energy density H might be calculated using the equation

$$\delta T = \frac{HR/4c}{Rh + h^2 + R^2/K^2} \tag{4}$$

where $K = \sqrt{3c/c_M}$ – dimensionless coefficient, which considers the ratio of the coefficients of thermal conductivity of the matrix and inclusion, $h = \sqrt{\alpha t_i}$ - thickness of the heated layer during the time of the pulse duration t_i. The radius of the most heated inclusion is equal to Kh, this value depends on the pulse duration. It turns out that the dependence of the maximum temperature of the heating on the inclusion radius in the inert matrix has the maximum, those position approximately corresponds to the minimum on the dependence of the critical initiation energy.

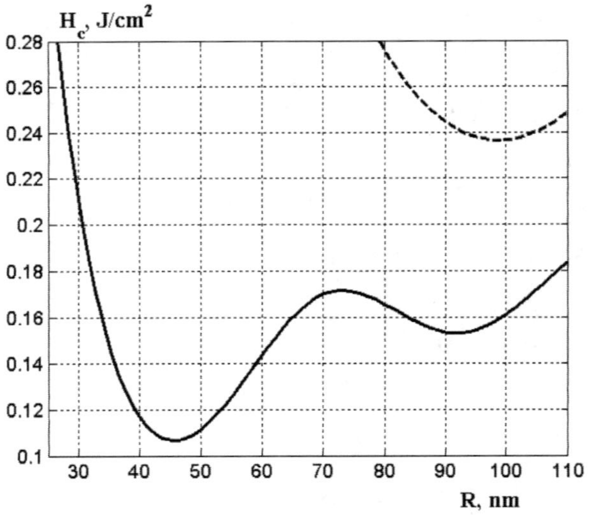

Figure 3. Dependence of the critical initiation energy density on the radius of the aluminium inclusions in the PETN matrix, calculated for the radiation wavelengths 532 nm (solid line) and 1064 nm (dashed line).

Critical energy density of the irradiation, which is expected to be found in an experimental research, is equal to the composition of the values presented on the pictures 1 and 2: $H_c(R) = H_1(R)/Q_{abs}(R,\lambda)$. Curve $H_c(R)$ for aluminium nanoparticles in PETN for the wavelengths 1064 and 532 nm presented on figure 3. Consideration of the absorptivity dependence on the wavelength leads to three main consequences. First of all, according to the experimental data [11] there is a spectral dependence of the critical energy density on the wavelength of the initiating pulse. Then, the minimum depth on the curve, describing the dependence of the critical energy density on the inclusion's radius, changes. The minimal energy density increases from 68.7 mJ/cm^2 (if the absorptivity is not taken into account) to 236.6 mJ/cm^2 (λ=1064 nm) and 106.6 mJ/cm^2 (λ=532 nm). This is due to the fact that the aluminium absorptivities are smaller than 1 for all particles' sizes and all researched (significant) wavelengths (figure. 1). And the last, the minimum's position changes from 79 nm (for Q_{abs} = 1) to 99 nm and 45 for the first and second harmonic of the ND:YAG laser correspondingly (figure. 3). So the minimum position is determined mainly by the optimal absorption conditions. Thus, the model predicts decrease of the critical energy density in case the initiating wavelength decreases.

The minimal critical energy densities calculated for the first and the second harmonic of the ND:YAG laser differ in 2.2 times, this result is in good agreement with the experiment (for the first harmonic – 1.15 J/cm^2, for the second harmonic – 0.7 J/cm^2) where they differ in 1.65 times.

This calculations allows to conclude that the RDX – aluminium composite is the perspective material to use in optic detonator especially for the second harmonic of the ND:YAG laser. This work was supported by Ministry of Education and Science of the Russian Federation (governmental project No. 2014/64).

Journal of Physics: Conference Series **552** (2014) 012037 doi:10.1088/1742-6596/552/1/012037

References

[1] Burkina R S, Morozova E Yu and Tsipilev V P 2011 Initiation of a reactive material by a radiation beam absorbed by optical heterogeneities of the material *Combust., Expl.,Shock Waves* **47(5)** 581–90

[2] Kriger V G and Kalenskii A V 1995 Initiation of heavy metal azides by pylse radiation *Chem. Phys. Rep.* **14(4)** 556-64

[3] Kriger V G and Kalenskii A V 1996 The effect of crystal size on initiation of decomposition of heavy metal azides by pulse radiation, *Chem. Phys. Rep.* **15 (3)** 351-58

[4] Kriger V G, Kalensky A V and Zvekov A A 2012 Relaxation of electronically excited products of solid-state reactions in the crystal lattice, *Russ. J. Phys. Chem. B* **48 (1)** 15–18

[5] Kalenskii A V, Bulusheva L G, Kriger V G and Mazalov L N 2000 Quantum chemical cluster modeling of boundary conditions for metal azides, *J. struct. Chem.* **41 (3)** 495-98

[6] Kalenskii A V, Anan'eva M V, Kriger V G and Zvekov A A 2014 Rate constant of capture of electron charge carriers on a screened repulsive center, *Russ. J. Phys. Chem. B* **8 (2)** 131–35

[7] Kriger V G, Kalenskii A V, Zvekov A A, Borovikova A P and Grishaeva E A 2012 Determining the width of the reaction wave front in the explosive decomposition of silver azide *Combust., Expl., Shock Waves* **48 (4)** 488-95

[8] Kriger V G, Kalenskii A V, Zvekov A A and Tsipilev V P 2009 Explosive decomposition of silver azide single crystals for various diameters of the irradiated area. *Combust., Expl., Shock Waves* **45 (6)** 729-31

[9] Aduev B P, Belokurov G M, Nurmukhametov D R and Nelyubina N V 2012 Photosensitive material based on PETN mixtures with aluminum nanoparticles *Combust., Expl., , Shock Waves* **48 (3)** 361–66

[10] Aduev B P, Nurmukhametov D R, Tsipilev V P and Furega R I 2013 Effect of ultrafine Al-C particle additives on the PETN sensitivity to radiation exposure *Combust., Expl., , Shock Waves* **49 (2)** 215–18

[11] Aduev B P, Nurmukhametov D R, Furega R I, Zvekov A A and Kalenskii A V 2013 Explosive decomposition of PETN with nanoaluminum additives under the Influence of pulsed laser radiation at different wavelengths *Russ. J. Phys. Chem. B* **7(4)** 453 – 56

[12] Kriger V G, Kalenskii A V, Zvekov A A 2010 Determining the onset of mechanical failure of silver azide crystals initiated by a laser pulse *Combust., Expl., , Shock Waves* **46 (1)** 60-63

[13] Kriger V G, Kalenskii A V, Zvekov A A, Anan'eva M V and Borovikova A P 2009 A Diffusion model of chain-branching reaction of the explosive decomposition of heavy metal azides *Russ. J. Phys. Chem. B* **3 (4)** 636–40

[14] Kriger V G, Kalenskii A V, Zvekov A A, Zykov I Yu and Nikitin A P 2013 Heat-transfer processes upon laser heating of inert-matrix-hosted inclusions *Thermophysics and Aeromechanics* **20 (3)** 367–74

[15] Kriger V G, Kalenskii A V, Zvekov A A, Zykov I Yu and Aduev B P 2012 Effect of Laser Absorption efficiency on the heating temperature of inclusions in transparent media *Combust., Shock Waves* **48 (6)** 705 – 8

[16] Kalenskii A V, Zvekov A A, Anan'eva M V, Zykov I Yu, Kriger V G and Aduev B P 2014 Laser wavelength influence on the critical density of initiation energy of energetic materials *Fizika Goreniya i Vzryva* **50 (3)** 98–104

[17] Zolotarev V M, Morozov V N and Smirnova E V 1984 *Optical Constants of Natural and Technical Media* (Leningrad.: Khimiya) p. 216

Journal of Physics: Conference Series **552** (2014) 012038 doi:10.1088/1742-6596/552/1/012038

Pulsed cathodoluminescence of LiF-Fe2O3 crystals

V I Korepanov*, P V Petikar and A A Kamrikova
Laser and Lighting Engineering Department, Institute of High Technology Physics, Tomsk polytechnic University, 30 Lenin Avenue, Tomsk 634050, Russia

E-mail: korepanov@tpu.ru

Abstract. We have conducted an in-depth study of pulsed cathodoluminescence of LiF-Fe_2O_3 crystal at temperatures 20 – 300 K and the time interval of 10 ns – 10 μs for the first time. It is shown that two types of luminescence centers can be formed on the basis of O^{2-} ion. The first center is excited by the doped impurity, and the second center is the part of the impurity complex. It is found that the processes of formation and accumulation of primary radiation defects affect the parameters and characteristics of the activator luminescence in LiF-Fe_2O_3 crystals.

1. Introduction

Lithium fluoride can be used as an efficient scintillator for systems to record astrophysical neutrinos and dark matter [1]. Under ionizing radiation, LiF crystals containing no activator impurity luminesce weakly and at low temperature only [2]. High yield of luminescence of lithium fluoride crystals is characteristic of lithium fluoride doped with metal oxides: LiF-WO_3, LiF-TiO_2 and others [1,3,4]. Previously, it was found that under electron irradiation the behavior of LiF-WO_3, LiF-TiO_2, LiF-Fe_2O_3 crystals is identical, and the patterns of pulsed cathodoluminescence (PCL) are similar [5,6]. For example, in all the crystals under electron irradiation, the reaction of O^{2-} center formation occurs as a result of destruction of the OH^- groups present in the crystal. A long-lasting buildup stage is characteristic of the activator luminescence in these crystals.

The behavior patterns of LiF-WO_3, LiF-TiO_2, LiF-Fe_2O_3 crystals under irradiation by electron beam are similar, apparently, due to the same set of defects present in the activator impurity in the crystal during its growth, for example, to compensate the excess charges. The main impurities are oxygen, OH^- groups and other hydrogen-oxygen defects [3]. The presence of OH^- is indicated by the reaction of transformation of $OH^- \rightarrow O_2^-$ and by IR absorption spectra, and atomic oxygen ions can be indicated by the absorption bands in the VUV spectral range. It is evident that these defects affect the scintillation properties and degradation characteristics of scintillators based on lithium fluoride in long-term exposure to radiation.

Previously, we have studied accumulation of O_2^- centers, F-, F- aggregate color centers in LiF-WO_3, LiF-TiO_2, LiF-Fe_2O_3 crystals under irradiation by electron beam [7]. However, the mechanism of luminescence excitation and the exact structure of the activator luminescence center have not been clarified so far. This data is of primary importance to improve characteristics of scintillation detectors. The aim of this research is to study the features of the activator luminescence in LiF-Fe_2O_3 crystals under irradiation by electron beam.

2. Technique

Content from this work may be used under the terms of the Creative Commons Attribution 3.0 licence. Any further distribution of this work must maintain attribution to the author(s) and the title of the work, journal citation and DOI.
Published under licence by IOP Publishing Ltd

PCL was excited by the electron beam generated by a nanosecond pulsed accelerator. The parameters of the electron beam were as follows: the duration of the electron current pulse was 2–10 ns; the probable current densities of the electron beam varied in the range of 0.1–1000 A/cm^2; the average electron energy was 200 – 250 keV. The spectral characteristics of the luminescence were recorded with the time resolution of 7 ns with MDR 204, PMT 97, 83 and oscilloscope Tektronix. The technique is described in detail in [8]. The stationary luminescence spectra were measured with Cary Eclipse spectrofluorometer at 300 K. The absorption spectra were measured with SF-256UVI spectrophotometer.

3. Research results

In LiF-WO_3 and LiF-TiO_2 crystals, the bands of activator absorption are found in the spectral range of 190–300 nm, and photoexcitation in this range leads to intense activator luminescence. The absorption in this spectral region in LiF-Fe_2O_3 crystal is not distinct. Excitation with the wavelength of 200 nm is known to cause photo-luminescence of O^{2-} oxygen centers (absorption band at 6.2 eV in LiF is reported in [9]). Optical absorption by oxygen centers can be observed in LiF- Fe_2O_3 crystals as well. Therefore, LiF-Fe_2O_3 can be used to study the photoluminescence of oxygen centers under excitation in the region of 6.0 eV.

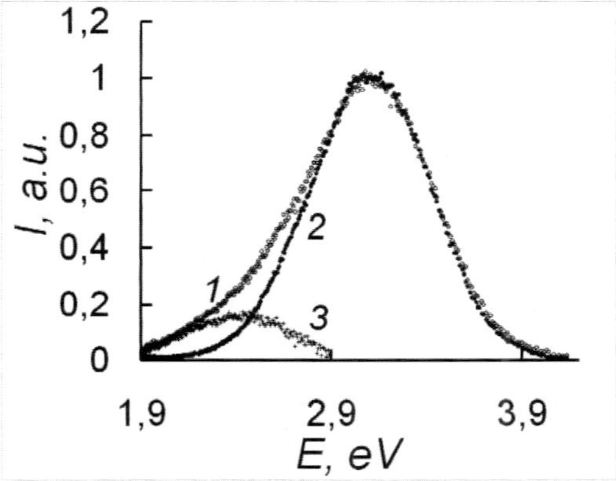

Figure 1. Photoluminescence spectra of LiF-Fe_2O_3 (1) and LiF-Li_2O (2) crystals under excitation with E = 6.05 eV at 300 K. 3 is a difference spectrum

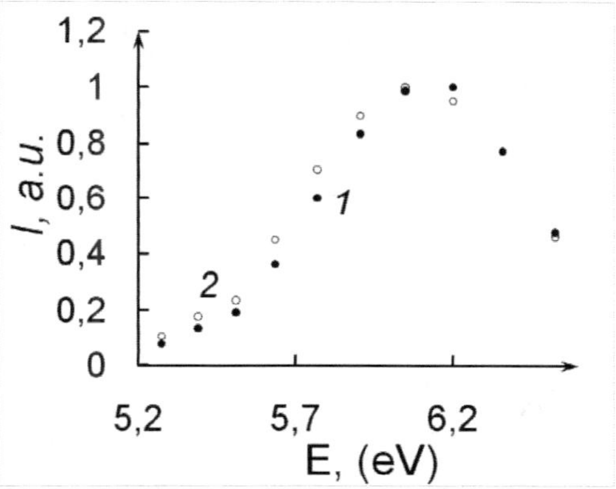

Figure 2. Excitation spectra of LiF-Li_2O (1) and LiF-Fe_2O_3 (2) crystals at 300 K.

Figure 1 shows the luminescence spectra of LiF-Li_2O and LiF-Fe_2O_3 crystals under photoexcitation by photons with the energy E = 6.05 eV. In lithium fluoride containing only oxygen, the luminescence is represented by one Gaussian band with the maximum at 3.1 eV and FWHM at 0,76 eV. This photoluminescence (PL) band is caused by (O^{2-} – V_a) centre, where V_a is an anion vacancy [3,4]. The luminescence excitation spectrum of this center measured at E = 3.1 eV is shown in figure 2 (curve 1). It consists of a band with the maximum at E = 6.0 eV.

Under excitation by photons with the energy E = 6.05 eV, the photoluminescence spectrum of LiF-Fe_2O_3 crystal is found to consist of two bands (figure 1, curve 2). The first band is identical to the luminescence band of O^{2-}oxygen ion, and the maximum of the second band is in the point where E = 3.1 eV. The excitation spectra of both luminescence bands are found to coincide figure 2 (curve 2).

We conducted an in-depth study of the spectral and kinetic characteristics of the activator PCL of LiF-Fe_2O_3 crystal in the temperature range of 20 – 300 K over the time interval of 10 ns – 10 ms. The kinetics of the activator luminescence decay after excitation by nanosecond electron pulse can be described as the sum of two exponentials.

$$I = I_{01}exp\,(-t/\tau_1) + I_{02}exp\,(-t/\tau_2)$$

At room temperature, at E = 3.0 eV: I_{01} =8 a.u., I_{02} = 1 a.u., τ_1 =1.9 μs, τ_2 = 74μs. Figure 3 (curves 1 and 2) show the normalized spectra of slow (decay within a millisecond time interval) and fast (decay within a microsecond time interval) components of PCL decay for LiF-Fe_2O_3 crystal at 300 K. The maxima of the luminescence bands in these spectra are at 2.65 eV and 2.9 eV. The bands of each component of PCL decay are asymmetrical and can be formally decomposed into two Gaussian components with the maxima in the regions of 3.05 eV and 2.6 eV, and FWHM equal to 0.47 eV and 0.51 eV. The results of this decomposition of the spectra at 300 K are shown in figure 3 (curves 3, 4, 5, 6). It is seen that the relations of the luminescence bands of the fast and slow decay components are different.

If the temperature of the sample decreases to 20 K, the PCL is shifted to the region of higher energies. At low temperatures, the spectrum can also be represented as the sum of two bands. The relation of these bands differs from that at 300 K, and the luminescence decay kinetics at 20 K is described only by one exponential with τ_1 = 3.4ms.

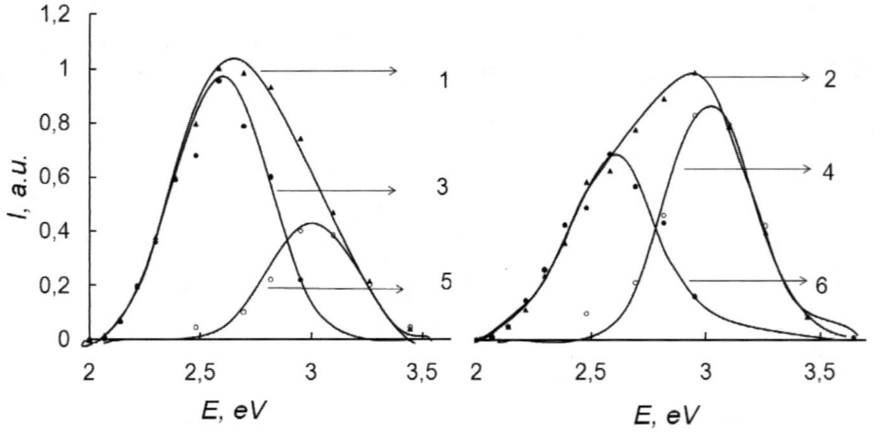

Figure 3. Normalized spectra of PCL for LiF-Fe_2O_3 crystal at 300 K. 1 is a slow decay component; 2 is a fast decay component; 3, 4, 5, 6 are results of decomposition into Gaussian components.

In case the temperature of the crystals under irradiation by electron beam changes, all the parameters of the kinetics of the activator luminescence decay change: decay time constant τ, intensity I, and the emitted light sum (S = I×τ). The dependences of the parameters of the fast and slow PCL decay components on temperature are different.

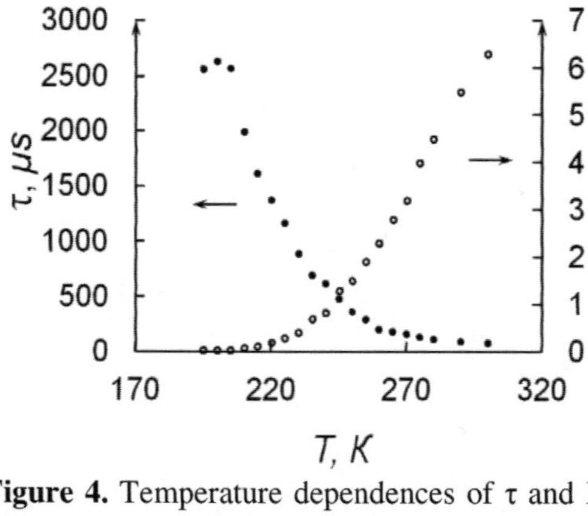

Figure 4. Temperature dependences of τ and I of the slow component of PCL decay for LiF-Fe_2O_3 crystal.

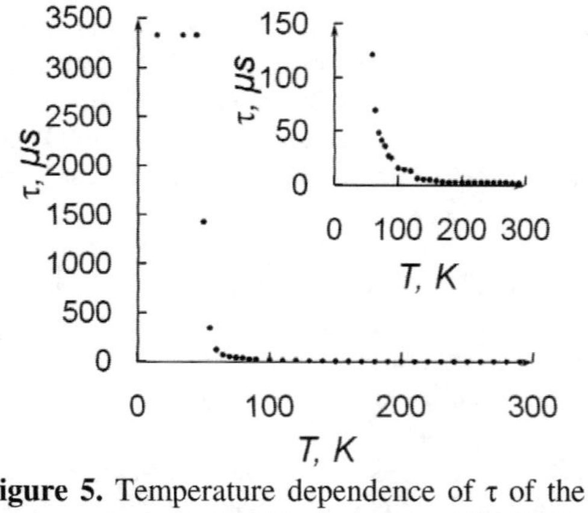

Figure 5. Temperature dependence of τ of the fast component of PCL decay for LiF-Fe_2O_3 crystal

Figure 4 shows the temperature dependence of τ and I for the luminescence of the slow decay component. In the temperature interval of 20 – 300 K τ of the slow decay component varies from 2.7 ms to 74 μs. The luminescence intensity in the range of 200 – 300 K increases. If temperature increases, the emitted light sum S increases starting from 200 K with a tendency to saturation at T > 250 K. From the results obtained it follows that at temperatures below 200 K, the slow decay component in PCL spectra is not observed due to strong decrease in the amplitude intensity. Therefore, at 15 K, both bands are represented by the fast component of the activator PCL decay only.

Figure 5 shows the temperature dependences of the parameters of the fast component of PCL decay for LiF-Fe_2O_3 crystal. It is seen that τ is constant at low temperatures, whereas at T > 50 K it sharply decreases. As temperature grows, τ keeps decreasing but at a slower rate. The second temperature point of significant change in τ falls within the range of 110 – 140 K (inset in figure 5). The luminescence intensity significantly increases at T > 50 K, and at temperatures above 120 K the intensity rapidly increases if temperature of the sample grows. The dependence of S = I×τ on temperature is shown in figure 6. This dependence reveals a number of characteristic temperature regions in which the emitted light sum S rapidly changes. These regions account for the temperature ranges of 40 – 60 K and 100 – 120 K.

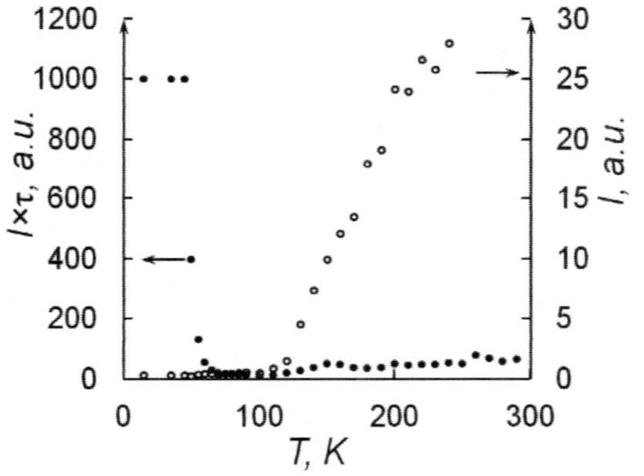

Figure 6. Temperature dependences of S = I×τ and I of the fast component of PCL decay for LiF-Fe_2O_3 crystal.

4. Discussion

PL spectra differ from PCL spectra in the positions of the maxima of bands and their FWHM. Therefore, different groups of luminescence bands are excited under photostimulated action and under irradiation by electron beam. These groups belong to different types of activator centers. Each group consists of two bands. The nature of each band can be substantiated by the following facts.

Only one luminescence band can be observed in the PL spectra of LiF- Li_2O crystal. As it has been shown in previous studies [3.4], this band may be due to the radiative transition in the (O^{2-}–V_a) center where V_a is anion vacancy. The band at 3.1 eV can be attributed to the emission of O^{2-} nearby the impurity. This is confirmed by the coincidence of the excitation spectra of the bands at 3.1 and 2.42 eV. The emission band in the spectral range 2 – 3 eV in crystals containing Mg impurity is described in [4]. Therefore, the two photoluminescence bands may belong to the radiative transitions in two types of (O^{2-} – V_a)-centers, weakly excited (3.1 eV) and strongly excited (2.4 eV) by iron impurity. The excited state of the center is always exposed to greater impact of the nearby defect. For this reason, the absorption spectra of the two types of (O^{2-} – V_a)-centers can be similar, whereas the emission spectra can be different.

The alternative interpretation is that both photoluminescence bands in LiF-Fe_2O_3 crystal belong to one type of the impurity (O^{2-} – V_a)-center, but to different electronic transitions. In this case, the impact of the activator center causes the second photoluminescence band. That is the iron impurity allows radiative electron transition at 2.4 eV. In both cases, excitation of the photoluminescence occurs as a result of the absorption of the energy by O^{2-} oxygen ion.

Both the PCL spectra of the activator luminescence and the PL spectra of LiF-Fe_2O_3 crystal consist of two luminescence bands. Under photoexcitation at 300 K, the parameters of these bands (peak position and FWHM) do not coincide with the parameters of the luminescence bands in ($O^{2-} - V_a$)-oxygen centers. It is evident that this is an activator intracenter luminescence. The electronic structure and nuclear configuration of the radiative state in this center are different from those of ($O^{2-} - V_a$)-center under photoexcitation. Therefore, two types of oxygen centers can be observed in the PL and PCL spectra of LiF-Fe_2O_3 crystal. The first type is ($O^{2-} - V_a$)-centre excited by impurity, and the second type is the complex which contains O^{2-} oxygen ion and activator impurity. The example of the second type of the luminescence center for LiF-TiO_2 crystal is reported in [4]. The cause of the two luminescence bands in PCL spectra may be the same as in the case of PL spectra, i.e. the existence of two types of nuclear configuration or two types of electronic transitions in the complex.

The analysis of the temperature dependences of the PCL decay parameters shows that abrupt changes of the values of these parameters are associated with characteristic temperature intervals of the formation and accumulation of color centers in pure lithium fluoride crystals. This distinguishes LiF-Fe_2O_3 crystal from LiF-WO_3, LiF-TiO_2 crystals. It is known that in pure crystals of lithium fluoride in the temperature range of 20 – 60 K, different types of the self-trapped excitons (STE) transform, the luminescence decays, and H-centers are delocalized [2]. In lithium fluoride crystal at temperatures close to 125 K, V_K - centers accumulated at low temperatures are delocalized. At 100 – 120 K, the efficiency of Frenkel defect formation grows. The characteristic points of the abrupt change of the decay time, the intensity of the emitted PCL light sum are observed in the same temperature range (figures 4–6).

The causes of the stage of luminescence buildup in LiF-WO_3, LiF-TiO_2, LiF-Fe_2O_3 crystals are crucial for understanding the mechanisms of PCL excitation. This stage indicates the inertial mechanism of energy transfer from the base of the scintillator (lithium fluoride) to the luminescence center. The results of our research suggest that LiF-Fe_2O_3 crystal is an appropriate system to study these processes. In contrast to LiF-WO_3, LiF-TiO_2 crystals, in this crystal we can find temperature regions within which we can observe change in the parameters of PCL and in the parameters of formation and accumulation of radiation defects.

5. Conclusion

1. It has been found that two types of luminescence centers can be formed on the base of O^{2-} ion: the center excited by the doped impurity; the center as a part of the impurity complex. Each of the two types of centers consists of two radiation bands.

2. The processes of formation and accumulation of primary radiation defects are found to affect the parameters and characteristics of the activator luminescence in LiF-Fe_2O_3 crystals. This allows an in-depth study of the mechanisms of energy transfer from LiF crystal to the activator radiative center.

Acknowledgement

This study was carried out under the state assignment of the Ministry of Education and Science of Russian Federation for 2014–2016 (research work № 8.2500.2014/K).

References

[1] Abdurashitov D, Gektin A, Nepomnyaschih A, Radzhabov E and Shiran N 2012 Possible applications of scintillation LiF crystals as detectors for dark matter particles *Proceedings of ISMART 2012* (Dubna)

[2] Lisitsyna L A, Grechkina T V, Korepanov V I and Lisitsyn V M 2001 Short-living primary radiation defects in LiF crystal *SSP* **43** 1613

[3] Egranov A V and Radzhabov E A 1992 Spectroscopy of hydrogen and oxygen impurity centers in alkali halide crystals *Novosibirsk*: *Nauka* 159

[4] Nepomnyaschih A I, Egranov A V and Radzhabov E A 1984 Color centers and luminescence of LiF crystals *Novosibirsk*: *Nauka* 113

[5] Korepanov V I, Lisitsyna L A, Lisitsyn V M, Timoshenko N N and Tupitsyna I A 2011 *Proceedings of Higher Education, Physics* **54** **1** 357-363

[6] Lisitsyna L A, Korepanov V I, Abdrahmetova A A, Timoshenko N N and Dauletbekova A K 2012 Impulse photo and cathodoluminescence crystals LiF, doped tungsten oxide *Optics and Spectroscopy* **112** **2** 200-206

[7] Lisitsyna L A, Korepanov V I, Lisitsyn V M, Petikar P V, Kasymkanova R N and Esilbaev D B 2012 Spectroscopic Properties of LiF Crystals Doped with W, Ti and Fe Oxides *Proceedings of Higher Education, Physics* **55** **11/3** 95-99

[8] Lisitsyn V M and Korepanov V I 2007 *Spectral measurements with a time resolution (Tomsk: TPU)* 97

[9] Lobanov B D, Maksimova N T, Tsirulnik P A, Schenina L I and Volkova N V 1984 *Optics and Spectroscopy* **56** **1** 172-174 (in Russian)

International Congress on Energy Fluxes and Radiation Effects (EFRE-2014) IOP Publishing
Journal of Physics: Conference Series **552** (2014) 012039 doi:10.1088/1742-6596/552/1/012039

Modern trends and development in high-dose luminescent measurements

V Kortov
Ural Federal University, Ekaterinburg, 62000 Russia

E-mail: vskortov@mail.ru

Abstract. Main application areas of high-dose dosimetry are described. The requirements to the materials for high-dose luminescent detectors are set. The examples of successful high-dose measurements using radiation-resistant phosphors are given. Viability of using materials with deep traps to detect intensive radiation flows is grounded. Characteristics of high-dose measurements using highly sensitive detectors TLD-500 (Al_2O_3:C) and LiF:Mg,Cu,P are discussed.

1. Introduction

The interest in high-dose measurements of ionizing radiations is increasingly growing due to development of radiation technologies. These technologies include radiation modification of the properties of various materials, medical products sterilization, disinfection of agricultural products, radiation control of defects in large machine parts and equipment [1]. All the above mentioned technologies employ intensive isotope radiation sources or charged particle accelerators providing exposure to the doses of several dozens of kGy. High-dose measurements are also used for radiation monitoring of nuclear power station (NPS) equipment and storages of spent nuclear fuel. High dose measurements for radiation resistance tests of the materials and equipment are of great importance as well. These served as the reasons for more extensive study for design and development of high-dose luminescent radiation detectors. The main advantages of luminescent dosimetry are a wide range of the measured doses, possibility to register different types of radiation (with low and high LET), high reproducibility of measurement results, automated dose readers.

The purpose of the paper is a brief analysis of the current status and modern trends in development of luminescent detectors for high-dose measurements.

2. Requirements to the materials for high-dose luminescent detectors

At present about twenty types of commercial luminescent detectors of radiation with the (10^{-6}-20) Gy measurement range are widely used for personal dosimetry and radiation monitoring of environment [2]. Still, the choice of luminescent detectors for high-dose dosimetry is limited. In the USA two types of thermoluminescent detectors (TLD): CaF_2:Mn (TLD-400) with the operating dose range (10^{-6}-10^{2}) Gy and $Li_2B_4O_7$:Mn (TLD-800) with the dose range (10^{-3}-10^{5}) Gy are used. In Russia TLD on the base of aluminophosphate (PST) and sodiumsilicate (TLD-K) glasses with the highest registered dose of (2-5) kGy are widely used [3].

A great number of storage materials are known to be promising for development of high-dose TLD as for these materials there is no tissue equivalence requirement. However, there are challenges

Content from this work may be used under the terms of the Creative Commons Attribution 3.0 licence. Any further distribution of this work must maintain attribution to the author(s) and the title of the work, journal citation and DOI.
Published under licence by IOP Publishing Ltd

to their usage. One of such challenges is saturation of the thermoluminescence (TL) yield at high-dose irradiation. In our experiments we were the first to measure TL yield in a wide range of gamma-radiation doses (10 mGy - 100 kGy) of the high-dose TLD-500 (Al_2O_3:C) detectors which we developed on the base of oxygen-deficient alumina single crystals [4]. Figure 1 shows a dose response of the main dosimetric peak at Tm=480 K of this detector in the given dose range. The linear dose response of the TL yield is shown to be limited to 10 Gy. With the further dose increase, TL intensity decreases to some extent. However, its medium value is only slightly different from the maximal one with the use of the dose of seven orders. It implies that the centers responsible for the main dosimetric peak in the TLD-500 detectors are not destroyed being exposed to high doses as aluminum oxide is a radiation-resistant material. However, the use of the main TL peak for dosimetric measurements is challenging due to saturation of its height at the doses higher than 10 Gy.

One of the main causes of the TL yield saturation at high-dose irradiation is known to be a limited capacity of electron traps which can be filled with free electrons formed at the exposure [5]. As a result, saturation of the TL yield appears with a particular dose related to filling of all the traps responsible for the main dosimetric peak. Thus, the material for a high-dose detector should have a high capacity of unfilled trapping centers prior to exposure.

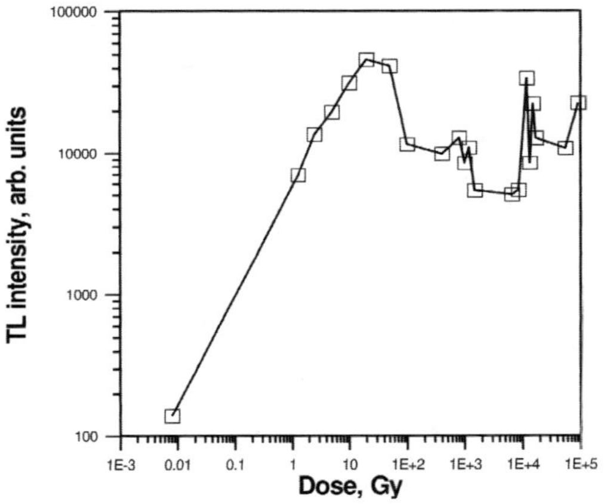

Figure 1. Dose response of TLD-500 (Al_2O_3:C) detectors at high dose γ-irradiation (Co-60).

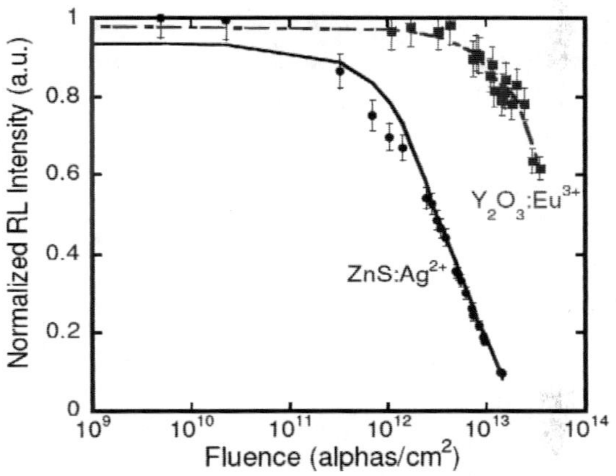

Figure 2. Normalized radioluminescence intensity as a function of alpha-particle fluence for ZnS:Ag^{2+} and Y_2O_3:Eu^{3+} [6].

Another challenge to the usage of phosphors for high dose measurements is the degradation of the luminescent centers. The degradation of the luminescent characteristics under intensive irradiation can be well seen when radioluminescence (RL) is registered. Figure 2 shows degradation of RL of the two well known phosphors exposed to α-particles [6]. Each of these phosphors has the α-particle fluence range with the constant RL intensity, then the intensity drops sharply with the fluence increase. The RL degradation kinetics proves that radiation resistance of the luminescent centers in Y_2O_3:Eu phosphors is much higher than in ZnS:Ag. Taking into account all the above mentioned, the main requirements to the material for a high-dose detector is made: the detector should be radiation-resistant and have a large capacity of the unfilled trapping centers of the charge carriers.

3. Radiation-resistant materials for high-dose dosimetry

Phosphors capable of keeping luminescent and dosimetric characteristics stable under high exposure doses belong to radiation resistance materials. Oxide dosimetric phosphors BeO and Al_2O_3 have the high radiation resistance and detectors based on the given oxides can be used for high-dose measurements [7,8]. Recently, dosimetric characteristics of the TL detectors based on yttrium orthoaluminate crystals ($YAlO_3$) with various dopants have been studied intensively [9,10]. Radiation-resistant oxides of aluminum and yttrium contained in $YAlO_3$ allow one to make detectors capable of registering doses up to (5-10) kGy. Natural silicates such as quartz, sand, topaz, onyx, etc., are also radiation-resistant materials [11-13]. The dose response of these materials was found to change linearly in the range from 10 Gy to several kGy. As an example, Figure 3 shows a dose response of topaz pellets irradiated with gamma-rays [12]. Dependence of radiation resistance on material structure was found. In particular, it has been shown that naturally disorder multiphase compounds, low-dimensional materials (glasses, ceramics), nanostructured materials, hetero-structures have high radiation resistance. These materials are disordered, therefore additional disorder caused by structural changes under irradiation does not affect significantly their characteristics in a wide dose range.

Nanostructured phosphors are the most suitable materials for high-dose detectors. They have big capacity of the trapping centers due to the surface of the centers and high radiation resistance. A great number of researches have been made by now [14,15]. They have found the common feature shared by nanostructured phosphors of various compositions: their dose response remains linear at the doses (1.5-2) orders higher than the ones in microcrystalline analogues [8]. Figure 4 shows TL response of LiF:Mg,Cu,P detectors exposed to gamma-rays of Cs-137 [14]. It is necessary to note that the commercial polycrystalline detector provides higher sensitivity, but its TL response curve begins flattening out at the dose of 20Gy. The dose response of the nanocrystalline phosphor is linear up to 20 kGy, and it can be used for measuring high doses.

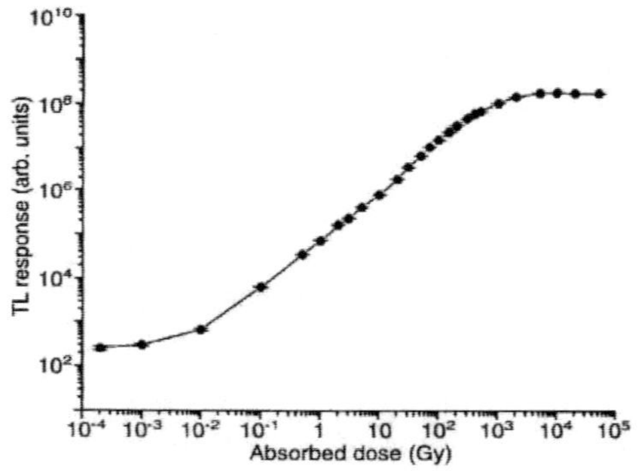

Figure 3. Dose response of topaz pellets irradiated with gamma rays (Co-60) [12].

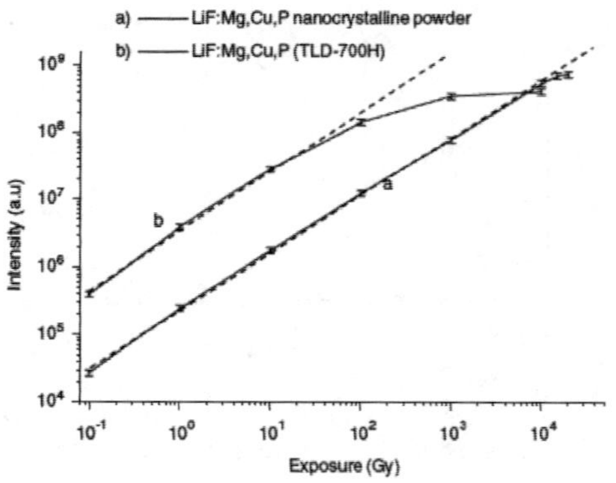

Figure 4. Comparison of dose response provided by nano- and commercial phosphor [14].

4. Use of deep traps for high-dose measurements

It is also possible to increase capacity of the traps of dosimetric phosphor using deep traps of the charge carriers. For instance, in oxygen-deficient aluminum oxide the deep traps are present both in a single crystal and in its nanostructured analogue. The energy depth of these traps is bigger than the one in the traps of the main dosimetric peak used for low-dose measurements. The nature of deep traps is not completely studied, they are considered to be formed by aggregates of the native and impurity defects [16]. Deep traps are found following high-temperature TL peaks after high-dose irradiation (Figure 5). Each of these peaks can be used for dosimetric measurements while heating the irradiated detector. For example, for the intensive TL peak at 603 K, the dose response changes linearly in the range of (1,5-80) кGy after exposure of the TLD-500 detector to an electron pulse beam from an accelerator [8]. Measurements of the high-temperature TL peaks appearing after beta-irradiation (Sr-90) of the TLD-500 detectors showed that the dose response in the majority of them does not exceed two orders [17]. It is connected with the fact that TL of the deep traps is also saturated with a particular dose. One can consider e the following process of deep traps filling (see Figure 5). All traps in the detector material begin to be filled under exposure. However, the probability of capturing the charge carriers decreases with the increase of the trap energy depth. Therefore, the traps of the main dosimetric TL peak (A) are filled in the beginning. With the dose near its saturation the first high-temperature TL peak (B) appears and grows with the dose increase. After its saturation the TL intensity of the second high-temperature peak (C) increases with the dose growth. Similarly, deeper traps responsible for other high-temperature TL peaks are filled. The task of registering high doses consists in measuring TL yield of all deep raps.

For this purpose the phototransfer luminescence (PTTL) [18] and photo-thermostimulation luminescence (PTSL) [19] methods are used. In both methods the detector exposed to a high dose is preliminary heated to (500-550) K to empty shallow traps. To register PTTL the detector is then exposed to intensive blue light for optical emptying the traps and for transfer of the electrons delocalized from the deep traps through the conduction band to the unfilled shallow traps. The following heating of the detector allows one to measure PTTL. Its intensity is proportional to a number of electrons which filled the deep traps under high-dose exposure. The disadvantage of the PTTL method is a long measuring procedure. Moreover, not all the deep traps might be emptied under photostimulation. The PTSL method makes it possible to eliminate these disadvantages. With the simultaneous photo-thermostimulation (thermally assisted OSL), the heating enhances electron energy in the traps and the photons of lower energy are required to empty the deep traps. The electron phototransfer to the shallow traps is similar to the one described for PTTL. These traps are thermoionized and the PTSL peaks are registered. The light sum of the peaks is proportional to the dose. The results of using the PTSL method while registering high doses by means of Al_2O_3:C detectors are presented in the work [20]. The PTSL peaks are located in the temperature range convenient for registration (up to 500 K), and the dose response is linear from 10^{-2} Gy to 10^{3} Gy.

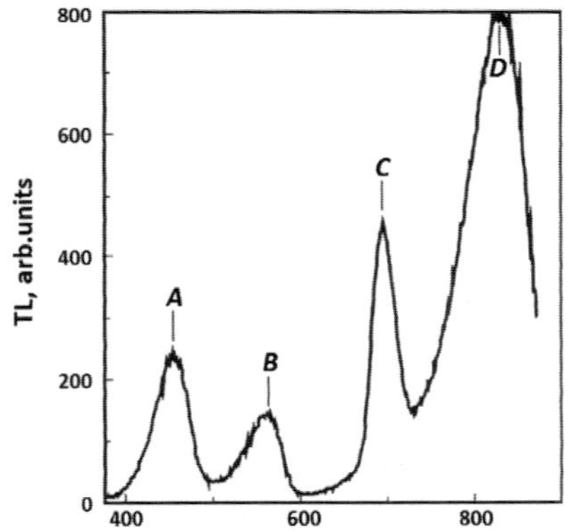

Figure 5. Glow curve of TLD-500 detector after nanosecond electron pulse irradiation (dose 75 kGy).

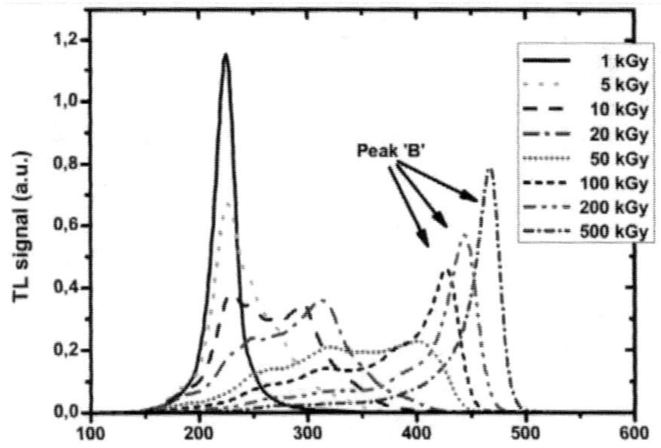

Figure 6. LiF:Mg,Cu,P glow-curves resulting from gamma irradiation (^{60}Co source) for the dose range (1–500) kGy [24].

Journal of Physics: Conference Series **552** (2014) 012039 doi:10.1088/1742-6596/552/1/012039

5. The possibility of registering high doses with highly sensitive radiation detectors

At present the most sensitive to radiation detectors, i.e. detectors capable of measuring very low doses (1mkGy and lower), are TLD-500 (Al_2O_3:C) and TLD-700H (LiF:Mg,Cu,P) detectors. Due to the necessity to measure high doses, the study of the behavior of these detectors under high-dose exposure has become important. The intensity of the main dosimetric peak of the detectors under study is saturated with the (10-20) Gy dose, as it was mentioned earlier. The exposure of the LiF:Mg,Cu,P detector to high doses causes decrease of the main TL peak intensity at Tm=220 °C with the simultaneous formation of new peaks. The Tm of these peaks is shifted to the range of high temperatures with the dose growth (Figure 6). At the dose about 50 kGy a stable TL peak (peak "B") begins forming with Tm=450 °C. The dose response of this peak remains linear up to 500 kGy [21]. The TL glow spectrum acquires new bands [22] but their nature is not studied. The obtained results give evidence of formation of new electron traps and recombination centers under high-dose irradiation. In our opinion, formation of a high-temperature dosimetric peak (peak "B") under high-dose irradiation provides evidence of appearance in the LiF:Mg,Cu,P detector of deep traps connected with radiation defects and their complexes.

The behavior of the TLD-500 (Al_2O_3:C) detector under high-dose irradiation is quite different from that of the LiF:Mg,Cu,P detector [23]. Figure 7 shows that the changes of the shape and temperature position of the main dosimetric peak are slight in a wide dose range. After exposure to nanosecond pulses of an electron accelerator, the saturation of the TL yield of this peak with the dose increase is observed up to 80 kGy (Figure 7). Then, a sublinear growth of the TL yield is registered with the dose increase up to 800 kGy. The measurements of photoluminescence (PL) of the detector exposed to a high dose show that a new wide band with the maximum at 550 nm appears in the PL spectrum. The intensity growth of this band is remarkable at the exposure doses of (70-80) кGy when the TL yield starts increasing. The performed analysis has shown that an new band in the PL spectrum is a superposition of glow bands of the aggregate F_2-type centers in different charge states. These centers are additional electron traps capable of capturing free electrons, thus, increasing the TL yield in the (80-800) kGy dose range.

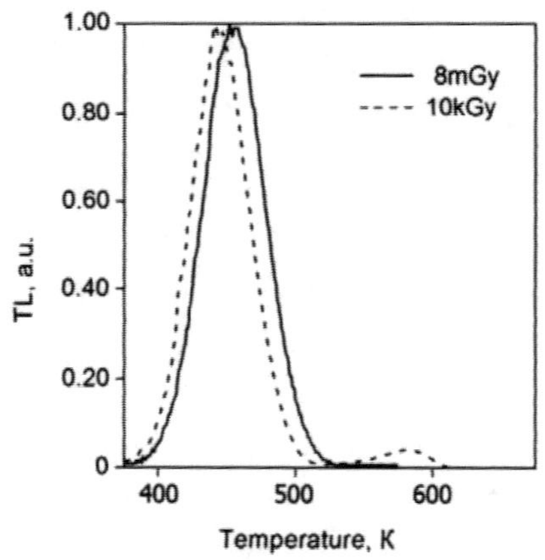

Figure 7. Normalized TL glow curves of TLD-500 detectors after low and high dose irradiation (^{60}Co).

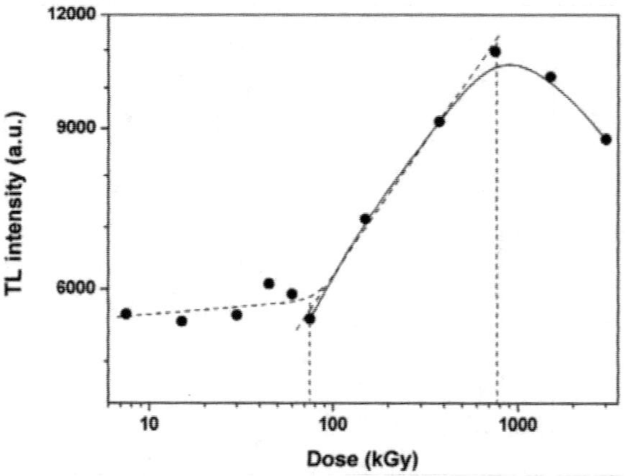

Figure 8. TL yield dependence in the main dosimetric peak at high-dose irradiation of the TLD-500 detector by an electron beam.

Thus, the possibility of registering high doses using highly sensitive detectors of radiation is shown. An increase of the traps capacity in a some dose range due to radiation disorders is a common feature.

The obtained results show the necessity of recovering the initial luminescent and dosimetric characteristics of the detectors exposed to a high dose. For his purpose, a special procedure of annealing with experimentally defined temperature and period is required. As in the LiF crystal structural changes affecting its luminescence occur at heating above 240 °C, development of the annealing procedure of the radiation disorders in this crystal is a real challenge. There are efforts to solve this problem [24]. However, LiF:Mg,Cu,P detectors can be used only for single measurements of high doses. Whereas for radiation-resistant TLD-500 (Al_2O_3:C) detectors, appropriate annealing conditions (900°C, 1 hour) have been found. Under these conditions the initial sensitivity, luminescent and dosimetric characteristics of such detectors recover. As a result, there is a possibility of multiple repeated uses of TLD-500 detectors for high-dose measurements [23].

6. Conclusions

The results obtained allow one to make some general conclusions on high-dose TL dosimetry current state and trends:

1. At present several successful researches on luminescent and dosimetric properties of some materials suitable for high-dose detector manufacture have been carried out. A number of such materials can be expected to increase, as there is no tissue equivalence reguirement for high-dose luminescent detectors.
2. The material for a high-dose detector should be radiation resistant and have significant trapping center capacity.
3. Low-dimensional phosphors, particularly nanostructured ceramics, are promising for creating high-dose luminescent detectors. Such materials have a large amount of charge carrier traps and high radiation resistance.
4. Selection of efficient phosphors containing deep traps involved in accumulating and storing dosimetric information is required.
5. It is necessary to develop an annealing procedure for every type of high-dose detectors to restore their initial properties and to maintain reproducibility when the dose is measured.
6. Development of dosimetric devices based on luminescence registration from the deep traps of the irradiated detectors by the PTTL and PTSL (TA-OSL) methods is a vital task.

References

[1] Kozlov Yu D, Stefanov I V, Ermolaev O P, Sidelnikov O P and Aleksandrov EN 2006 *High Technologies with the Use of Ionizing Radiation Sources in Industry* (Moscow:Energy-Atom Press) p 714
[2] Kortov V S 2007 *Radiat. Meas.* **42** 576
[3] Aluker N L, Artamonov A S, Bakulin Yu P, Danilovich E N, Krysanova O U, Riskina R V and Sogoyan AV 2006 *Nuclear Sience and Ingineering Issue: Series "Physics of Radiation Effect on Radioelectronic Devices"* № 1-3 86
[4] Akselrod M, Kortov V, Kravetsky D, Gotlib V 1990 *Radiat. Prot. Dosim.* **33** 119
[5] Chen R, Lo D and Lawless J 2006 *Radiat. Prot. Dosim.* **119** 33
[6] Cress C D 2008 *Effect of Ionizing Radiation on Nanomaterials and III-V Semiconductors* (Rochester Institute of Technology: PhD Dissertation) p 155
[7] Kortov V S, Milman I I and Lesz J 1995 *Sci. Rep. Techn. Univ.* Opole **15** 31
[8] Kortov V and Ustyantsev Yu 2013 *Radiat. Meas.* **56** 299
[9] Zhydachevskii Ya, Suchocki A, Berkowski M, Bilski P and Warchol S 2010 *Radiat. Meas.* **45** 516
[10] Premkumar H B, Sunitha DV, Nagablushana H, Sharma SC, Nagabhushana BM, Shivakumara C and Rao J L 2014 *Journal of Alloys and Compounds* **601** 75
[11] Teixeira M I and Caldas L V E 2004 *Nucl. Instr. and Methods in Phys. Res.* **B218** 194
[12] Souza D N, Lima J F, Valerio M E G and Caldas V E 2002 *Radiat. Prot. Dosim.* **100** 413
[13] Teixeira M I and Caldas V E 2012 *Appl. Radiat. Isot.* **70** 1417
[14] Salah N, Sahare P D and Rupasov AA 2007 *J. Lumines.* **124** 357

[15] Salah N, Habib S S, Khan Z H and Lochab S P 2008 *Radiat. Meas.* **43** 236
[16] Kortov V S, Milman I I and Nikiforov S V 1999 *Radiat. Prot. Dosim.* **84** 35
[17] Lo D, Lawless and Chen R 2006 *Radiat. Prot. Dosim.* **119** № 1-4 71
[18] Alexander C S and McKeever S W S 1998 *J. Phys.D: Appl. Phys.* **31** 2908
[19] Polymeris G S, Rapitis S, Afouxenidis D, Tsirliganis N G and Kitis G 2010 *Radiat. Meas.* **45** 519
[20] Soni A, Mishra D R, Bhatt B C, Gupta S.K, Rawat N S, Kulkarni M S and Sharma D N 2012 *Radiat. Meas.* **47** 111
[21] Bilski P, Obryk B, Olko P, Mandowska E, Mandowski A and Kim J 2008 *Radiat. Meas.* **43** 315
[22] Mandowska E, Bilski P, Obryk B, Mandowski A, Olko P and Kim J 2010 *Radiat. Meas.* **45** 579
[23] Kortov V S, Zvonarev S V, Pustovarov V A and Slesarev A I 2014 *Radiat. Meas.* **61** 74
[24] Obryk B, Skowronska K, Sas-Bieniarz A, Stolarczyk L and Bilski P 2013 *Radiat. Meas.* **56** 171

Kinetics of the cathodoluminescence flash in crystals with competing trapping centres

V Lisitsyn[1*], L Lisitsyna[2], Z Karipbaev[3], D Valiev[1]
[1]National Research Tomsk Polytechnic University, Tomsk, Russia
[2]Tomsk State University of Architecture and Building, Tomsk, Russia
[3]L.N. Gumilyov Eurasian National University, Astana, Kazakhstan

E-mail: lisitsyn@tpu.ru

Abstract. The present work aims to bring proof of possibility of the existence of the process to increase the number of luminescence centers in the radiative state. That apparently has the effect of the buildup of luminescence excited by a pulsed electron beam. To demonstrate the possible influence of competing centers on the decay kinetics of luminescence materials were selected on the basis of crystal $YLiF_4$ and $YLiF_4$: Nd^{3+}. Found that the kinetics of flash luminescence when excited $LiYF_4$: Nd crystals pulse electron beam at room temperature in the initial stage of a flare-up of up to 300 ns was observed. The model of the exchange of electronic excitations between the centers of the emission and capture is discussed.

1. Introduction

Exposure of wide-band gap materials to short time pulses radiation initiates a luminescence flash which then decays in time. In the kinetics of some materials under certain experimental conditions the luminescence intensity is found to increase at the initial period after excitation in time exceeding the duration of the excitation pulse. The effect of the luminescence buildup under excitation by a pulsed electron beam was observed in [1-3]. The luminescence buildup at the initial period after exposure may be caused by increase in the amount of luminescence centers in a radiative state due to transfer of energy from the trapping centers excited by radiation pulse. The present research sets out to prove the possibility of this process and to develop the model of the luminescence buildup in the crystal with competing luminescence and trapping centers.

2. Test samples and experimental technique

The shape of the kinetic curves of the pulsed cathodoluminescence flash depends on a number of factors: the type of the material, the temperature of the sample under excitation, the history of the sample, the excitation power, and etc. To demonstrate the possible effect of the competing centers on the luminescence decay kinetics we chose $YLiF_4$-based materials. In these materials, at temperatures close to room temperature the cathodoluminescence buildup is well manifested after exposure to radiation pulse. Undoped "pure" $YLiF_4$ crystals and those doped with Nd^{3+} grown in INCROM Ltd were studied. The test samples were cleaved from a single crystal at an angle of 45° to the crystal optical axis C and then polished.

The samples were excited by pulsed electron beams with the following characteristics: FWHM of 10 ns pulse duration, the average electron energy of 250 keV, and the energy density of the excitation pulse equal to 15 mJ/cm^2. The pulsed cathodoluminescence (PCL) of the crystals was investigated in the spectral range of 2.0–4.0 eV at temperatures ranging from 80–300 K, and the kinetics of the luminescence flash in the time interval of $1 \cdot 10^{-8}$–$1 \cdot 10^{-4}$ s after the end of the pulse action. To record the luminescence and change in optical absorption we used PMT-106, PMT-84-6, a digital oscilloscope Gwinstek GDS-2204, Tektronix TDS-2014 and monochromators MDR-206 and MDR-3.

Content from this work may be used under the terms of the Creative Commons Attribution 3.0 licence. Any further distribution of this work must maintain attribution to the author(s) and the title of the work, journal citation and DOI.
Published under licence by IOP Publishing Ltd

3. Experimental results

The structure of the PCL spectra of pure $YLiF_4$ crystals, and those doped with impurities is complex [4, 5]. Two groups of bands are distinct in the PCL spectrum: in the ranges of 3.7 ... 5.5 and 2.0 ... 3.7 eV. The characteristics of the two groups of bands differ due to different luminescence centers. The long-wave region of the spectrum is caused by uncontrolled impurities which occur in the crystal in its growing. Intentional doping with impurities suppresses the luminescence of the uncontrollable impurities and changes the spectrum.

Figure 1 shows the PCL spectra in the region of 2.0 ... 3.7 eV measured after 10 and 200 ns after the end of the excitation pulse for $YLiF_4$ crystal doped with 0.7 mol% Nd at 300 K. The spectrum measured instantaneously after the end of the excitation pulse is found to be a superposition of the spectra caused by the uncontrollably entered impurity and intentionally doped neodymium. In time after the excitation, the radiative spectrum caused by the uncontrolled impurity disappears and after 200 ns its contribution is negligible. The luminescence caused by the doped neodymium with the bands characteristic of this chemical element becomes dominant. The shape of the spectrum changes and the intensity of the neodymium luminescence increases in time.

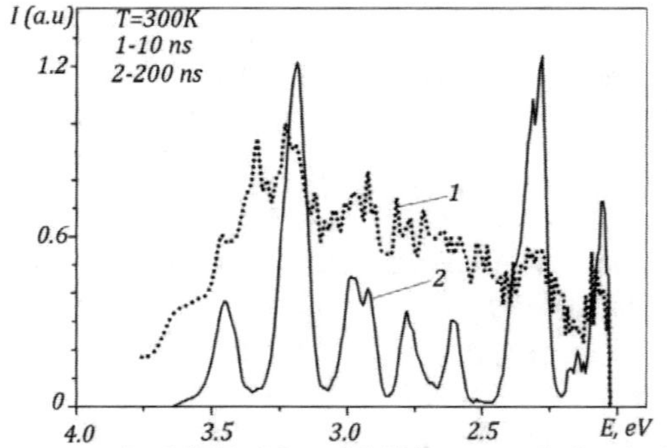

Figure 1. Luminescence spectra for $LiYF_4$-0.7 mol% Nd measured after 1–10 ns and 2–200 ns at T = 300 K

Figure 2 shows the kinetic curves of the change in the luminescence intensity at 2.3 and 2.5 eV. One of the main lines of the neodymium luminescence is found at 2.3 eV, and the greatestspectrum notching accounts for 2.5 eV. In this region, intense luminescence after 10 ns is caused by the luminescence of the uncontrolled impurity. The research results show a substantial difference in the kinetic curves.

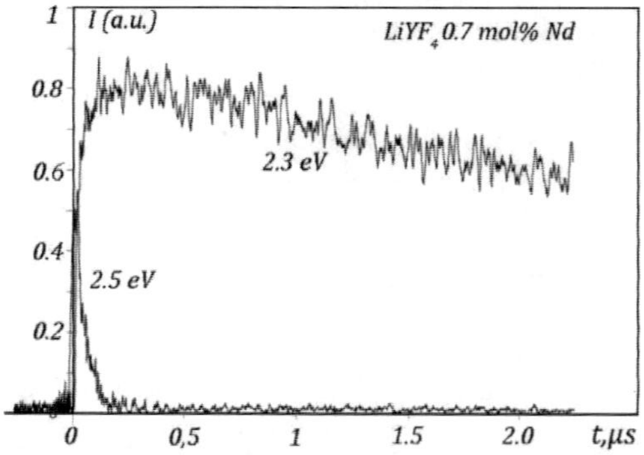

Figure 2. Kinetic curves of PCL initiated by pulsed electron beam in the $YLiF_4$-0.7 mol% Nd crystal at 2.5 eV and 2.3 eV at 300 K

At 2.5 eV, the luminescence decays after 200 ns, whereas at 2.3 eV, the luminescence grows in the range of 200–300 ns, and in the microsecond time range, the luminescence decays. We can conclude that the excitation energy is transferred in time from the luminescence centers responsible for the luminescence at 2.5 eV to the neodymium centers responsible for the luminescence at 2.3 eV.

3.1 Model of the luminescence flash buildup kinetics

Consider the model of the process shown in Figure 3. Assume that the crystal contains two competing centers of excitation energy trapping. In the energy scheme of the process, E_1, E_2, I_1 and I_2 are the energies and intensities of transitions in C_1 and C_2 centers, and ΔE_1 and ΔE_2 are the energy barriers for the exchange of the electron excitation between the centers.

Assume that the characteristics of the luminescence centers (trapping centers) are as follows:

1. The radiation efficiency of the luminescence centers C_1 is high (high light yield). The radiation efficiency of the trapping (luminescence) centers C_2 is low or transition to the ground state is nonradiative.
2. The luminescence centers C_1 appear to be deeper electron excitation traps than C_2 centers. However, the cross section of electron excitation trapping by C_2 centers is larger than that by C_1 centers.

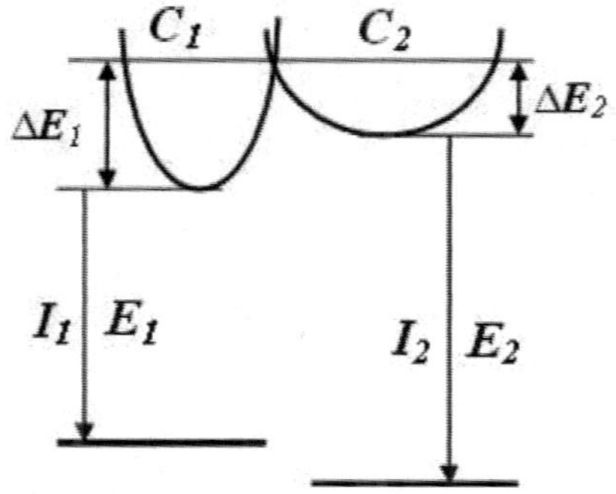

Figure 3. Energy scheme of the model of the competing centers excitation relaxation.

3. Exchange of electron excitations between the luminescence centers is considered to be possible. Since, according to the model ΔE_2, the activation energy of the transition of electron excitation from C_2 to C_1 is less than ΔE_1, the activation energy of the transition of C_1 to C_2, the transfer of the electron excitation energy will occur mainly from C_2 to C_1.
4. The lifetime τ_2 of C_2 luminescence centers in an excited state is much shorter than the lifetime τ_1 of C_1 centers.

Recombination in the radiative center C_1 is followed by generation of radiation with the intensity of I_1. The radiation intensity and the power of radiation from the unit of the crystal volume are determined by the recombination rate:

$$I_1 = k_{11}\frac{dN_1^*}{dt}, \qquad (1)$$

where k_{11} is the radiation energy yield, N_1^* is the concentration of the excited centers per time t disappearing within time Δt. The concentration of C_1 centers in an excited state is determined by the concentration of the luminescence centers entered in the crystal and the density of the absorbed excitation flux energy. The concentration of the luminescence centers in phosphor crystals is typically more than 10^{18} cm^3. The density of the absorbed energy of the excitation pulse is typically less than 1 J. At high pulse energies, mechanical fracture of the crystal occurs. In case the probability of the electron excitation energy transfer to the luminescence center is k_{21}, the concentration of the luminescence centers excited by instantaneous energy pulse will be:

$$N_1^* = k_{21}N_1 p \qquad (2)$$

where P is the density of the absorbed energy of the excitation pulse, N_1 is the concentration of the entered luminescence centers.

The luminescence centers in an excited state spontaneously transfer to the ground state with the lifetime of τ_1:

$$N_1^* = k_{21} N_1 p \cdot e^{-\frac{t}{\tau_1}}, \tag{3}$$

In this case the radiation intensity caused by the luminescence centers is:

$$I_1 = k_{11} k_{21} N_1 p \tau_1 \cdot e^{-\frac{t}{\tau_1}}, \tag{4}$$

According to the model, the electron excitations can be trapped by a different luminescence center (trapping center). The concentration of C_2 centers in an excited state and relaxation of the excitation in C_2center can be written by analogy with (3.4):

$$N_2^*(t) = k_{22} N_2 p \cdot e^{-\frac{t}{\tau_2}}, \tag{5}$$

$$I_2 = k_{12} k_{22} N_2 p \tau_2 \cdot e^{-\frac{t}{\tau_2}}, \tag{6}$$

The case with k_{12}=0 means that relaxation is nonradiative.

The model assumes that the electron excitations can be exchanged between the luminescence centers. Since the depth of the trap under condition of C_2 is less, the transition of electron excitations occurs mainly from $C_2$to C_1. In this case the number of the excited states of C_1 centers is to increase in time by the value proportional to the decrease in the concentration $N_2^*(t)$. Assume that one part $N_{21}^*(t)$ of a number of the excited states $N_2^*(t)$ recombines at time t radiatively (or nonradiatively) during transition to the ground state. The other part of the excited states $N_{22}^*(t)$ (electrons, excitons and etc.) jumps barrier ΔE_2 under certain conditions and after that it is found in C_1 center. The change in the balance of the centers in an excited state in time can be written as:

$$N_2^*(t) = N_{21}^*(t) + N_{22}^*(t), \tag{7}$$

The number of the excited states which can jump the barrier at time t is determined by the expression:

$$N_{22}^*(t) = k_{30} \cdot e^{-\frac{\Delta E_2}{kT}} N_2^*(t), \tag{8}$$

where is the probability of the electron excitation transfer from C_2center to C_1 center.

Assume that the law of variation in time $N_{22}^*(t)$ and $N_2^*(t)$ is described by an exponential. This assumption is satisfied if one of the processes of relaxation of the excited C_2 centers (to the ground state or transition to C_1 state) is dominant. Consequently:

$$N_{22}^*(t) = k_{30} e^{-\frac{\Delta E_2}{kT}} \cdot k_{22} \cdot N_2 \cdot p \cdot e^{-\frac{t}{\tau_2}} \tag{9}$$

Transition of electron excitations from C_2 state to C_1 state C_2 causes increase in the concentration of centers $N_1^*(t)$ at time t by the value:

$$\Delta N_{1add}^*(t) = \int_0^t N_{22}^*(t) dt = k_{30} e^{-\frac{\Delta E_2}{kT}} \cdot k_{22} \cdot N_2 \cdot p \cdot (1 - e^{-\frac{t}{\tau_2}}) \tag{10}$$

The total concentration of the excited C_1 centers at time t with account of recombination in these centers in time τ_1 (with $\tau_1 > \tau_2$) becomes equal to:

$$N_{1n}^*(t) = N_1^*(t) + N_{1add}^*(t) = k_{21} \cdot N_1 \cdot p \cdot e^{-\frac{t}{\tau_1}} + k_{30} e^{-\frac{\Delta E}{kT}} \cdot k_{22} \cdot N_2 \cdot p \cdot (1 - e^{-\frac{t}{\tau_2}}) \cdot e^{-\frac{t}{\tau_2}} \tag{11}$$

Hence, the kinetics of the luminescence flash in centers C_1 is to be described by the expression:

$$I_{1\Pi} = k_{11} \cdot k_{21} \cdot N_1 \cdot p \cdot \tau_1 \cdot e^{-\frac{t}{\tau_1}} + k_{11} k_{22} k_{30} \cdot e^{-\frac{\Delta E_2}{kT}} \cdot N_2 \cdot p \cdot (\tau_1 \cdot e^{-\frac{t}{\tau_1}} - \tau_2 \cdot e^{-\frac{t}{\tau_2}}) \tag{12}$$

Figure 4 shows, as an example, the calculation results for the kinetic curve of the cathodoluminescence flash initiated by instantaneous excitation pulse at different ratios of the characteristic relaxation times of C_1 and C_2 centers.

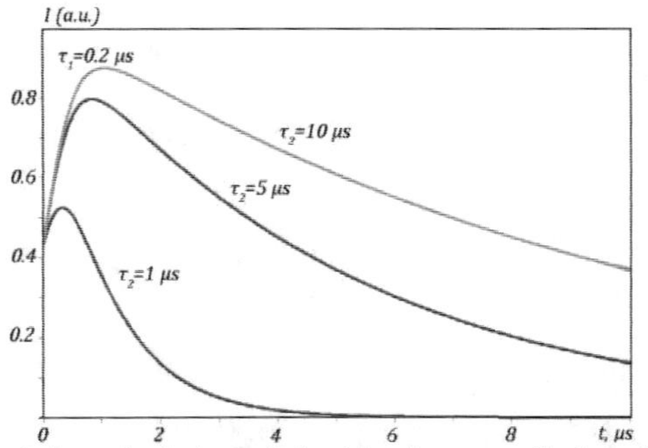

Figure 4. The results of the calculation of the kinetic curve flash cathodoluminescence initiated instantaneous excitation pulse at different ratios of the characteristic relaxation times

4. Conclusion

In the kinetics of the luminescence flash in $LiYF_4$: Nd crystals excited by nanosecond electron beam at room temperature, the buildup is observed at the initial stage up to 300 ns. The buildup of PCL can be attributed to the transfer of the excitation energy stored in the trapping centers to the luminescence centers. A model of the exchange of electron excitations between the luminescence and trapping centers is suggested. According to the model, the buildup stage becomes possible if a number of conditions have been fulfilled. The buildup is possible only for certain ratios of the characteristic times of excitation relaxation in competing centers, concentration of these centers, the sample temperature and excitation power. Typically, the excitation pulse energy is not instantaneous in relation to the time resolution of the measuring system. The temporal resolution of the device is to be adjusted to the chosen pulse duration to obtain the maximum relevant measurement. In case the pulse shape is known and takes the form of $p(t)$, substitution of the function $p(t)$ in (12) makes it possible to take into account the impact of the pulse shape on the luminescence flash kinetics. Moreover, it allows us to obtain information with a high temporal resolution by means of mathematical processing.

Acknowledgement
The research is financially supported by the Ministry of Education and Science of the Russian Federation, project “Science”, 3.1329.2014 (2152)

References

[1] LisitsynaL A, OleshkoV I, PutintsevaS N and LisitsynV M2008*Optics and Spectrosc.* **105** 531–537.
[2] LisitsynaL A, KorepanovV I, Abdrakhmetova A A, Timoshenko N N, Dauletbekova A K 2012*Optics and Spectrosc.* **112**175–181.
[3] Yakovlev V, Trefilova L, Meleshko A 2009*J. of Lum.* **129**790–796
[4] Chinkov E P, KrasnousovI V, LisitsynaL A, ReiterovV M1990*Izv. Latviyskoy AN***5** 60–64.
[5] LisitsynV M, Bikhert Y V, Lisitsyna L A, Dauletbekova A K, Reiterov V M, Karipbaev Z T 2014*Advanced Mater. Res.* **880**13–18.

Journal of Physics: Conference Series **552** (2014) 012041 doi:10.1088/1742-6596/552/1/012041

Substitutional impurities influence on hydrogen sorption properties of titanium and zirconium

T I Spiridonova[1], E V Tuch[2], A V Bakulin[2,3], S E Kulkova[2,3*]
[1] National Research Tomsk Polytechnic University, pr. Lenina 30, Tomsk, 634050, Russia
[2] Institute of Strength Physics and Materials Science, pr. Akademichesky 2/4, Tomsk, 634021, Russia,
[3] National Research Tomsk State University, pr. Lenina 36, Tomsk, 634050, Russia

E-mail: kulkova@ms.tsc.ru

Abstract. The influence of substitutional impurities on the hydrogen absorption energy in Ti and Zr is investigated by the projector augmented wave method within the density functional theory. Obtained results demonstrate that simple metal atoms are repulsive to H, whereas transition metal atoms attract H and provide traps for H. The relationship between the interaction energy, the lattice distortion and the electronic structure of metals is analyzed.

1. Introduction

Metals of group IVB such as titanium and zirconium are widely used for technological applications in aerospace, marine, chemical, etc. industries. Both metals have high corrosion resistance and good biocompatibility with human tissue that provides their widespread applications in medicine for dental and bone implants, various fixation devices and medical instruments. Besides zirconium and its alloys are very important construction materials for nuclear reactors. The hydrogen is the main factor of degradation of metals and their alloys under working condition. It is known that hydrogen does not dissolve appreciable in metals at room temperature and normal pressure whereas their hydrides can be easily formed. The appearance of hydride phase results in brittle behavior of the metal. This phenomenon is called as hydrogen embrittlement [1]. The increase of temperature leads to better dissolution of hydrogen in the metal matrix and therefore, in particular, titanium exhibits a ductile behavior. It is believed that the alloying allows improving the hydrogen embrittlement and restrain the hydride phase formation. It was reported in [2-4] that substitutional impurities can influence on the terminal solubility of hydrogen in metals due to a trapping of hydrogen. The detailed investigation of influence of simple and transition metals on the solubility of hydrogen in titanium was studied in [5] using the first principles calculations. It was shown that the different mechanisms can be responsible for hydrogen interaction with alloying metals. In particular, the repulsive interaction between simple metals (SM) and hydrogen is due to the hybridization of their orbitals whereas the interaction of H with transition metals (TM) is attributed to the atomic size effect [5]. However, there are contradictory results about preferential position of hydrogen even in pure α-Ti [5, 6]. Thus, the goal of present paper is to study the preferential position for H absorption in both Ti and Zr as well as the influence of substitutional impurities on the hydrogen absorption properties.

2. Calculation method

Content from this work may be used under the terms of the Creative Commons Attribution 3.0 licence. Any further distribution of this work must maintain attribution to the author(s) and the title of the work, journal citation and DOI.
Published under licence by IOP Publishing Ltd

The calculations of the electronic structure of IV group metals in the hexagonal phase was carried out by projector augmented wave method [7] implemented in VASP code [8] within the generalized gradient approximation for the exchange-correlation functional. The model of the supercell constructed from a 2×2×1 replication of the conventional hexagonal unite cell, containing 8 atoms of metal, are shown in Fig. 1. The optimization of supercell was performed by keeping its shape, i.e. the supercell was relaxed by changing the lattice constant homogeneously with a ratio of c/a fixed at the experimental values 1.588 and 1.593 for Ti and Zr, respectively. The octahedral (O) and tetrahedral (T) sites for hydrogen atom, shown in figure 1, were considered. One of metal atom which is nearest to H site, was replaced by a metal impurity atom as shown in figure 1. In this case an additional optimization of the atomic positions (after volume optimization) until the forces at each atom ~0.01 eV/Å, was performed. The total energies were converged up to 10^{-5} eV. The Γ-centered k-points grid 7×7×7 was applied for the integration over the Brillouin zone. The energy cut-off for plane waves was set as 400 eV.

The binding (E_b) and absorption (E_{abs}) energies were calculated using the following equation:

$$E_b = E(N-1, X, \mathrm{H}) - E(N-1, X) - E(\mathrm{H}) \quad (1)$$

$$E_{abs} = E(N-1, X, \mathrm{H}) - E(N-1, X) - 1/2E(\mathrm{H}_2), \quad (2)$$

where $E(N-1, X)$ and $E(N-1, X, \mathrm{H})$ are total energies of system with alloying metal (X) without and with hydrogen (H), $E(\mathrm{H})$ and $E(\mathrm{H}_2)$ is the total energies of hydrogen atom and molecule, respectively. The factor 2 takes into account the presence of one H atom inside supercell.

The interaction energy between impurity atom and H was calculated as

$$\Delta E = [E(N-1, X, \mathrm{H}) + E(N)] - [E(N-1, X) + E(N, \mathrm{H})] \quad (3)$$

where $E(N)$, $E(N, \mathrm{H})$ are the total energies of supercell with N sites and that with hydrogen.

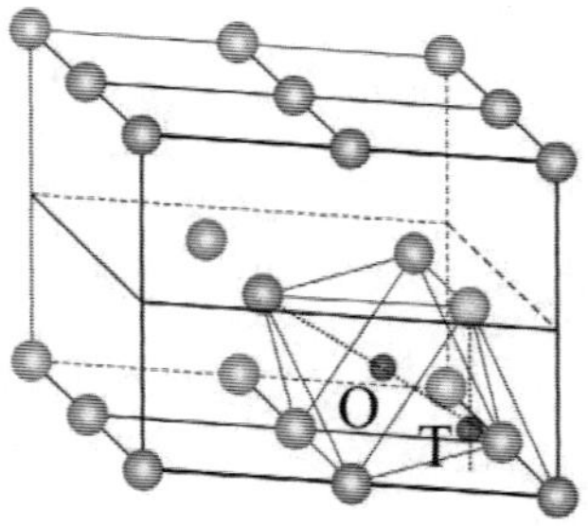

Figure 1. Atomic structure of IVB group metal supercell 2×2×1 with hydrogen at the octahedral and tetrahedral sites (small red balls) and impurity metal (green ball) instead one host atom.

3. Results and discussion

The calculated binding and absorption energies of hydrogen at both O and T sites are given in table 1. It is necessary to note that greater absolute value of the binding or absorption energy means better bonding or dissolving hydrogen in the metal matrix. As can be seen from table 1 the hydrogen prefers to occupy the octahedral site in the hexagonal lattice in case of both considered metals. The difference between H energies at the O and T sites decreases if the relaxation of the atomic positions due to H incorporation in metal matrix was taken into account. The obtained difference in the total energies of Ti with H at the O and T is only 0.16 eV in our calculations that is one order less than the difference (E_{tot}(T)–E_{tot}(O)) in [5] where the preference of the T site for hydrogen absorption was found within linear muffin-tin orbital method (LMTO-ASA).

Figure 2 presents the change of the hydrogen absorption energy at both O and T sites if one of Ti atom nearest to the hydrogen position was replaced by simple or 3d transition metals. It is seen that substitutional impurities do not influence on the obtained trend in Ti whereas some impurities such as Sc, V, Cr, Ni can change the preference of considered positions for H absorption in Zr. Figure 2 demonstrates that the transition metal impurities of the middle of 3*d*-period lead to increase of the hydrogen dissolution in both Ti and Zr matrix.

Table 1. Binding and absorption energies for H at both O ad T sites calculated in 2×2×1 supercell keeping its shape with volume change and with relaxed atomic positions (in brackets).

Metal	E_b (eV)		E_{abs} (eV)	
	Octahedral site	Tetrahedral site	Octahedral site	Tetrahedral site
Ti	−2.66 (−2.71)	−2.42 (−2.55)	−0.38 (−0.44)	−0.15 (−0.28)
Zr	−2.64 (−2.68)	−2.59 (−2.66)	−0.37 (−0.41)	−0.32 (−0.40)

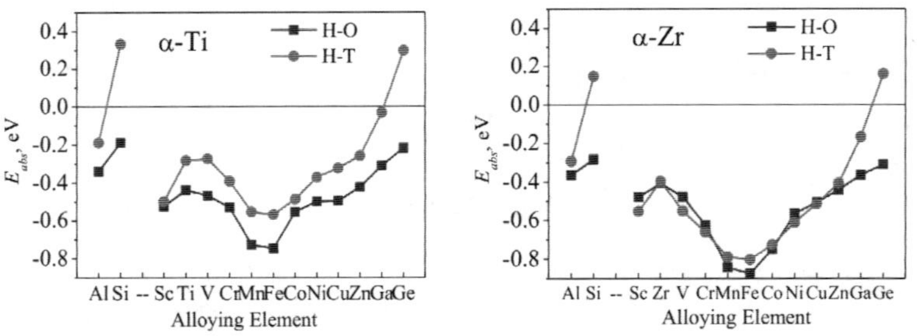

Figure 2. The absorption energies of H at both O and T sites in dependence on substitutional impurity.

The interaction energy between alloying element and hydrogen occupying O and T sites are given in figure 3. Our calculations show that for simple metals ΔE is positive irrespective of the H site in the metal matrix. The positive values of the interaction energy suppose that considered SM (Al, Si, Ga, Ge) are repulsive to the hydrogen atom. It is seen from figure 3 that ΔE(T) is larger than ΔE(O) because the distance between H and SM is shorter when H occupies the T site than the O-site. In general, hydrogen cannot be trapped at both O and T sites with neighboring simple metals in Ti and Zr as well. However, most transition metals have negative ΔE that means attractive interaction between TM and H at both interstitials whereas the attractive interaction only at the T-site was found in [5]. Consequently, both interstitials having TM as neighbor are trapping sites for H. It is known that the trapping of hydrogen by substitutional impurities was discussed in [2-4]. In particular, the hydrogen trapping by impurities was attributed to size difference between the substitutional impurity and host atoms. It was supposed in [4] that the dissolution of hydrogen reduces the strain in metal matrix due to undersized impurity atom. In order to this effect was more pronounced the H atom should be closer to the impurity atom that suggests the trapping of H. The oversized impurity atom may expand the interstitial sites that makes them more preferential for H occupancy and leads also to trapping of H as was suggested in [2, 3]. It is necessary to note that in both models only structural factors were taken into account.

The figures 3c,d show the change of volume of both Ti and Zr caused by substitutional impurities and hydrogen. In our calculations H expands the equilibrium volume of Ti (Zr) lattice by 1.46-3.32%, (0.96-2.50%), whereas TM metal with atomic radius less than that of host metal reduces this volume as shown in figure 3c,d. The change of volume is slightly larger in case of H at the tetrahedral site than that for H at the O site that can be indication of the preference of the octahedral site for H absorption. In the presence of H the lattice contraction is reduced due to undersized impurity atom. As can be seen from figure 3c,d the volume change due to TM demonstrates a parabolic behavior as in [5] and reaches a maximum change at Fe, Co. These results coincide well to the interaction energies presented in

Journal of Physics: Conference Series **552** (2014) 012041 doi:10.1088/1742-6596/552/1/012041

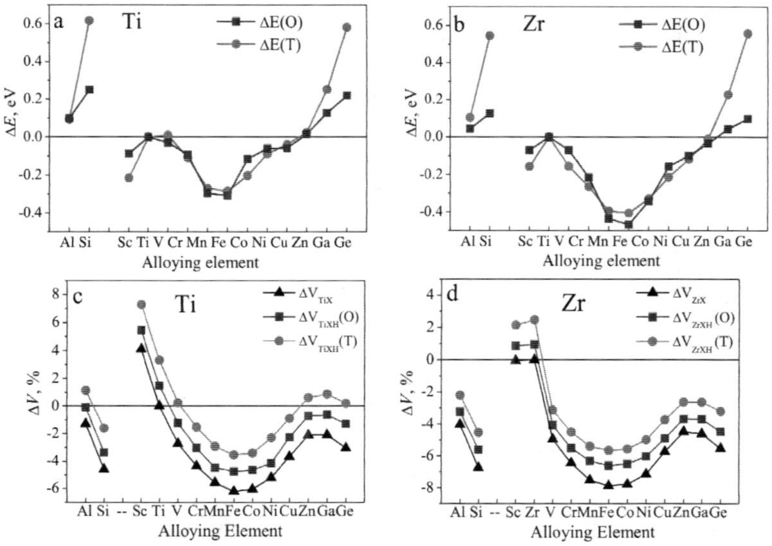

Figure 3. Interaction energy between hydrogen and alloying element (a,b) and the change of equilibrium volume relative to that of pure Ti (c) or Zr (d).

figure 3a,b. Among considered transition metals only Sc has the radius larger that Ti but almost the same as Zr. So, the presence of Sc in Ti matrix increases its volume as seen in figure 3c. Obtained results reveal the attractive H-Sc interaction at both H sites inside metal matrix whereas the repulsive behavior was found in [5] for H at the tetrahedral site. The similar behavior was obtained in our calculation without relaxation of atomic positions around Sc. Thus, it is very important in theoretical calculations to take into account the local strain and the lattice distortion due to substitutional impurity and H absorption. It is necessary to emphasize that the lattice distortion can depend strongly on the concentration of alloying metal and hydrogen. In general, our calculations of the interaction energies between TM and H support also as in [5] the strain field model suggested in [4].

Several words should be said about the electronic structure of system in the presence of H and substitutional impurities. Figure 4 presents the total densities of states (DOS) of Ti with impurity atom (V, Cu) and the system with both V(Cu) and H at the octahedral site. It can be seen that low-lying states are appeared at –7 eV in the DOS curve for system with hydrogen. These states are formed due to interaction of H *s*-state with *s*,*d*-states of host and impurity atoms. These hybridized states represent the covalent bonding between H and TM. It is seen in figure 4 that the total DOS changes only slightly due to the presence of H in Ti-X-H system. The same can be revealed from the local DOS curves of impurity atom and H comparing them in Ti-X-H, Ti-H and Ti-X systems. In this context we support the conclusion of Ref. [5] that the electronic structure may be not the main factor for understanding the microscopic nature of H trapping in metals doped by TM. It should be noted that mentioned above trends are valid for Zr also.

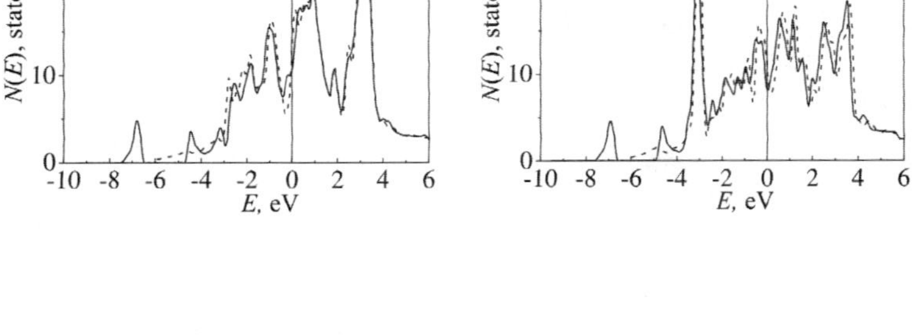

Figure 4. Total densities of states of Ti-X (dashed lines) and Ti-X-O (solid lines) system for H in the octahedral site.

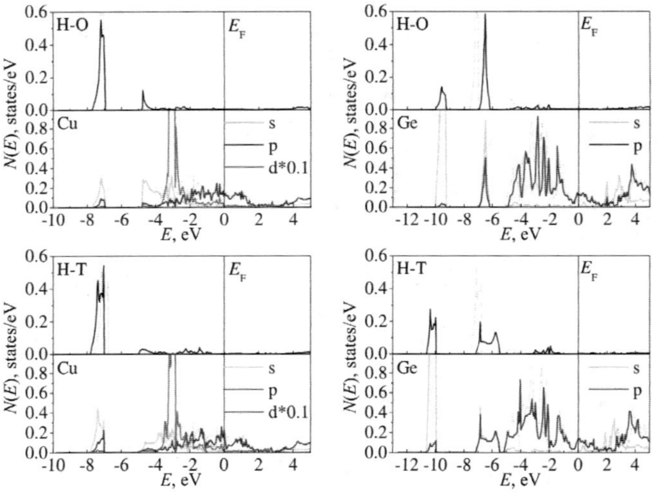

Figure 5. Local DOS of H in the O (top) and T (bottom) sites and Cu (left) and Ge (right) in Ti-X-H systems. DOS of H in Ti-H and that of TM or SM in Ti-X are given by dots.

In case of simple metals in Ti or Zr matrix there are low-lying states also. Moreover they are common feature of simple metal and also present in Ti-X system due to hybridization of SM *s,p*-states with *s,d*-states of Ti (Zr). In system with hydrogen (Ti-X-H) these low lying states are split into two bands. It is clearly can be seen in the local DOS of H (figure 5a,b) that there is bonding (at –10 eV) and antibonding (at –6.5 eV) states with gap between them in case of H at the O and T-sites. In spite of the hybridization between SM and H states contributes to the binding energy, however the contribution to kinetic energy due to the Pauli principle, is larger that leads to a positive ΔE and to repulsive interaction between the simple metal and hydrogen in Ti-X-H system.

4. Conclusion

The first principles investigation of the influence of substitutional impurities on the H absorption in α-Ti and α-Zr was performed. The relationship between impurity trapping of hydrogen and lattice distortion as well as electronic structure was analyzed. It was shown that H prefers to occupy the octahedral position in both metals, however the difference in absorption energies for O and T sites in Zr is lower than that in Ti. The substitutional impurities do not influence on the obtained trend in Ti whereas the T site may be more preferred for H absorption in Zr in case of some metals such as Sc, V, Cr and Ni. The calculations of the interaction energy show that simple metals are repulsive to H whereas transition metals are attractive to H and may provide its trapping. Different factors are responsible for the SM-H and TM-H interactions: the electronic factor may be a main reason of the repulsive SM-H interaction while the size effect can be responsible for TM-H interaction.

References

[1] Bernstein I M and Thompson A W 1974 *Hydrogen in Metals* (OH: American Society for Metals).
[2] Westlake D G and Miller J F 1979 *J. Less-Common Met.* **65** 139.
[3] Miller J F and Westlake D G 1980 *Trans. Jpn. Inst. Met. (Sappl.)* **21** 153.
[4] Matsumoto T 1977 *J. Phys. Soc. Jpn.* **42** 1583.
[5] Hu Q M, Xu D S, Yang R, Li D and Wu W T 2002 *Phys. Rev. B* **66** 064201.
[6] Han X L, Wang Q, Sun D L, Sun T and Guo Q 2009 *Int. J. Hydrogen Energy* **34** 3983.
[7] Blöchl P E 1994 *Phys. Rev. B* **50** 17953.
[8] Kresse G P, Hafner J 1994 *Phys. Rev. B* **49** 14251.

The role of deep centers in formation of luminescent and dosimetric properties of wide-gap materials

S V Nikiforov*, V S Kortov
Ural Federal University, Ekaterinburg, Russia

E-mail: s.v.nikiforov@urfu.ru

Abstract. The direct and indirect methods of experimental detection of deep traps in wide-gap insulators are described. The experimentally observed effects of influence of deep traps with different nature on luminescent and dosimetric properties of materials are analyzed. It is established that the most wide-spread and well-studied effects are the sensitization and superlinearity of dose response. They are interpreted in terms of the kinetic model of competitive electron traps. Taking into account the temperature dependence of capture probability by deep traps in this model allows one to explain some new effects associated with luminescence thermal quenching. The luminescence model of Al_2O_3:C single crystal is described. In this model the temperature dependence of competitive interaction between the main and deep traps is caused by thermal ionization of excited states of F-centers.

1. Introduction

The mechanisms and features of thermoluminescence (TL) in different materials are studied in terms of necessity of solution of fundamental and applied tasks associated with its application in radiation dosimetry, geology, archaeology and temperature sensing. In a simple TL model only two localized levels are taken into account: an electron trap and recombination center. The development of this model is an account of the competing effects of deep trapping centers. The deep centers are traps with energy depth more than in the main dosimetric ones which remain thermally stable and not emptied during usual TL registration. The available literature information about the role of deep centers in the TL mechanism in different materials has an odd and non-systemized character. In this regard, the aim of this work is to reveal general regularities and mechanisms of effect of deep traps on luminescent and dosimetric properties of wide-gap materials on the base of the analysis of available experimental data and kinetic models of TL.

2. Experimental methods of deep traps detection

Experimental difficulties in deep traps detection by direct measurement of high-temperature TL peaks are due to the presence of significant thermal radiation of the heating element at T>400 °C. In some cases it is possible to observe the TL peaks associated with deep traps after high-dose irradiation of the material. The high temperature TL peaks were registered after excitation of LiF:Mg,Ti; LiF:Mg,Cu,P and Al_2O_3:C by different types of ionizing radiation [1-3]. In Al_2O_3:C intensive TL peaks of deep traps at T=300-600 °C are also observed after excitation by unfiltered UV-radiation of a mercury lamp at high temperatures [4]. In this case the traps are filled by optical ionization of F-centers. The

Content from this work may be used under the terms of the Creative Commons Attribution 3.0 licence. Any further distribution of this work must maintain attribution to the author(s) and the title of the work, journal citation and DOI.
Published under licence by IOP Publishing Ltd

probability of this process increases with the temperature growth. The concentration of F-centers in Al_2O_3:C is high (10^{17} cm^{-3}), resulting in high efficiency of deep traps filling in this way.

The indirect methods of deep traps detection include the observation of the phototransferred thermoluminescence (PTTL) peaks. For registration of a PTTL signal a sample is pre-irradiated by ionizing radiation which fills the deep traps. After thermal empting of the main traps the optical transfer of charge carriers from deep traps to shallow ones occurs. The main shallow traps give PTTL peaks after further thermal stimulation. PTTL is experimentally observed in many materials: SiO_2; LiF:Mg,Ti; CaF_2:Mn; Al_2O_3:C [5-8]. A simple kinetic PTTL model includes two electron traps (main and deep) and a recombination center. The development of these concepts is a model which takes into account optical bleaching of the main traps TL occurring simultaneously with the empting of deep centers [5]. The model has allowed one to explain non-monotonic dependence of the PTTL intensity versus photostimulation time observed experimentally, in particular, in Al_2O_3:C and SiO_2 [5,9]. There are other indirect methods to prove the existence of deep trapping centers in dosimetric materials based on the detection of the effects of sensitization and superlinearity of dose response. Experimental manifestations and mechanisms of these effects will be discussed below.

3. Models of competing interaction of trapping centers involving deep traps in wide-gap insulators

3.1. The model of competitive electron traps

The model contains the main TL-active trap, deep trap and recombination center. Excitation of phosphor leads to the formation of the electron-hole pairs. Resulting free charge carriers can be captured at localized levels. During the subsequent thermal stimulation the captured electrons are released from active traps in the conduction band. Then free electrons can recombine with holes at the luminescent centre accompanied by generation of photons. Another possibility is to capture electrons by a deep trap. The situation when charge carriers released into the conduction band from one center can be captured by another trap is considered to be the interaction of these traps. This is the so-called interactive kinetics of TL [10].

The most important practical application of the model considered is the explanation of the effect of superlinearity of dose response. The term "superlinearity" means faster than linear dependence on the dose. The effect of superlinearity is observed in different dosimetric materials [11-13] and can occur in the whole range of registered doses, and in some parts of dose dependences. In theoretical works [14-16] the analyzed model is considered in two stages of TL signal formation: excitation and heating. It is shown that the gradual filling of competitive deep traps with increasing doses reduces the probability of capturing electrons and increases the number of recombination acts at the luminescent centre that causes superlinearity. Other authors considered two types of active traps in terms of the model of competing traps. They are interactive (interacting with the deep traps) and non-interactive [17]. This modified model was used to calculate superlinearity of TL dose response of peak 5 in LiF:Mg,Ti [18].

The effect of increasing the sensitivity of phosphor to radiation (sensitization effect) is associated with the superlinearity of dose response. The sensitivity is TL response to low test dose. Sensitization in repeated irradiation-heating cycles is observed, in particular, for the TL peak at 110 °C in quartz. It can be described in terms of the model of competing electron traps [19]. As a rule, the effect of sensitization is observed in the same dose range as the superlinearity and characterized by a similar mechanism associated with competing capture of charge carriers by deep traps.

We modified the model under study by consideration of the temperature dependence of the probability of capturing by deep traps, which was observed experimentally for Al_2O_3:C single crystals [4]. Alongside with the phenomena of superlinearity and sensitization, this model has allowed us to explain new experimentally observed effects [4,13], in particular, the influence of the heating rate on the degree of dose responses superlinearity (figure 1).

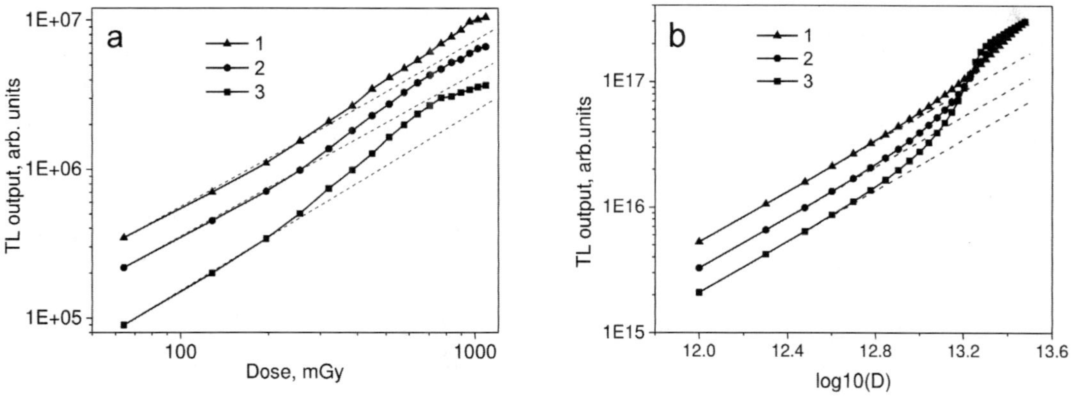

Figure 1. Experimental (a) [13] and calculated (b) [20] dose dependences of the main TL peak in Al_2O_3:C at heating rates of 0.5 (1), 2(2) and 6 K/s

3.2. The models taking into account the deep hole traps

The kinetic model is analyzed [21] which takes into account the competition between the recombination centers, which may be regarded as deep hole traps. It is shown that the competition between the hole centers at a excitation stage gives rise to the effect of dose response superlinearity. In addition, this model explains the dose-rate dependence of the TL intensity and anomalous fading [22,23].

A more complex model which includes two electron and two hole centers is described in literature [24]. This model explains the appearance of non-monotonic dose dependence, experimentally observed in some phosphors (LiF:Mg,Ti; Al_2O_3:C; natural quartz) [25-27]. In this case, the TL output increases with a growing dose, reaching the maximum, and then decreases. An important conclusion made by the authors is that non-monotonic dose dependence in the range of high doses can be caused not only by radiation destruction of active centers but their competitive interaction.

Deep traps capable of capturing the charge carriers with a different sign, are identified in Al_2O_3:C single crystals. In particular, the sensitization/desensitization effects after annealing of the samples preliminary irradiated by UV-radiation [28] and a pulsed electron beam [3] are analyzed. It was found that at temperature ranges of 350-500 °C and 650-750 °C mainly electron traps are emptied. The hole traps are active at T=500-650 °C and T>700 °C. The obtained results agree in the main with those of other authors [26,29] who used other ways of deep centers filling.

According to the nature of the deep traps in Al_2O_3:C kinetic models are proposed which take into account the competing capture of electrons and their non-radiative recombination with holes on the electron and hole deep traps respectively. In literature [30,31] a model which includes one electron and one hole deep trap, is analyzed at the stage of UV and ionizing radiation excitation, as well as thermal stimulation. It is found that this model provides a good quantitative description of the dose dependence of the TL output in the main peak and concentration of luminescent centers. In our paper [28] it is shown that the additional account of the temperature dependence of the probability of capturing by deep electron traps allows one to describe not only the effects of changes in sensitivity, but also the influence of the occupancy of deep traps of different nature on the dependence of TL output in dosimetric peak on a heating rate (figure 2). A significant role of deep traps in heating-rate effect in Al_2O_3 samples with oxygen vacancies created not only by thermal but also radiative coloration, as well as for nanostructured modification was further established [32,33]. In our paper [34] the model was supplemented by taking into account the hole traps contributing to the broadening of the main TL peak at 180 °C. In terms of the modified model the sensitization/desensitization effects were analyzed in Al_2O_3:C samples with different half-width of the main TL peak.

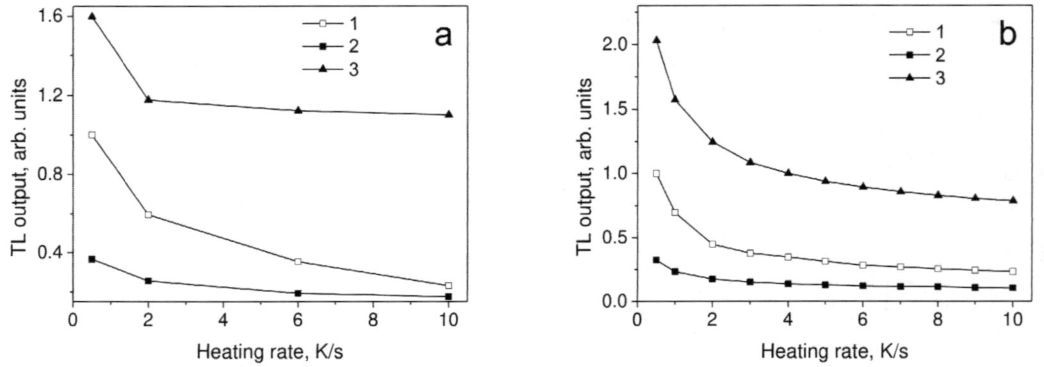

Figure 2. Experimental (a) and calculated (b) dependence of the TL output in the main peak on the heating rate for different states of deep traps [28]: 1 - deep traps are emptied; 2 – deep hole traps are partially occupied; 3 - deep electron traps are partially occupied

3.3. A model of thermal quenching of luminescence in Al_2O_3:C single crystals taking into account thermal ionization of F-centers

The disadvantage of previously considered kinetic models describing the TL mechanism in wide-gap materials is lack of consideration of the energy structures of the particular defects (traps and luminescent centers). We proposed the model [35] which describes the mechanism of luminescence in Al_2O_3:C single crystals. This model, along with competing capture of charge carriers by deep traps, takes into account the thermal ionization of excited states of luminescent centers (F-centers) (figure 3).

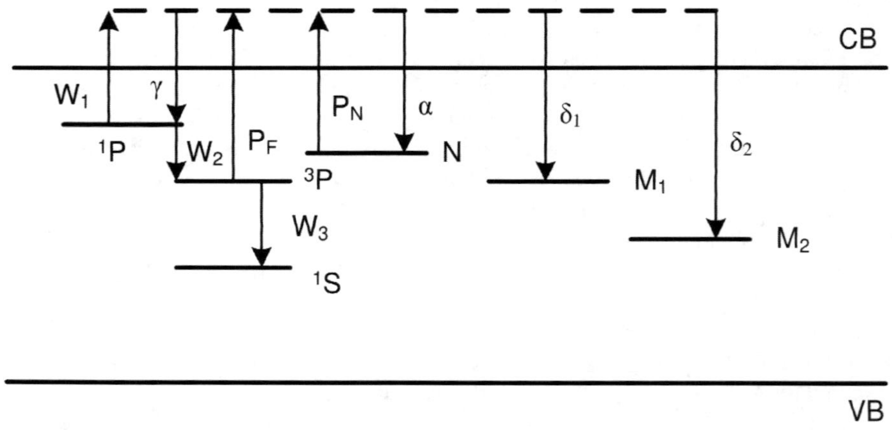

Figure 3. The energy band diagram of the luminescence model in Al_2O_3:C taking into account the thermal ionization of F-centers [35]

In this figure N is the main (dosimetric) trap, M_1, M_2 are deep electron traps, ^1S - ground state, ^1P, ^3P - excited levels of F-center. The ^1P level is near the bottom of the conduction band, so the probability of W_1 transition is assumed to be independent on temperature. Thermal ionization of the excited state (^3P) corresponds to the P_F transition, probability of which increases exponentially during heating of the sample. The ionization of F-center results in formation of F^+-center which after electron capture (transition γ) could once again become an excited F-center. Luminescence of F-center (3.0 eV) corresponds to the W_3 transition. During the excitation by ionizing radiation free charge carriers are generated which are captured by the traps. During the subsequent thermal stimulation the electrons released from trap N are captured by F^+-center, forming an excited F-center. Then free electrons formed by ionization of excited (^3P) state of F-center may be captured by the deep traps. Thus, in this model the main and deep trapping centers are considered in an interactive relationship.

The simulation results show [35] that the process of thermal ionization of excited states of F-centers leads to a decrease in the share of radiative transitions inside the centre and the increase in the probability of electron capturing by deep traps. These effects cause thermal quenching of photo - and radioluminescence in the crystals under study. In addition, the shape of the calculated quenching curves depends on the degree of deep traps filling (figure 4), which was observed experimentally [36]. The proposed model also explains the described earlier regularities of influence of deep electron traps occupancy on TL sensitivity and the degree of its dependence on the heating rate, as well as on the shape of the main TL peak [35]. Later this model and its modifications were used to analyze the effects of optically stimulated luminescence and thermally stimulated conductivity in Al_2O_3:C [37].

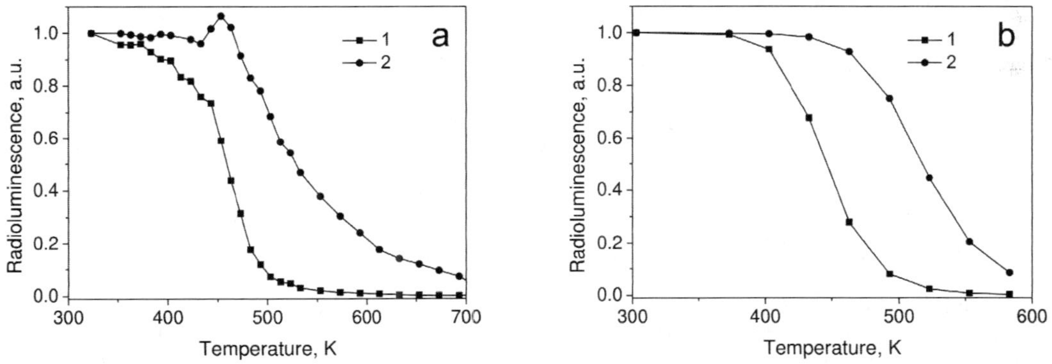

Figure 4. Experimental (a) and calculated (b) thermal quenching curves of F-centers radioluminescence in Al_2O_3:C for two states of deep traps occupancy [35]: 1 - deep traps are emptied; 2 – deep electron traps are partially occupied

4. Conclusions

Deep trapping centers in wide-gap insulators can be experimentally detected by the method of direct observation of the high-temperature TL peaks as well as by a number of indirect methods. Competing influence of deep electron traps causes the effects of sensitization and superlinearity of TL dose response. Taking into account the effects of hole centres allows one to explain a number of new effects, such as non-monotonic behaviour of the TL output with increasing dose, its dose-rate dependence, and anomalous fading. The nature of deep traps which cause sensitization/desensitization effects in Al_2O_3:C single crystals was analyzed. Consideration of the temperature dependence of the capture probability by deep traps allows one to explain new effects in this material. These effects are associated with the influence of deep centers occupancy on luminescence quenching, the dependence of TL output and degree of superlinearity of dose response on heating rate. The model describing the luminescence mechanism in Al_2O_3:C with regard to the process of thermal ionisation of excited states of F-centers is proposed.

References

[1] Bilski P, Obryk B and Stuglik Z 2010 *Radiat. Meas.* **45** 576
[2] Lo D, Lawless J L and Chen R 2006 *Radiat. Prot. Dosim.* **119** 71
[3] Nikiforov S V, Kortov V S, Zvonarev S V, Moiseykin E V and Kazantseva M G 2014 *Radiat. Meas. (in press)* http://dx.doi.org/10.1016/j.radmeas.2014.02.010.
[4] Milman I I, Kortov V S and Nikiforov S V 1998 *Radiat. Meas.* **29** 401
[5] Alexander C S, Morris M F and McKeever S W S 1997 *Radiat. Meas.* **27** 153
[6] Sunta C M and Watanabe S J 1976 *J. Phys. D: Appl. Phys.* **9** 1271
[7] Allen P and McKeever S W S 1990 *Radiat. Prot. Dosim.* **33** 19
[8] Kortov V S, Milman I I, Moiseykin E V, Nikiforov S V and Ovchinnikov M M 2006 *Radiat. Prot. Dosim.* **119** 41
[9] Walker F D, Colyott L E, Agersnap Larsen N and McKeever S W S 1996 *Radiat. Meas.* **26** 711
[10] McKeever S W S and Chen R 1997 *Radiat. Meas.* **27** 625
[11] Zimmerman J 1971 *J.Phys C:Solid State Phys.* **4** 3265
[12] Yang X B, Xu J, Li H J, Bi Q Y, Cheng Y and Tang Q 2009 *J. Appl. Phys.* **106** 033105
[13] Kortov V S, Milman I I, Nikiforov S V, Moiseykin E V and Kondrashov S V 2007 *Radiat. Meas.* **42** 590
[14] Bowman S. G E and Chen R 1979 *J. Lumin.* **18/19** 345.
[15] Kristianpoller N, Chen R and Israeli M 1974 *J. Phys. D: Appl. Phys.* **7** 1063
[16] Chen R and Fogel C M 1993 *Radiat. Prot. Dosim.* **47** 23
[17] Sunta C M, Yoshimura E M and Okuno E 1994 *J. Phys. D: Appl. Phys.* **27** 852
[18] Sunta C M, Okuno E, Lima J F and Yoshimura E M 1994 *J. Phys. D: Appl. Phys.* **27** 2636
[19] Chen R, Fogel G and Kristianpoller N 1994 *Radiat. Meas.* **23** 277
[20] Nikiforov S V, Kortov V S and Kazantseva M G 2014 *Phys. of the Solid State* **56** 554
[21] Chen R, Fogel G and Lee C K 1996 *Radiat. Prot. Dosim.* **65** 63
[22] Chen R and Leung P L 2000 *J. Phys. D: Appl. Phys.* **33** 846
[23] Chen R and Hag-Yahya A 1997 *Radiat. Meas.* **27** 205
[24] Chen R, Lo D and Lawless J L 2006 *Radiat. Prot. Dosim.* **119** 33
[25] Horowitz Y S, Moscovitch M and Wilt M 1986 *Nucl. Instrum. Methods in Phys. Res. A* **244** 556
[26] Yukihara E G, Whitley V H, Polf J C, Klein D M, McKeever S W S, Akselrod A E and Akselrod M S 2003 *Radiat. Meas.* **37** 627
[27] Ichikawa Y 1968 *Jpn. J. Appl Phys.* **7** 220
[28] Nikiforov S V and Kortov V S 2010 *Radiat. Meas.* **45** 527
[29] Akselrod M S and Gorelova E A 1993 *Nucl. Tracks Radiat. Meas.* **21** 143
[30] Pagonis V, Chen R and Lawless J L 2007 *Radiat. Meas.* **42** 198
[31] Pagonis V, Chen R and Lawless J L 2008 *Radiat. Meas.* **43** 175
[32] Kortov V S, Nikiforov S V, Milman I I and Moiseykin E V 2004 *Radiat. Meas.* **38** 451
[33] Kortov V S, Nikiforov S V, Moiseykin E V, Vokhmintsev A S and Simanov A G 2013 *Phys. of the Solid State* **55** 2088
[34] Nikiforov S V, Kortov V S, Nosal A A and Moiseykin E V 2011 *Phys. of the Solid State* **53** 2141
[35] Nikiforov S V, Milman I I and Kortov V S 2001 *Radiat. Meas.* **33** 547
[36] Kortov V S, Milman I I and Nikiforov S V 1999 *Radiat. Prot. Dosim.* **84** 35
[37] Pagonis V, Chen R, Maddrey J W and Sapp B 2011 *J. Luminescence* **131** 1086

Photoluminescence of nanosized Zn2SiO4:Mn depending upon preparation method

K A Petrovykh[1, 2], V S Kortov[1], A A Rempel[1, 2]
[1]Ural Federal University named after the first President of Russia B.N. Yeltsin, 19 Mira Str., Ekaterinburg, 620002, Russia
[2]Institute of Solid State Chemistry, Ural Branch of the Russian Academy of Sciences, 91 Pervomaiskaya Str., Ekaterinburg, 620990, Russia

E-mail: kspetrovyh@mail.ru

Abstract. Nanosized Zn_2SiO_4:Mn powders were prepared by two different methods: a high-energy ball-milling of microcrystalline powder (so-called "top-down") and a sol-gel method ("bottom-up"). It was shown that it is possible to obtain particles of 30±10 nm by means of the ball-milling. A particle size of the Zn_2SiO_4:Mn synthesized by the sol-gel method ranged from 20 to 110 nm. It was found all samples exhibit photoluminescence (PL) in the green spectral region with a maximum emission wavelength from 515 to 520 nm. A nanopowder obtained by the ball-milling showed a significant decrease of the PL intensity comparing with bulk material. The PL intensity of the samples prepared by sol-gel method is much higher than that of ball-milled Zn_2SiO_4:Mn.

1. Introduction

Ability to manage the physical, mechanical, optical and other properties by changing the linear particle size causes a great interest in nanomaterials [1]. Manganese Mn^{2+} activated zinc orthosilicate, Zn_2SiO_4:Mn, is a well-known phosphor, which exhibits an intense green luminescence under UV light or electron beam. Decreasing of the particles to nanosize improves such important characteristics as the luminescence quantum efficiency, radiation resistance and adhesion to the substrate [2, 3]. Therefore, nanosized Zn_2SiO_4:Mn phosphor is perspective for a modern plasma and field emission panels, lights and other devices.

Currently, a lot of methods allowed to obtain nanomaterials on the "top-down" and "bottom-up" principles are described in literature. The first group includes methods wherein nanoparticles are obtained by milling of powder with an initial particle size of a few micrometers. The second group is nanoparticles formed from atoms and molecules during chemical reaction [4]. Each of the existing techniques has its own advantages and features. Thus, an important task is to identify the optimal method provided Zn_2SiO_4:Mn with the certain particle size and morphology and optical properties. In this work two methods of synthesis of nanosized Zn_2SiO_4:Mn were selected: high-energy ball-milling ("top-down") and sol-gel technology ("bottom-up").

The disintegration of the material by means of a high-energy ball-milling in mills of various designs is one of the simplest and most effective methods of the nanoscale achieving. This way allows obtaining the nanopowder with particles having a minimal size of about 20 nm from coarse-grained material [5, 6]. Since the process involves a high energy, it is important to study the influence of milling duration on the fluorescent properties of the material. In turn, the sol-gel technology makes

Content from this work may be used under the terms of the Creative Commons Attribution 3.0 licence. Any further distribution of this work must maintain attribution to the author(s) and the title of the work, journal citation and DOI.
Published under licence by IOP Publishing Ltd

it possible to obtain multicomponent materials homogeneous on a molecular level and change the activator concentration in a wide range [7, 8].

Accordingly with the above, the aim of the present work was to reveal and study the PL features of the nanosized Zn_2SiO_4:Mn obtained by the two fundamentally different methods of nanotechnology.

2. Samples and experimental techniques

The Zn_2SiO_4:Mn microcrystalline powder with the activator concentration of 1.1 at. % and an average particle size of 3 μm was taken for disintegration. The ball-milling of the coarse-grained powder within 15, 30, 60, 120 and 240 min was carried out in a planetary ball mill *Retsch PM* 200. Isopropyl alcohol was used as a milling liquid. At the end of the process it was removed by drying. In addition, the nanopowders obtained by milling longer than 60 min were annealed in air at 300 °C for two hours to relax the residual stresses and partially remove the adsorbents. Other ball-milling parameters were described in details in [9].

Zinc and manganese chlorides and tetraethoxysilane were used as precursors for preparation of Zn_2SiO_4:Mn sol. The activator Mn^{2+} concentration was 1.0 at. %. Gelation occurred at room temperature, after that the gel was dried to remove the liquid phase. Since the resulting material does not possess an ordered crystalline structure, powder subjected to annealing to obtain a crystalline modification of α-Zn_2SiO_4. Annealing was occurred in air at temperature range from 500 to 1200 °C in steps of 100 °C and hold for two hours at each temperature.

The Zn_2SiO_4:Mn nanopowders obtained by methods described above have been characterized by the crystal structure, phase composition and particle size. X-ray diffraction studies of coarse-grained and disintegrated Zn_2SiO_4:Mn revealed that the process duration does not affect the crystal structure of the samples. Significant broadening of the diffraction peaks, indicating a particle size decrease, is observed after 15 min of milling. It should be noted that impurity phases have not been detected in the samples. Crystallinity and phase composition of the Zn_2SiO_4:Mn obtained by sol-gel method, on the contrary, essentially depend on the annealing temperature. As expected an increasing of the crystallinity degree of the material, gradually decreasing of the amorphous phase amount, and increasing of α-Zn_2SiO_4 content were observed with increasing of annealing temperature. The impurity phases of ZnO and SiO_2 were detected in the samples (Table 1). The average particle size of the material was determined by the broadening of the diffraction peaks by Williamson-Hall method. As seen from table 1, the amount of nanoparticles in the disintegrated powder increased significantly with increasing of the process duration. The minimal particle size was 30 ± 10 nm. In Zn_2SiO_4:Mn obtained by sol-gel method a gradual coarsening of the particles occurs with increasing of annealing temperature.

Table 1. Particle size and crystallinity of Zn_2SiO_4:Mn produced by ball-milling and sol-gel methods

Ball-milling		
Milling time, *t* (min)	Particle size, *D* (nm)	Relative volume fraction of nanoparticles, V_{nano} %
15	70 ± 20	40
60	50 ± 10	80
120	30 ± 10	100
240	30 ± 10	100
Sol-gel		
Annealing temperature, *T* (°C)	Particle size, *D* (nm)	Relative volume fraction of crystalline phases, V_{cryst} %
As dried at 220	20 ± 10	Amorphous
600	70 ± 15	Zn_2SiO_4 – 94, ZnO – 6

800	85 ± 15	Zn_2SiO_4 – 95, ZnO – 5
1000	90 ± 10	Zn_2SiO_4 – 97, ZnO – 3
1200	110 ± 10	Zn_2SiO_4 – 82.5, SiO_2 – 14.7, ZnO – 2.8

PL of obtained samples was studied using the *Perkin Elmer LS55* spectrometer. The spectra registration was carried out in the mode of phosphorescence, the delay time was 1 ms, the wavelength of the exciting radiation $\lambda_{ex} = 250$ nm. Monochromator slit width was chosen depending on the luminescence intensity of the samples.

3. Results and discussion

Optical properties of an ion-activator are mainly determined by the crystal structure of the host matrix. α-Zinc orthosilicate crystallizes in the trigonal system (space group *R*-3*m*) and its structure is formed by $[SiO_4]^{4-}$ and $[ZnO_4]^{6-}$tetrahedra. When Mn^{2+} ions (electronic configuration $3d^5$) are injected into the matrix, they occupy positions of zinc Zn^{2+} in the lattice. The splitting of 3*d*-energy sublevels of ion activator occurs under the influence of a weak crystal field. Excitation by UV light leads to the transfer of charge carriers from the ground state of Mn^{2+} ($^6A_1$) to the conduction band (CB) of Zn_2SiO_4:Mn. From this, the electrons move to an excited state of the Mn^{2+} ion ($^4T_1$) by nonradiative transition. Relaxation of electrons to the ground state ($^4T_1 \rightarrow {}^6A_1$) causes intense luminescence with emission maximum in the range from 510 to 520 nm.

Figure 1 shows the excitation PLE (*a*) and emission PL (*b*) spectra of microcrystalline and disintegrated Zn_2SiO_4:Mn powders. The excitation spectrum of a microcrystalline powder has a broad band in the range from 220 to 360 nm with several maxima near 245, 260, 280 and 340 nm. According to [10], the first three bands belong to the electronic transition from the ground state ($^6A_1$) of Mn^{2+} to the CB of the host. An intense band with a maximum at 340 nm corresponds to the *d-d* transition of the activator ion. As seen in figure 1, *a*, after milling the shape of the spectrum does not change, but the intensity decreases significantly.

The emission spectrum of Zn_2SiO_4:Mn shows that the microcrystalline material has the highest intensity of luminescence (figure 1, *b*). As mentioned above, the intensity and position of the PL bands greatly influenced by the crystalline environment of the activator ion. Since the disintegration of microcrystalline Zn_2SiO_4:Mn does not affect the crystal structure, the main reason of PL quenching can be considered the presence of strong deformation destruction of particles that leads to the appearing of the dislocation network. Residual strain may change the crystalline field near Mn^{2+} ions, leading to decomposition of the matrix-activator solid solution. The PL quenching may occur due to diffuse redistribution of activator ions to dislocations arising during milling process. [11].

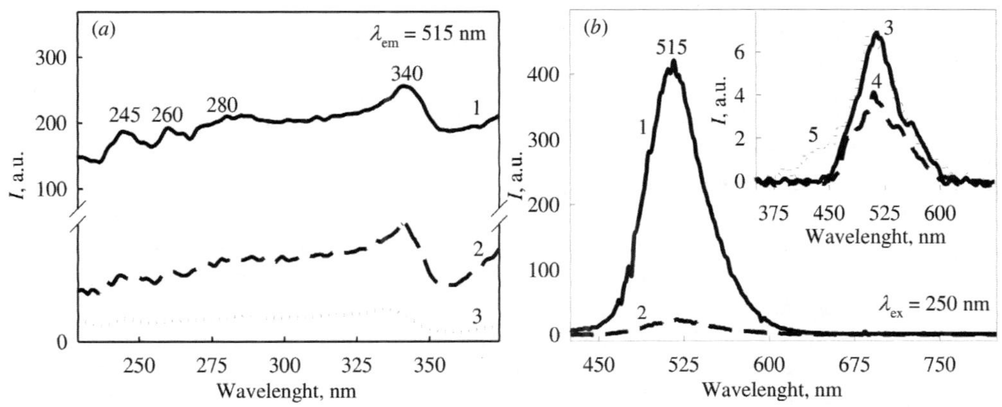

Figure 1. PLE (*a*) and PL (*b*) spectra of coarse-grained Zn_2SiO_4:Mn powder (1) and nanopowder

after ball-milling within 15 min (2); 120 min (3); 240 min (4); sample (4) annealed at 1100 °C – (5).

Since annealing of disintegrated Zn_2SiO_4:Mn at 300 °C does not influence the PL spectral characteristics, samples were further annealed at 1100 °C. The inset in figure 1, *b* shows that high-temperature annealing promotes partial recovery of the PL (curves (4) and (5)), but the intensity of the emission is still much lower than that of the bulk powder. Maximum of the emission wavelength of the samples is at 517 ± 3 nm, that corresponds to an electronic transition $^4T_1 \rightarrow {}^6A_1$.

In turn, Zn_2SiO_4:Mn obtained by sol-gel method also shows the dependence of the PL properties upon the preparation conditions. The emission of the material is not observed up to the annealing at 600 °C. Figure 2 shows that the intensity of the bands in the excitation and emission spectrum increases significantly with increasing of annealing temperature. The change is associated with crystallinity improving of the material and the gradual increase of the Zn_2SiO_4:Mn phase content. In this case, one can say of a homogeneous distribution of the activator ions in the host and the formation of a solid solution of Mn^{2+}-Zn_2SiO_4. The excitation spectra of the samples are presented by a broad band in the range from 220 to 300 nm with three maxima at 230, 250 and 260 nm (figure 2, *a*). The last two peaks indicate the transfer of energy from the ground level of Mn^{2+} ion to the CB of the host. It can be assumed that the presence of the band with maximum at 230 nm and a further sharp intensity increase toward short wavelengths indicate the absorbance of the Zn_2SiO_4 matrix itself [12, 13].

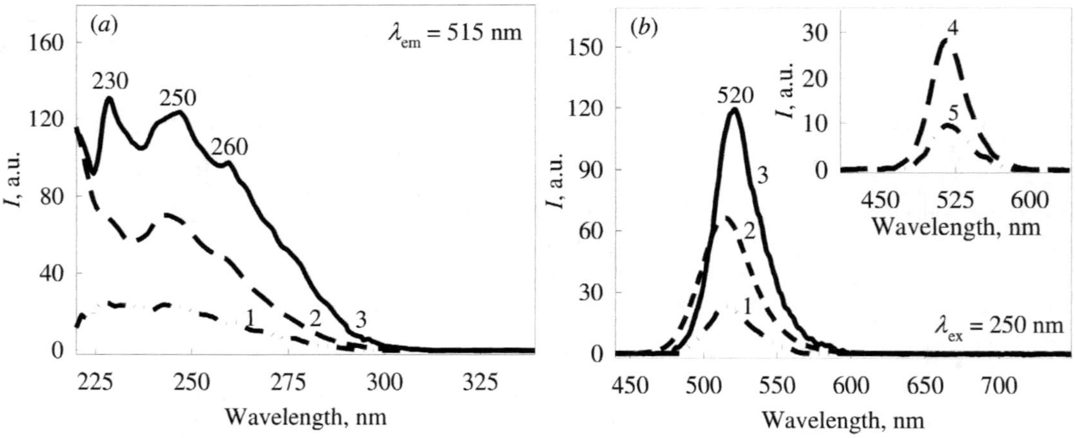

Figure 2. PLE (*a*) and PL (*b*) spectra of Zn_2SiO_4:Mn synthesized by sol-gel method and annealed at 600 °C (1), 700 °C (2); 900 °C (3); 1100 °C (4); 1200 °C (5).

Samples prepared by the sol-gel technology, exhibit only a narrower band emission with a maximum at 518 ± 2 nm (figure 2, *b*). It is important to note that after annealing at temperatures above 1000 °C there is a significant decrease in the intensity of luminescence. Explanation for this would be the changing of the coordination of the activator ions. In Zn_2SiO_4:Mn annealed at temperatures up to 1000 °C a gradual increase in the symmetry of the crystal field around Mn^{2+} ions occurred, that led to an increase in the PL intensity. On the other hand, the absence of SiO_2 impurity phase is observed at higher temperatures (table 1). Due to that the tetrahedral coordination of the activator ions can be significantly distorted, resulting to the appearance of nonradiative recombination centers. Similar phenomenon was reported in [14].

4. Conclusion

In this work Zn_2SiO_4:Mn nanopowders were prepared by means of "top-down" and "bottom-up" methods. The samples were characterized by the particle size and crystal structure. PL properties of the Zn_2SiO_4:Mn were studied depending on the preparation conditions: milling duration and annealing temperature. Despite the fact that the disintegrated material is characterized by small particle size, high energy process causes significant deformation of the crystal lattice and degradation of the PL. The additional high-temperature annealing does not ensure recovery of the emission intensity. Thus, while using high-energy milling it is possible to achieve a narrow size dispersion of nanoparticles, this method is not effective enough for production of studying phosphor.

The PL of Zn_2SiO_4:Mn obtained by the sol-gel process has an intense band in the green region of the visible spectrum. Dependence of the luminescence intensity on the annealing temperature is observed. In addition, one of the features is a coarsening of nanoparticles during annealing. The sol-gel method can be considered more promising to obtain the luminescent nanomaterials than ball-milling. It is needed further improvement of the sol-gel method for the synthesis of nanosized phosphors with narrow particle size dispersion, good crystallinity and high luminescence intensity.

Acknowledgments

This work was partially supported by the Ural Branch of the Russian Academy of Sciences (project no. 12-*P*-234-2003, Program of the Presidium of the RAS no. 24 Fundamental Issues in Technologies of Nanostructures and Nanomaterials).

References

[1] Andrievsky R A, Glezer A M 1999 *Physics of Metals and Metallography*. **88** 45
[2] Feldman C, Justel T, Ronda C R and Schmidt P J 2003 *Adv. Funct. Mater.* **13** 511
[3] Yen W M, Shionoya S and Yamamoto H 2006 *Phosphor Handbook* (Boca Raton: CRC Press)
[4] Rempel A A 2007 *Rus. Chem. Rev.* **76** 474
[5] Kurlov A S, Gusev A I 2011 *Technical Physics* **81** 76
[6] Valeeva A A, Schroettner H, Rempel A A 2011 *Inorganic Mat.* **47** 464
[7] Takesue M et al 2009 *Progr. Cryst. Growth and Charact. Mater.* **55** 98
[8] Cho T H, Chang H J 2003 *Ceram. Int.* **29** 611
[9] Petrovykh K A, Rempel A A, Kortov V S, Valeeva A A and Zvonarev S V 2013 *Inorganic Mat.* **49** 1099
[10] Orgel L E 1955 *J. Chem. Phys.* **23** 1004
[11] Michailov M M, Vladimirov V M and Vlasov V A 2000 *Bulletin TPU* **303** 191
[12] Su F, Ma B, Ding K, Li G, Wang S, Chen W, Joly A G, McCready D E 2006 *J. Lum.* **116** 117
[13] Chang H, Park H D, Sohn K S and Lee J D 1999 *J. Korean Phys. Soc.* **34** 545
[14] Yan J, Ji Z, Xi J, Wang C, Du J and Zhao S 2006 *Thin Solid Films* **515** 1877

Journal of Physics: Conference Series **552** (2014) 012044 doi:10.1088/1742-6596/552/1/012044

Initiation of furazanotetrazinedioxide and mixes on its basis by high-current electron beam

V I Oleshko[1,*], V E Zarko[2], V V Lysyk[1], V P Tsipilev[1] and P I Kalmykov[3]
[1]National Research Tomsk Polytechnic University, Tomsk, Russia
[2]Institute of Chemical Kinetics and Combustion SB RAS, Novosibirsk, Russia
[3]Federal Research & Production Center ALTAI, Biysk, Russia

E-mail: oleshko@tpu.ru

Abstract. The results of study of physicochemical processes developing in the samples of furazanotetrazinedioxide (FTDO) and its mixes with dinitrodiazapentane (DNP) upon irradiation by the high-current electron beam with the energy density varied in the range of 0.05–60 J/cm^2 are presented. Pre-explosion processes taking place in materials under examination at below threshold modes of excitation are studied. Electron beam threshold energy densities leading to explosive decomposition of FTDO and FTDO/DNP mixes are determined. Noticeable effect of the electron beam energy density on kinetic characteristics of explosive decomposition process of FTDO is discovered. Spectra of the products of FTDO explosive decomposition are measured at explosion initiation in the atmosphere.

1. Introduction

The endothermic substance furazanotetrazinedioxide (FTDO, $C_2N_6O_3$) with high enthalpy of formation (4200 kJ/kg) is of great interest as a perspective high-energy material [1]. However, its practical use is complicated due to the high sensitivity to the mechanical impact which is comparable to sensitivity of lead azide. For resolving this problem it is suggested to crystallize FTDO in a mix with dinitrodiazapentane (DNP, $C_3H_8N_4O_4$) [2]. For the safe work with FTDO it is necessary to know not only its sensitivity to mechanical impact but its sensitivity to other energetic impacts including high-current electron beams (HCEB) action.

In the present work the results of experimental investigations of physicochemical processes developing in FTDO and its mixes with DNP (with the FTDO content of 75 weight %) under irradiation by electron beam with varied energy density are presented. Researches in this direction will allow more complete determination of explosive conversion mechanism of FTDO and mixes on its basis.

2. Experiment

FTDO samples were pressed with a pressure of 400 MPa up to density of 1.44 g/cm^3 in the form of tablet with 3 mm diameter and 250–450 μm thickness. Samples of FTDO/DNP mixes of 450–800 μm thickness were prepared by crystallization of corresponding melt at the substrate temperature of 20°C. Explosive decomposition of FTDO was initiated by electron beam in atmosphere. FTDO/DNP samples were initiated in vacuum chamber at the pressure of ~10^{-3} Torr. Nanosecond electron accelerator based on pulsed high-voltage generator GIN-600 (average electron energy ~250 keV, duration of current pulse at FWHM ~15 ns) was used as a source of excitation (initiation). The energy

Content from this work may be used under the terms of the Creative Commons Attribution 3.0 licence. Any further distribution of this work must maintain attribution to the author(s) and the title of the work, journal citation and DOI.
Published under licence by IOP Publishing Ltd

density H of electron beam ejected into the atmosphere was varied in the range of 0.05–0.28 J/cm^2 by changing the distance between output foil of electron accelerator and sample and it was measured by radiation-chemical method. Upon irradiation of samples in a vacuum chamber the H value was varied in the range of 0.05–60 J/cm^2 by changing the cathode-anode distance and using diaphragms.

Schemes of initiation of energy materials explosion by electron beam are given in [4]. FTDO sample under examination was fixed on a sample holder placed in atmosphere and was excited by electron beam through the anode aperture and a 30 μm thick aluminum foil that separates vacuum diode from atmosphere. Maximum energy density of HCEB ejected into the atmosphere was observed in near-foil region and reached 0.28 J/cm^2. Increasing distance between foil and sample led to decrease of HCEB energy density that reached 0.05 J/cm^2. Sample holders were prepared from different metals (aluminum, duraluminum, copper and etc.). An obstacle (20 mm × 10 mm × 2 mm aluminum plate) was placed at distance L from explosive's surface. It was stopping the plasma jet formed during FTDO explosion. The purpose of using an obstacle was increasing intensity of explosive glowing and dimensional localization of plasma torch. Energy density of electron beam H was 0.05, 1.0 and 60 J/cm^2 during experiments of FTDO/DNP mixes initiation in vacuum chamber. High H values were obtained in the regime of electron beam self-focusing and were determined by the formation of erosion spots on the surfaces of metals with different sublimation temperatures. Kinetic characteristics of FTDO cathodoluminescence and glowing of the products of explosive decomposition were registered by measurement system based on grating monochromator MDR-23, PMT-84 and oscilloscope Tektronix DPO 3034 paired with PC. The resulting time resolution of measurement system was ~15 ns. Integral spectra of explosion were registered by means of a fiber-optical spectrometer AvaSpec - DUAL. Spectral range of measurements was 480–980 nm, spectral resolution ~1.5 nm. Measurements were carried out at initial temperature of 300 K.

3. Experimental results

3.1. Luminescence, destruction and explosive decomposition of FTDO samples upon excitation by electron beam in atmosphere.

Electron beam impact on FTDO samples at energy density varied in the range of $0.05 \leq H \leq 0.28$ J/cm^2 has been studied. A sample under examination was fixed on metal sample holder and was excited by electron beam.

Cathodoluminescence. It has been revealed that irradiation of samples by electron beam with the value of energy density H ~0.05 J/cm^2 is accompanied by pulsed cathodoluminescence (CL). CL spectrum consists of a wide emission band with maximum at $\lambda = 580$ nm. The band's full width at half maximum is ~0.36 eV, decay time of luminescence $\tau \leq 10$ ns.

Kinetics of explosive glowing. Explosive decomposition of samples is observed at energy density of electron beam $H \geq 0.09$ J/cm^2. The value of threshold energy for FTDO is close to the threshold energy of initiation for heavy metal azides (HMA) and amounts ~0.1 J/cm^2. Kinetics of FTDO explosive decomposition represents two pulses following one after another [4]. The first short peak corresponds to sample cathodoluminescence and can be used as a reference representing irradiation instant of sample. The second long-time peak corresponds to glowing of products of explosive decomposition. At H^* ~0.09 J/cm^2 glowing of explosion products arises with a delay of t_{ind} ~1200 ns after irradiation instant. An increase of electron beam energy density within the range of $0{,}09 \leq H \leq 0{,}25$ J/cm^2 leads to monotonous reduction of the induction period from 1200 to 30 ns and simultaneous reduction of duration of explosive luminescence pulse. Duration of induction period versus HCEB energy density relation is given on figure 1.

Dimensional effect. During the study a dimensional effect was discovered. This effect is that the explosion initiation probability depends on FTDO sample thick (figure 2). At $H = 0.25$ J/cm^2 FTDO samples with thickness of 320 μm had initiation probability $P = 1$. Meanwhile the samples with thickness less than 280 μm didn't explode ($P = 0$) but they were only destructed into separate segments.

Spectra of the products of explosive decomposition. It has been found that two types of glowing are distinguished in spectra of explosion plasma of FTDO: line spectrum of atoms being a part of explosive material and a sample holder and continuous spectrum representing glowing of dense low-temperature plasma. Line and continuous spectra intensity ratio depends on a mass of explosive, sample holder material and conditions of plasma torch expansion (free expanding into vacuum and interaction with a solid obstacle). Emission spectra of products of explosive decomposition of FTDO samples fixed on aluminum and magnesium alloy (MA18) holder were measured by means of fiber-optical spectrometer AvaSpec - DUAL. They are presented on figure 3 a and b respectively.

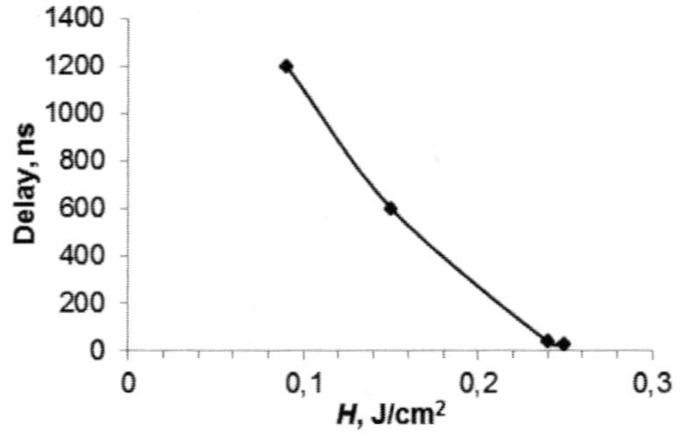

Figure 1. Effect of density of electron beam energy on duration of induction period of FTDO

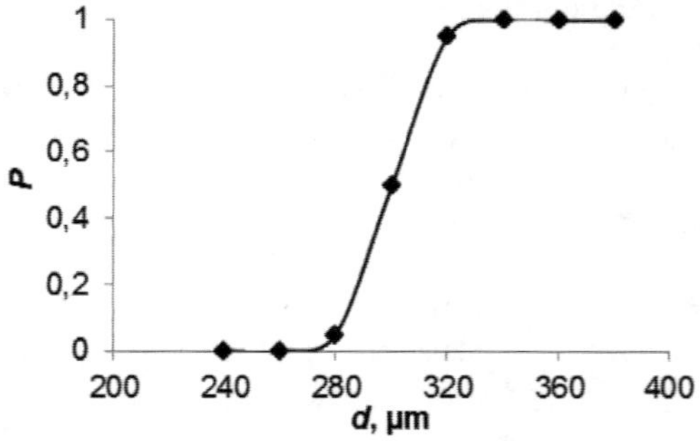

Figure 2. Dependency of the explosion initiation probability on the thickness of FTDO samples excited by HCEB (H=0.25 J/cm^2)

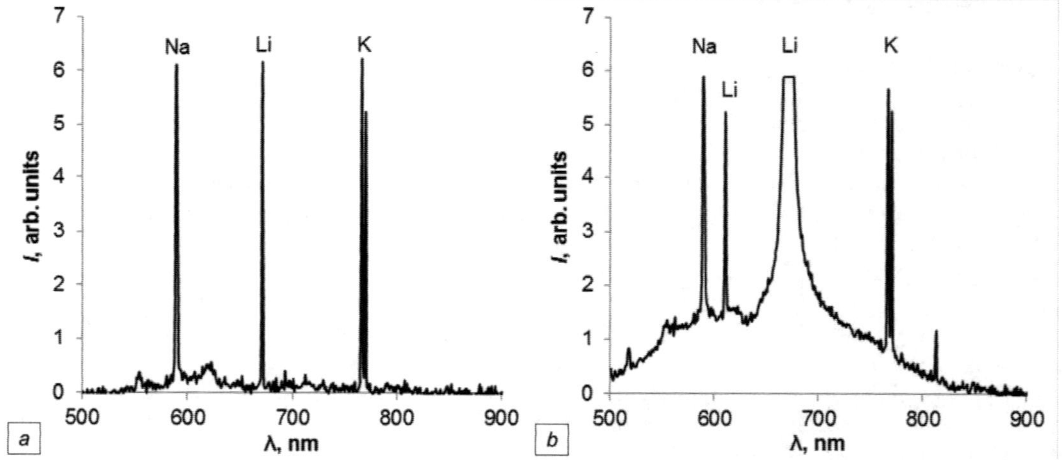

Figure 3. Spectra of explosive glowing of FTDO placed on aluminum (a) and magnesium alloy MA18 (b) holders at initiation by electron beam.

It can be seen that the most intensive lines in FTDO explosion spectra in visible region refer to atomic lines of alkali metals: Na I: 588.99; 589.59; Li I: 670.78; K I: 766.49; 769.89 nm. Two atomic lines Na I 588.99 & 589.99 nm are registered as one line due to insufficient spectral resolution of AvaSpec - DUAL spectrometer. A comparison of spectra obtained during explosion of FTDO samples on two different holders shows significant grow of Li I line (670.78 nm) intensity and appearance of additional Li I line (610.36 nm on figure 3, b). This evidences that the elements from holder get into plasma torch of explosion (MA18 alloy contains 10% of lithium). Analysis of experimental data shows that alkali metals enter into FTDO during the process of its producing.

3.2. Kinetics of explosive decomposition of FTDO/DNP (75/25) samples at excitation by electron beam in vacuum.

Explosive decomposition of desensitized FTDO/DNP (75/25) samples are observed only at the maximum energy density of HCEB (H=60 J/cm^2). At $H \geq 0.1$ J/cm^2 only destruction of irradiated surface takes place. And the effects that are typical for explosive decomposition are not observed.

Slow combustion and low-speed detonation of FTDO/DNP samples at $H = 0.1$ and 1.0 J/cm^2 have not been registered. Irradiation of FTDO/DNP samples by electron beam with H~1.0 J/cm^2 doesn't lead to their explosive decomposition. However size of produced fragments shrinks to dozens of micrometers (a powder is formed at sample location). Only a pulsed cathodoluminescence has been observed at the instance of excitation pulse. Irradiation of FTDO/DNP samples at H^* ~60 J/cm^2 leads to detonation. The high rate of chemical reaction expansion through the sample (~9 km/s) and scabbing from the backside of metal plate carrying an energetic material sample are the facts proving this. Oscillograms of explosive glowing of FTDO/DNP (75/25) during free expansion of explosion plasma are identical to oscillograms of explosive glowing of FTDO at $H > 0.25$ J/cm^2. That is the glowing of products of explosive decomposition of FTDO/DNP (75/25) appears with a delay of ~20–30 ns after sample irradiation instant, and duration of pulse of explosive glowing does not exceed 400 ns.

4. Discussion

Qualitative description of low-threshold initiation of FTDO samples and high-threshold initiation of FTDO/DNP (75/25) mixes on physical level can be provided in the context of thermal mechanism of process developing. This description is the following. Decomposition rate of FTDO is high and during the beam action this process supersedes gas-dynamic unloading from irradiated volume framed by beam diameter and depth of electron penetration (about 0.4 mm). So, conditions of process development are close to adiabatic, especially at high impact levels. This can explain both low energy threshold of initiation and reduction of explosion delay and duration of explosive glowing with increasing H. An alternative scenario occurs in case of FTDO/DNP (75/25) mixes. At relatively low excitation levels (H ~1 J/cm^2) characteristic time of gas-dynamic unloading is less than characteristic time of heat generation from chemical reaction in heating foci. This is proved by observed processes of samples fracture and expansion of unreacted substance. It is obvious that initiation of desensitized compound is possible only at high intensities of beam energy input which accumulates necessary heat in decomposition foci. In other words it is possible at high values of H. It is necessary to note that the value $H = 60$ J/cm^2 used in these experiments certainly exceeds threshold of initiation energy which will be refined in further experiments.

Analyzing explosive conversion of pressed FTDO samples it should be kept in mind that approximate estimate of substance heating gives very low values of surface layer temperature. If we take H^* equal to 0.1 J/cm^2, take into account FTDO heat capacity that is equal to 1 J/g·K and its density of 1.5 g/cm^3 and take into account that electrons are being absorbed in 0.04 cm thick layer, all this will give average layer heating of 2.2 K. That is to say, here is the same situation as with the electron initiation of HMA where thresholds of explosive decomposition are laying in the range of 0.1 J/cm^2 [5]. In [5, 6] a model of "hot centers" formation in volume of EM during localization of energy evolved as a result of electric breakdown was suggested for description of HMA behavior. At sufficiently high rates of the reaction of chemical decomposition the process may develop out of the region of energy release of electron beam and expand on the whole mass of explosive.

Analyse of results of FTDO initiation by electron beam proves that initiation model can be based on electronic discharge model published in [5–9]. Particularly the evidence of this is dimensional effect discovered in the presented work. Namely with decreasing of FTDO sample thickness to 280 μm a part of high-energy electrons traverse the sample and are not absorbed. As a result, intensity of electric field related to negative spatial charge of electron beam injected into the sample decreases and becomes insufficient for initiation of electric breakdown.

5. Conclusion

Initiation of FTDO and crystalized FTDO/DNP (75/25) mix by high-current electron beam has been studied. It has been revealed that luminescence and samples fracture are observed at under-threshold densities of electron beam. Initiation thresholds of explosive decomposition have been measured for materials under examination. It has been established that the sensitivity of FTDO to the action of a pulse of electron beam is close to the sensitivity of HMA, and the sensitivity of FTDO/DNP mix is close to the sensitivity of PETN. In contrast to HMA, experiments with FTDO showed strong dependence of ignition delay time and duration of a pulse of explosive glowing on energy density H of electron beam. With an increase of H from 0.1 to 1.0 J/cm^2, the velocity of chemical decomposition increases by more than an order and reaches ~9 km/s. Dimensional effect proving electric-discharge mechanism of FTDO initiation by electron beam has been discovered. It has been shown that behavior of FTDO and FTDO/DNP (75/25) mixes under HCEB irradiation can be described in terms of models of electron beam initiation of HMA and PETN respectively. The results of experiments evidence possibility of using electron beam initiation method for rapid testing of varying mixes.

Acknowledgments

This work was performed with the financial support of Russian Ministry of Education and Science, government order "Science", theme 3.1329.2014.

References

[1] Kiselev V G, Grican N P, Zarko V E et al. 2007 *Comb., Expl. and Shock Waves* **43** 77–81
[2] Kalmykov P I, Zarko V E, Sidelnikov A A et al. 2011 *Russian J. Appl. Chem.* **84** 248–55
[3] Zarko V E, Kvasov A A, Simonenko V N et al. 2011 *42nd Intern. Conf. ICT* (Karlsruhe, Germany) p 23-1–23-9
[4] Oleshko V I, Zarko V E, Lysyk V V et al. 2013 *Russian Phys. J.* **56** 63–8
[5] Oleshko V I, Damamme G, Malys D and Lisitsyn V M 2009 *Tech. Phys. Lett.* **35** 55–61
[6] Oleshko V I, Lisitsyn V M, Damamme G and Malys D 2010 *Nuclear Instruments and Methods in Physics Research* **19** 3265–8
[7] Oleshko V I, Korepanov V I, Lisitsyn V M and Tsipilev V P 2007 *Zababakhin Scientific Readings: Proceedings of 9th Int. Conf.* p 58
[8] Oleshko V I, Korepanov V I, Lisitsyn V M and Tsipilev V P 2007 Comb., Expl. and Shock Waves **43** 87–9
[9] Oleshko V I, Lisitsyn V M, Scripin A S and Tsipilev V P 2012 *Tech. Phys. Lett.* **38** 37–43

Journal of Physics: Conference Series **552** (2014) 012045 doi:10.1088/1742-6596/552/1/012045

Inhomogeneous change of temperature of ionic crystals under the action of a pulsed electron beam

S A Stepanov*, V F Shtan`ko and E P Chinkov

National Research Tomsk Polytechnic University, 30 Lenin Avenue, Tomsk 634050, Russia

E-mail: stepanovsa@tpu.ru

Abstract. This paper presents the results of mathematical modeling of temperature change in ionic crystals in an area of dissipation of energy of an electron beam. Fluence is varied from 0.1 to 1.2 J/cm^2. Electron beam pulse duration was 24 ns, and the maximum electron energy of 280 KeV. The initial temperature of the crystals varied from 20 to 400 K. The calculation is performed considering the inhibitory effect of electrons in the field of space charge formed by a vacuum gap 1000 micron before the irradiated surface.

1. Introduction

Using pulsed electron beams allows generating short-lived defect concentration sufficient for detection and research of the mechanisms of defect formation in solids. However, the impact on ionic crystals pulsed electron beams has some specific features compared with low-intensive sources of radiation. In an area of energy dissipation of pulsed electron beam is formed strong electric field and the temperature rises along with the generation of non-equilibrium electron-hole pairs, which affects the efficiency of formation and decay of an electron-hole excitation and also on the postradiation processes [1]. Effect of temperature increase in an area of energy dissipation is not enough studied, making it difficult to correct account of this factor in the analysis of research results.

Effect of temperature increase in the area of energy dissipation is not enough researched, making it difficult to correct account of this factor in the analysis of research results.

2. Method of calculating

In the calculations used spectrum of a pulsed electron beam [2], which was determined experimentally by attenuation in thin foils, as described in [3]. To calculate the distribution of thermalized electrons in the sample and the absorbed energy density applied a variation "enlarged" collisions of Monte Carlo method using the angular distribution of Moliere-Bethe [2]. Ionization losses are determined by the formula Bethe-Bloch and the average ionization potential is defined by the formula Sternheimer [4]. Distributions of charge density are determined by the algorithm presented in [2]. To estimate the cooling time of the heated region can use the relation given in [5]:

$$t_b << D/s_l << D^2/\chi$$

where χ – coefficient of thermal diffusivity of the target material; D – the characteristic size of the heated region; t_b – heating time. Variable D^2/χ it makes sense characteristic cooling time of the heated region. Condition means that during pulsed electron beam (PEB) irradiation, the temperature field varies weakly and determined by the distribution of the radiation energy.

Content from this work may be used under the terms of the Creative Commons Attribution 3.0 licence. Any further distribution of this work must maintain attribution to the author(s) and the title of the work, journal citation and DOI.
Published under licence by IOP Publishing Ltd

3. Results of calculation

Simulation of the temperature distribution over the thickness of the irradiated material (Z) was performed for ionic crystals LiF, CaF_2, NaCl and KBr. Results of calculations shown in figure 1.

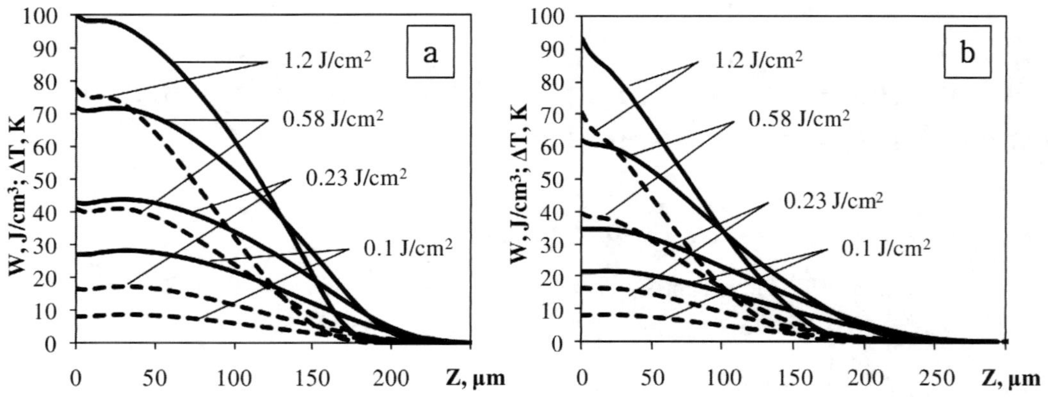

Figure 1. Calculated forms of energy release (W, dashed curves) and change of temperature of crystal (ΔT) at the end of PEB in crystals CaF_2 (a) and KBr (b) at initial temperature of the crystal (T_M) 20 K

From the results of calculations that at low initial temperature of the crystal at the end of irradiation the greatest temperature increase of the crystal (ΔT) takes place in the subsurface layer and then decreases monotonically with increasing distance from the irradiated surface. The magnitude of such a large temperature change is due, firstly, the low specific heat capacity of the material at low initial temperatures. Secondly, with increasing fluence PEB increased influence of field of space charge in the vacuum gap, which leads to large energy in the subsurface layer of the crystal and thus to greater heating of his. [6]. The totality of these processes leads to the fact that the form $\Delta T(Z)$ is different from the form of energy release $W(Z)$. These regularities are common to all of the crystals.

For a detailed consideration of the time variation of the temperature and energy release in the layers of crystal with increasing distance from the irradiated surface was performed calculations with allowance the spectral-temporal parameters of PEB, leading to spatiotemporal dependence energy release in the sample. Results of calculations shown in figure 2. It is seen that with increasing distance from the irradiated surface total time of energy release in the layer is reduced. Spatiotemporal function of energy release defines complicated dependence of change of temperature in the layers with the irradiation time. With increasing distance from the irradiated surface is characteristic reduction of heating time of layer. But the form of $\Delta T_Z(t)$ differs from the form of $W_Z(t)$, which is attributed to variation of the specific heat of the crystal layer with time, as it increases its temperature.

Consider this effect the example of the crystal CaF_2 (figure 2a). At start time, the crystal had a temperature of 20 K, after irradiation PEB with duration of 1 ns, for example, in the surface layer (figure 2a, layer «0 µm») temperature will amount to about 22 K. Changing the temperature of the layer will increase its specific heat and thus to further reduce the growth rate of temperature of the layer against the background increasing energy release in the layer. So from 5 to 13 ns (figure 2a, layer «0 µm») ΔT_Z becomes almost constant though W_Z is continuously growing. This effect is characteristic throughout the irradiated volume of the crystal (i.e. for all layers). After some time (for CaF_2 after 14 ns and KBr after 12 ns) been a sharp rise W_Z and synchronous growth ΔT_Z in subsurface layers (figure 2, layer «0 µm»), due to the inhibitory effect of the field of space charge in the vacuum gap [6].

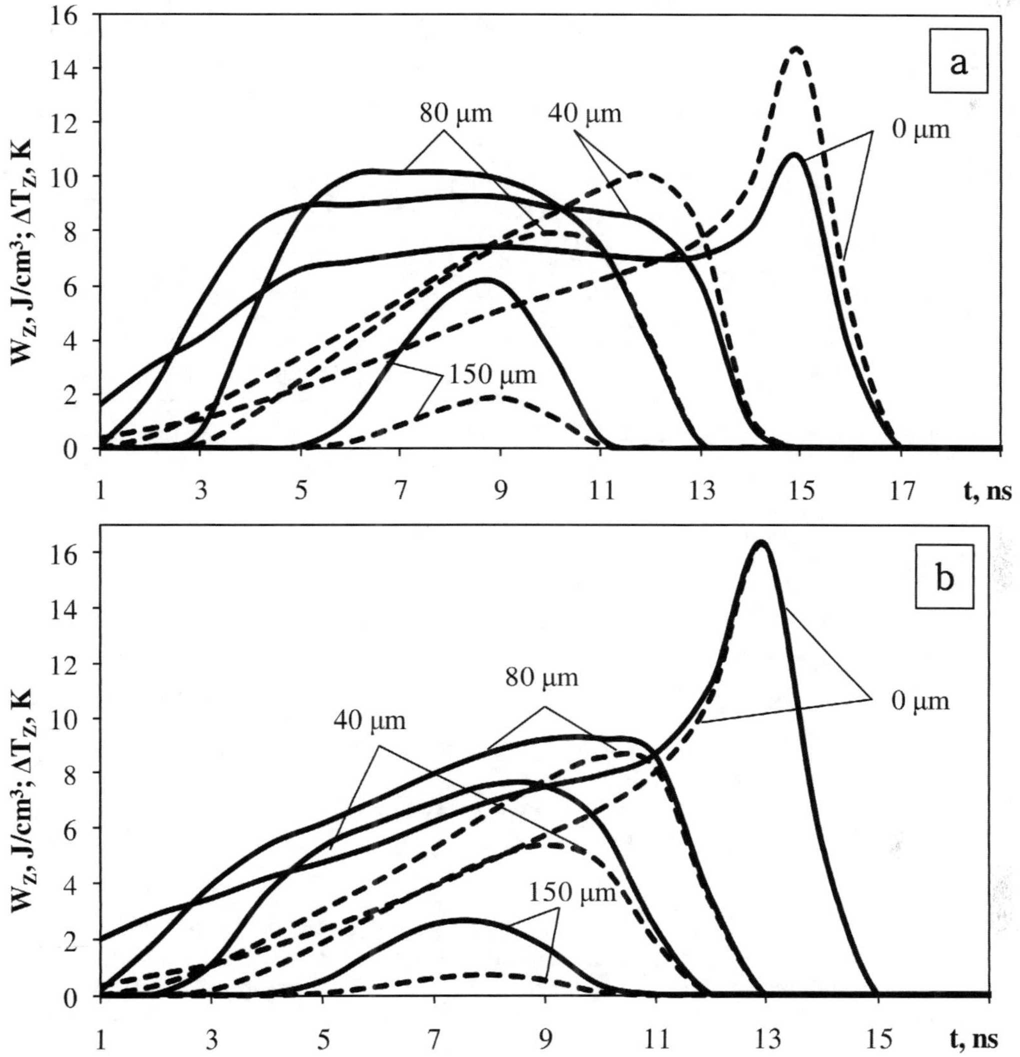

Figure 2. Calculation of changes temperature of layers (ΔT_Z) and the time dependence of energy release (W_Z, dashed curves) in crystals CaF_2 (a) и KBr (b) at T_M 20K and a fluence of 1.2 J/cm^2 (layer «0 μm» corresponds to the irradiated surface of the crystal)

From the results it follows that ΔT_Z are maximized for subsurface region of the crystal. In Figure 3 shows the calculated dependence of ΔT of subsurface layers from T_M in crystals CaF_2, KBr, NaCl and LiF at the end of irradiation PEB. In these layers (up to 75 microns from the irradiated surface) energy release is almost 50 percent of a total energy release throughout the irradiated volume of the crystal (figure 1). On a plot of dependency $\Delta T(T_M)$ where $T_M < 80$ K the determining factor of high heating for all of the crystals is a low specific heat capacity of materials. With an increase of T_M more than 80 K values of $\Delta T(T_M)$ are practically constant. In this area the high temperature is mainly due to high energy release in the subsurface region of the crystal due to presence a field of space charge in the vacuum gap. So at a fluence of PEB equal to 0.58 in KBr temperature reaches an average of 35 K, in NaCl ΔT – an average of 20 K, в CaF_2 ΔT – an average of 18 K, в LiF ΔT – an average of 10 K.

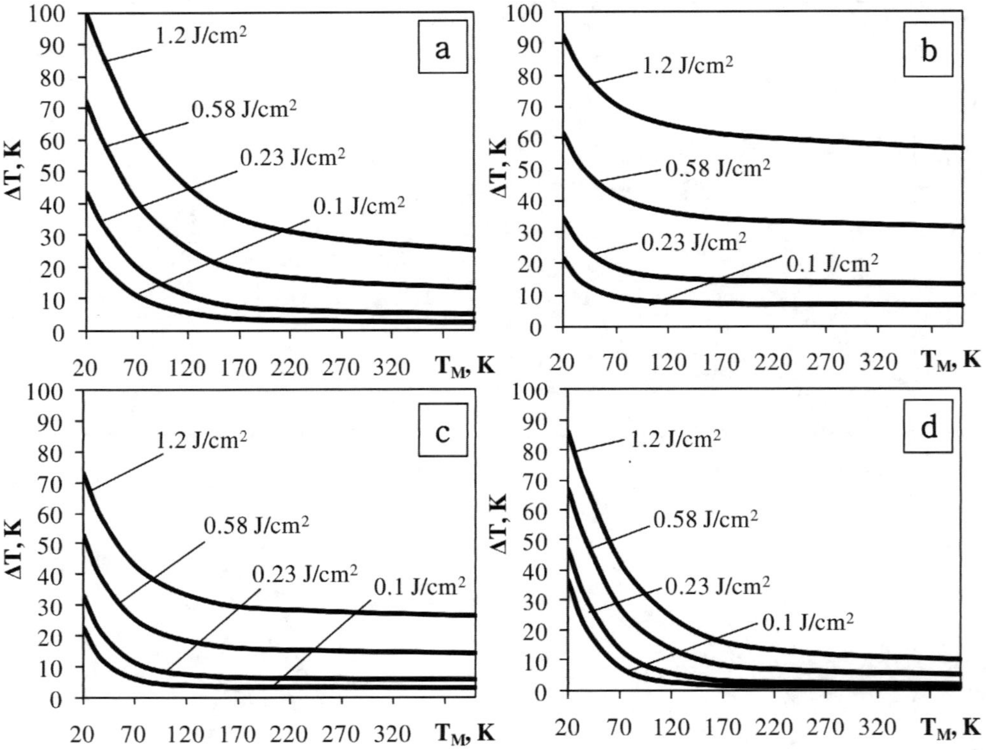

Figure 3. Calculated dependence ΔT of the maximum of energy release from T_M in crystals CaF_2(a), KBr (b), NaCl (c), LiF (d) at the end of PEB

4. Conclusion

Studying the temperature dependences of the spectra of pulsed cathodoluminescence, values of decay component, as well as the temperature dependence of the yield of products of irradiation, an important factor is the temperature of the material in which the measurements are carried out. Consequently, as shown in this paper, the presence of spatiotemporal inhomogeneity of temperature of the crystal, as well as a significant heating of the material during irradiation PEB, even at low fluence, will make a significant distortion in the measurement results. This effect of inhomogeneous change of a temperature must be taken into account in the research of the temperature dependence of the yield of products of irradiation and kinetic processes postradiational.

Acknowledgement
This work was supported by the Ministry of Education of the Russian Federation, government contract "Science", theme 3.1329.2014 (2152).

References

[1] Shtanko V F, Tolmachev V M and Glybin V G 1998 *Tech. Phys.* **68** 53
[2] Shtanko V F, Tolmachev V M, Chinkov E P and Stepanov S A 2012 *Tech. Phys.* **57** 225
[3] Krastelev E G, Meskhi G O and Yablokov B N 1976 *Prib. Tekh. Eksp.* **3** 39
[4] Sternheimer R M 1966 *Phys. Rev.* **145** 247
[5] Zalyubovskii I I, Kalinichenko A I and Lazurik V T 1986 *Introduction to Radiation Acoustics* (Kharkov: Vysshaya Shkola) p 165

[6] Shtanko V F, Chinkov E P, Stepanov S A and Obukhova E E 2011 *Izv. Vyssh. Uchebn. Zaved. Fiz.* **11/3** 179

Journal of Physics: Conference Series **552** (2014) 012046 doi:10.1088/1742-6596/552/1/012046

Electrical and photoelectric properties of polycrystalline diamond films deposited from an abnormal glow discharge

F V Konusov, A V Kabyshev, S A Linnik, A V Gaydaychuk and G E Remnev
National Research Tomsk Polytechnic University, 30, Lenin Avenue, 634050, Tomsk, Russia

E-mail: konusov@hvd.tpu.ru

Abstract. Electrical and photovoltaic properties of polycrystalline diamond films (PDF), deposited from the abnormal glow discharge were analysed. Features of charge carrier transfer in PDF are determined by continuous energy distribution in the band gap of defect states of different nature. Dominated n-type conductivity activation component of electrical conduction and photoconductivity is complemented by a hopping mechanism with the participation of states near the Fermi level with a density $5.6 \cdot 10^{17}$–$2.1 \cdot 10^{21}$ eV^{-1} cm^{-3}. Activation transfer is realized in the exchange of charge carriers between the allowed bands and donor levels with the activation energy 0.007–0.21 eV, which are sparsely populated and have wide variation in their parameters. Trapping centers and carriers recombination are heterogeneously distributed in grain boundaries. Under lighting, the state density increases 3-5 times and probable jump length decreases by 1-3 nm. Spectral distribution centers of photosensitivity are correlated with the distribution of deep-level defects, determining the absorption spectra.

1. Introduction

Unique chemical, mechanical, electrical, thermal, optical and photoelectrical properties of a diamond contribute to its wide application in high-frequency, high-temperature electronics and semiconductor manufacturing of a special purpose [1–5]. Due to the limited possibility of using diamond single crystals and epitaxial diamond films, because of their high cost in the most appropriate instruments are the polycrystalline diamond films (PDF), which are obtained by well-studied methods of the vapor deposition [1-8]. PDF are successfully used for the production of the stable radiation-resistant UV detectors and ionizing radiation, as well as for laser and photodiode structures [1,4]. Dielectric PDF are used for these purposes (electrical conduction ($\sigma = 10^{-15}$–10^{-11} $S \cdot cm^{-1}$). Depending on the characteristics of PDF polycrystalline structure, the content of alloying impurity atoms and defects in the PDF crystal structure, the electrical characteristics, as well as the mechanism of transport and type of charge carriers vary widely ($\sigma = 10^{-6}$–10^{3} $S \cdot cm^{-1}$) [1,4–15]. It stimulates the study of the peculiarities of polycrystalline films electric and photoelectric properties, and the investigation of the mechanisms determining these properties and the influence of deposition conditions on the change in the characteristics of the film material.

The aim of the work is to determine the electrical and photovoltaic characteristics of the dominant transfer mechanism, charge carriers and the energy spectrum of the defect levels in PDF, deposited from plasma abnormal glow discharge.

Content from this work may be used under the terms of the Creative Commons Attribution 3.0 licence. Any further distribution of this work must maintain attribution to the author(s) and the title of the work, journal citation and DOI.
Published under licence by IOP Publishing Ltd

2. Experimental

Methods of film deposition from abnormal glow discharge and the study of their structural features are described in detail in [2]. Dark surface conductivity σ and photoconductivity $\Delta\sigma_{ph}=\sigma_{ph}-\sigma$ (σ_{ph} conductivity at lighting), photosensitive $K(h\nu)=\Delta\sigma_{ph}/\sigma$ were measured at a constant voltage across the electrodes U=0.01–300 V, at temperature T=300–700 K and photon energy hν=1.5–4.0 eV. The electrodes were deposited on the surface by rubbing the soft graphite. Temperature dependences of σ, $\Delta\sigma_{ph}(T)$ are approximated by the equation for the activation mechanism

$$\sigma_a(T) = \sigma_0 \times e^{(-\varepsilon_\sigma / k\cdot T)}, \qquad (1)$$

where σ_0 – pre–exponential factor, ε_σ – activation energy, k– Boltzmann constant and equation for the hopping mechanism of transport between localized states (LS) near the Fermi level E_F in the model of Mott

$$\sigma_p(T) = \sigma_0' \times e^{(-(T_0/T)^{0.25})}, \qquad (2)$$

where σ'_0 – pre–exponential factor, T_0–activation energy [16–18]. The density of states $N(E_F)$ was calculated from T_0 according to [16–18]. The sign of the dominant charge carriers was determined by the method [16].

3. Result and discussion

Current voltage characteristic (CVC) I(U), field dependence of K(U) and carrier transfer characteristics ε_σ, σ_0, T_0, σ'_0, $N(E_F)$ vary in thickness and deposition plane, due to the inhomogeneity of PDF properties (figures 1, 2). Quantitative differences in the parameters σ and $\Delta\sigma_{ph}$ are recorded from the growth smooth and the opposite rough sides (figures 1, 3). Conductivity from the growth side is higher than from the side of the substrate and the ratio between K is opposite (figure 1). In vacuum σ in 1.1–1.3 times lower than in the air, due to the weak influence of the absorption currents. Only physical adsorption of molecules of atmospheric gases on the surface of PDF affects the change of σ(T), so stable chemical bindings are not formed. CVC coincidence at U>0 and U<0 indicates a weak influence of space charge cloud on σ and $\Delta\sigma_{ph}$. The annealing of the films before T_{ann}<600 K stabilizes the properties. CVC PDF with $\sigma=10^{-10}$–10^{-4} S is ohmic (figure 1). These films have a low, but stable photosensitivity to heat and change U (K=0.005–0.1), as compared with dielectric PDF (K=0.1–1.0). In PDF with $\sigma=10^{-14}$–10^{-11} S deviation from the linear law is fixed, CVC obeys the equation $I\propto\exp(\beta U)$, where β=0.02–0.05 (figure 1, curves 4 and 5)), which indicates the current effect, limited by the space charge, as in [8]. In CVC nonlinearity makes the spatial distribution of the energy barrier that exists on the border with electrodes and grain boundaries [6–8]. Variation in PDF electrical parameters may be caused by inhomogeneity in the distribution of crystallites on size and growth defects in their entirety and on the boundaries or by the influence of hydrogen impurity in the surface layers [4, 11–14].

Dependencies σ, $\Delta\sigma_{ph}(T)$ in the interval ΔT=300–600 K are determined by thermally stimulated electron exchange between the shallow donor levels with energy ε_σ=0.007–0.21 eV and the conduction band (CB). For $\Delta\sigma_{ph}$ value ε_σ at 0.03–0.05 eV is lower than for σ by reducing the population of LS upon excitation by 5–10 times. There is usual quenching $\Delta\sigma_{ph}(T)$ at T≥340 K. N–type σ and $\Delta\sigma_{ph}$ are dominated in PDF layers. It cannot be excluded the influence on CVC, σ, $\Delta\sigma_{ph}(T)$ and K(U, hν) the exchange of holes between LS defects and the valence band (VB) [13]. The parameters ε_σ, σ_0 indicate the effect on the transfer of hopping conductivity, which is confirmed by approximation σ(T) in the interval ΔT=300–600 K (figure 2). LS density calculated from σ, $\Delta\sigma_{ph}(T)$ is large $N(E_F)=5.6\cdot10^{17}$–$1.5\cdot10^{21}$ $eV^{-1}cm^{-3}$ in PDF with ε_σ=0.007–0.15 eV. In PDF LS density, exponentially distributed near E_F consist $N(E_F)=(4–8)\cdot10^{19}$ [17, 18] and 10^{15}–10^{18} $eV^{-1}cm^{-3}$ [19]. The magnitude of the most probable jump length is calculated according to [16–18] for the values of $N(E_F)$, is R=1.1–8.1 nm and under

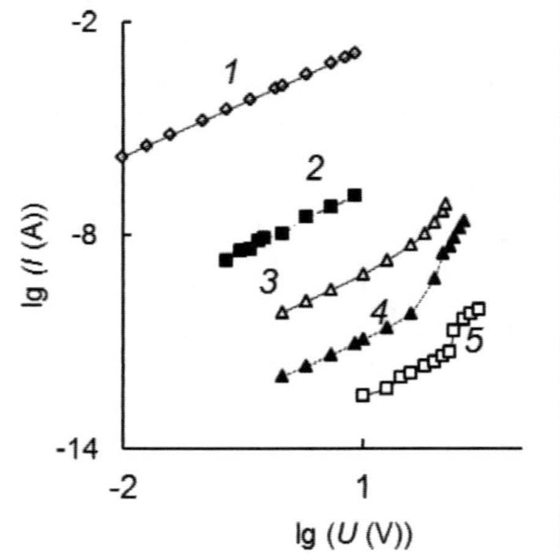

Figure 1. CVC of PDF from the growth side (curves 1 and 2) and from the substrate side (curve 3–5) in the center (curves 1,3) and on edge deposition (curves 2, 4 and 5).

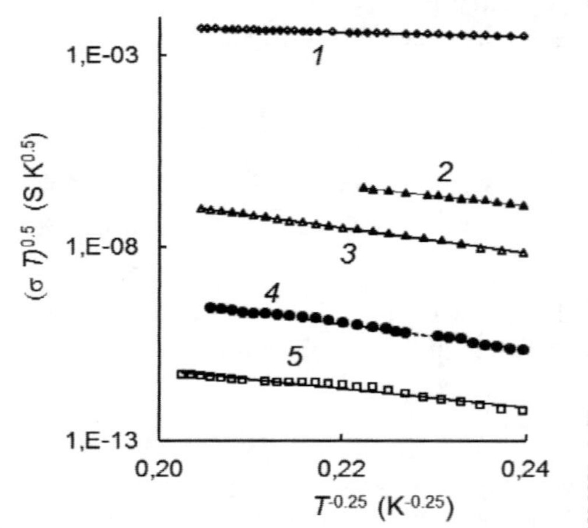

Figure 2. Temperature dependences of conductivity of PDF from the growth side (curves 1 and 2) and from the substrate side (curves 3–5) and in the center (curves 1 and 3) and on edge deposition (curves 2, 4 and 5).

lighting was reduced to 1.0–6.0 nm in line with the increase in the LS density of 3–5 times for PDF with $\sigma<10^{-10}$ S.

Dependencies between parameters σ_h, σ_a and K allow establishing the predominance of one of the transfer mechanisms. Decrease σ_0 with the increasing ε_σ indicates the predominance of σ_a (figure 3, curve 1). This interrelationship is confirmed by $N(E_F)(\varepsilon_\sigma)$ is typical for the uncrystalline semiconductors with a high content of defects [16–19]. By analogy with [6–13, 16–19], we can conclude that PDF hopping conductivity on LS near the Fermi level dominates over the activation in the range of T=300–400 K. At T=400–600 K is determined by the electrical conductivity activation $\sigma_a(T)$ with LS defects interference on grain boundaries and distributed in band gap (BG) in the area of "tails" of the allowed bands. Charge carriers are excited through mobility gap in the area of delocalized states. At T>600 K, the influence of the barrier between the crystallites is increases [6, 8]. The largest contribution to $\Delta\sigma_{ph}(T)$ gives the activation component σ_a, as confirmed by K(T, U, ε_σ) (figure 3, curve 2).

Shape of the spectrum K(hν) is similar to the shape of PDF absorption spectra (figure 4) [1, 4]. There is a constant value of photosensitivity in the interval hν=1.2–2.0 eV. In the interval hν=2.0–3.2 eV there is an increase of photosensitivity with further decrease in hν≥3.2 eV. A similar change of K(hν) is recorded in PDF at hν=1.8–2.9 eV [4]. In [12], the threshold at 2.6 eV is determined by electronic transitions between the VB and LS defects in diamond thin oxidized layer after hydrogenization. Such a spectrum K(hν) in UNCD films is induced with transitions in LS tails zones and LS of inclusions of graphite–like phase [1]. Growth in K at hν>3.4 eV is typical for most structures based on PDF in detectors of ultraviolet irradiation [4]. The oxygen from the air adsorbed on growth defects may have a passivating effect on the increase in photosensitivity of surface of deposited PDF in ultraviolet band.

Accumulation of optical concentration levels is associated with decreasing of LS density near the Fermi level which are involved in hopping transfer, as it is shown by its reduction from $N(E_F)=10^{19}$ to $5\cdot10^{17}$ $eV^{-1}\cdot cm^{-3}$ an increase of the absorption coefficient from $\alpha=2\cdot10^4$ to $5\cdot10^4$ cm^{-1}. Reduction of density responsible for $\sigma_h(T)$ states $N(E_F)$ is also correlated with an increase in the Urbach energy E_U levels which determine the characteristics of the optical absorption in the intervals hν=1.15–1.3 eV

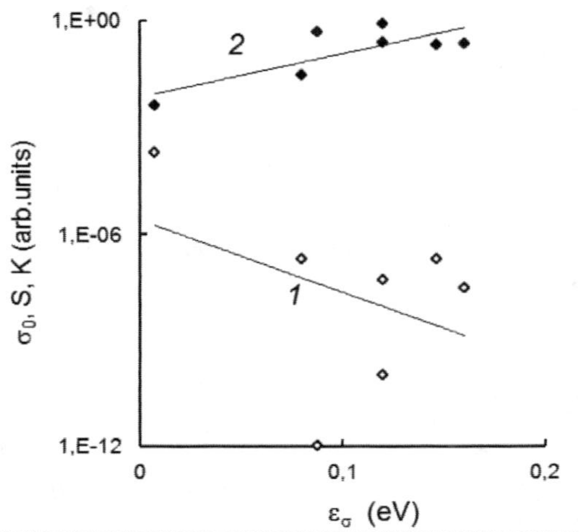

Figure 3. Interrelation between activation energy of conductivity ε_σ, pre-exponential factor σ_0 (curve 1) and photoconductivity K (curve 2) in PDF

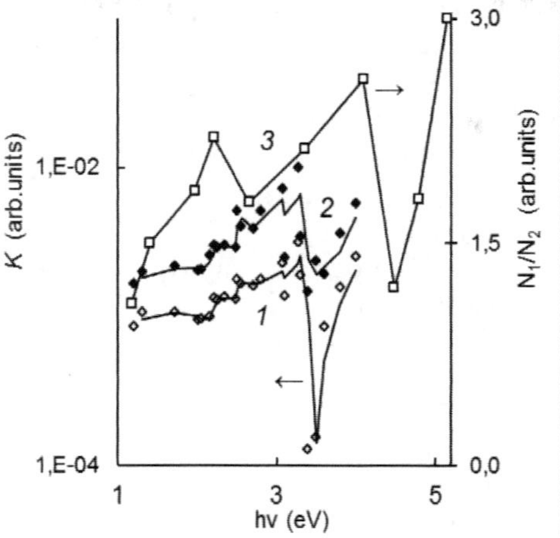

Figure 4. Spectral dependencies K(hν) from the substrate side in the center (curve 1), on edge deposition (curve 2) and ration of the concentration of absorption centers from the substrate side N_1 to its value from the growth side N_2 (curve 3).

and 1.6–3.0 eV. Growth of its value from E_U=0.07 to 0.15 eV with the decreasing of conductivity from $\sigma=10^{-4}$ to 10^{-11} S and the increasing of the activation energy from ε_σ=0.007 to 0.16 eV is determined by influence on the properties of the disorder induced with defects in the crystalline phase material due to the presence of disordered graphite-like phase in the films. As a result, exponential distribution of LS on energy is formed in BG [8, 17, 19].

4. Conclusion

Electrical and photoelectric characteristics, mechanism of charge transfer in the deposited PDF are caused by the defects, the levels of which are continuously distributed across the energy in BG. Activation component of dark and photoconductivity of n-type is realized in the exchange of charge carriers between the conduction band and shallow donor levels with the activation energy $\varepsilon\leq0.21$ eV. Conductivity of the activation type is complemented by a hopping mechanism, involving localized states in the band gap near the Fermi level. Density of states, where the hopping conduction occurs, is significant $N(E_F)=5.6\cdot10^{17}–2.1\cdot10^{21}$ eV^{-1} cm^{-3}, its variation over the surface and the film thickness is determined by the content of growth defects, heterogeneously distributed in grain boundaries. Thermal quenching of photoconductivity is appeared. Under lighting LS density increases 3–5 times, and the probable jump length of photoconductivity decreases on values 1–3 nm. The spectral distribution of photosensitivity levels is correlated with the spectrum of deep levels of the optical absorption defects, which are induced in BG defects.

Acknowledgments
The paper has been completed within megaproject of National Research Tomsk Polytechnic University «Materials for Extreme Conditions».

References

[1] Williams O A 2011 *Diamond and Relat. Mater.* **20** 621
[2] Linnik S A, Gaydaychuk A V 2013 *Diamond and Relat. Mater.* **32** 43
[3] Linnik S A, Gaydaychuk A V 2012 Tech. Phys. Lett. **38** 258

[4] Polyakov V I, Rukovishnikov A I, Avdeeva L A, Kun'kova Z E, Varnin V P, Teremetskaya I G and Ralchenko V G 2006 *Diamond and Relat. Mater.* **15** 1972
[5] Chiquito A J, Berengue O M, Diagonel E, Galzerani J C 2007 *J. Appl. Phys.* **101** 033714
[6] Kopylov P G, Lotonov A M, Apolonsky I A and Obraztzov A N 2009 *Vestn. of MSU.Ser.3(in russian)* 2 54
[7] Trucchi D M, Cappelli E, Conte G, Mattei G, Gramaccioni C and Ascarelli P 2005 *Diamond and Relat. Mater.* **14** 575
[8] Conte G, Rossi MC, Spaziani F and Arcangeli R 2005 *Diamond and Relat. Mater.* **14** 570
[9] Hikavyy A, Clauws P, Maes J, Moshchalkov V V, Butler J E, Feygelson T, Williams O A, Daenen M and Haenen K 2006 *Phys. status solidi* **A 203** 3021
[10] Muret P and Saby Ch 2004 *Semicond. Sci. and Technol.* **19**1
[11] Hubik P, Mares J J, Kozak H and Kromka A 2012 *Diamond and Relat. Mater.* **34** 63
[12] Stallhofer M, Seifert M, Hauf M V, Abstreiter G, Stutzmann M, Garrido J A and Holleitner A W 2010 *Appl. Phys. Lett.* **97** 111107
[13] Yutaka Itoh, Yu Sumikawa, Hitoshi Umezawa and Hiroshi Kawarada 2006 *Appl. Phys. Lett.* **89** 203503
[14] Chaudhary A, Welch J O and Jackman R B 2010 *Appl. Phys. Lett.* **96** 242903
[15] Ri Sung-Gi, Takeuchi Daisuke, Kato Hiromitsu, Ogura Masahiko, Makino Toshiharu, Yamasaki Satoshi, Okushi H, Rezek B and Nebel Ch E 2005 *Appl. Phys. Lett.* **87** 262107
[16] Kabyshev A V, Konusov F V, Remnev G E and Pavlov S K 2013 *Proc. of the universities. Physics (In Russian)* **56** 30
[17] Achatz P, Williams O A, Bruno P, Gruen D M, Garrido A and Stutzmann M 2006 *Phys. Rev.* **B 74** 155429
[18] Gan L, Bolker A, Saguy C, Kalish R, Tan D L, Tay B K, Gruen D and Bruno P 2009 *Diamond and Relat. Mater.* **18** 1118
[19] Nath S and Wilson J I B 1996 *Diamond and Relat. Mater.* **5** 65

Calculation of the spatial distribution of defects and cascade-probability functions in the materials

A I Kupchishin[1] *, A A Kupchishin[2], E V Shmygalev[2], T A Shmygaleva[2], K B Tlebaev[2]
[1] Al-Farabi Kazakh National University, al-Farabi av.71, Almaty, 050012, the Republic of Kazakhstan
[2] Abai Kazakh National Pedagogical University, Dostyk av.13, Almaty, 050010, the Republic of Kazakhstan

E-mail: ankupchishin@mail.ru

Abstract. In this article we carried out the calculations of the depth distribution of implanted ions of arsenic and indium, loss of energy and cascade-probability functions in silicon. Comparison of the calculations with the experimental data is in the satisfactory agreement. The computer simulation and analysis of the characteristics of ions depending on the depth of penetration and the number of interactions were carried out.

1. Introduction

It is known that the high-energy particles produce in solids, as well as in semiconductor materials, a number of consecutive interactions with electrons and atoms condensed matter. As a result the energy of the primary and secondary radiation is transferred to this substance that causes the radiation-chemical transformations in the material by the ionization and excitation of atoms and molecules. In addition we observe the cascade of secondary particles (primary sputtered atoms) having a sufficiently high kinetic energy to generate secondary sputtered atoms and molecules. It should be noted that the charged particles (electrons, ions, etc.) produce in condensed media direct ionization of atoms and molecules, and the photon and neutron radiations produce indirect ionization due to the interaction of this radiation with the medium, whereby the charged particles are formed.

Further, electrons and atoms of a material, absorbing the energy of the primary interactions become ancestors of a large number of further consecutive interconnected events at the atomic and molecular level, which ultimately leads to a change in the structure and various properties of the irradiated material. Consequently, after irradiation we can observe the irreversible changes in the structure and properties of materials accompanied by a set of effects that are characterized by dose or integral effects, since the degree of their activity depends primarily on the energy and intensity of the radiation and absorbed dose. As a rule, radiation effects occur at the atomic and molecular level in the form of various radiation-induced defects.

The most typical type of radiation defects in the structure of solids are Frenkel pairs in the form of interstitial atoms and vacancies in the crystal lattice. In this case, the colliding neutral or charged particles interact elastically with the nucleus (the target) and shift their its field of the nodes of crystal lattice [1-4]. At the elastic interaction the total kinetic energy of the displaced atom and the primary particle remains unchanged. The primary knock-on atom with kinetic energy sufficient for subsequent interactions with the atoms of the crystal lattice, may cause further displacement. The

Content from this work may be used under the terms of the Creative Commons Attribution 3.0 licence. Any further distribution of this work must maintain attribution to the author(s) and the title of the work, journal citation and DOI.
Published under licence by IOP Publishing Ltd

regularities of formation of the secondary displaced atoms are considered in various models and theories of cascades. Moreover, different types of ionizing radiation have different ability of formation of secondary etc. displacements [1 - 3].

Despite the large amount of works in this field, there are little calculations and analysis of the characteristics of interactions and the depth of observation. In this paper, we carried out the calculations of the spatial distributions of defects, loss of energy, concentration of knocked-on atoms and cascade-probability functions in silicon irradiated by ions of arsenic and indium.

2. Calculation procedure

The computer simulation of the spatial distributions of implanted ions was carried out using the standard program Srim.

The calculations of cascade-probability functions (CPF) were made by the formula (2):

$$\psi_n(h,E_0)=\frac{1}{n!\lambda_0^{\ n}}\left(\frac{E_0}{E_0-kh}\right)^{-l}exp\left(\frac{h}{\lambda_0}\right)*\left[\frac{ln\left(\frac{E_0}{E_0-kh}\right)}{ak}-h\right]^n, \qquad (1)$$

where h is the depth of the particle detection, n is a number of interactions, E_0 is the energy, K is the r specific particle energy loss factor, σ_0, a, l are the approximation coefficients.

The CPF has the following physical meaning: it is the probability that the primary particle with energy E reaches a depth h after n interactions.

For ions unlike electrons, protons and α-particles dependence of the approximation coefficient of cross section on energy is represented as:

$$\sigma(h)=\sigma_0\left(\frac{1}{a(E_0-kh)}-1\right). \qquad (2)$$

In finding the approximation parameters of (1) and (2) we used the following (Table 1):

1. The cross section of atom-atom interactions were calculated by the Rutherford formula.
2. The observation depths and other parameters were found out using tables of parameters of the spatial distribution of ion-implanted impurities [3].

2. Results and discussion

As an example, Figures 1 - 4 show the distribution of implanted ions of arsenic and indium and energy loss in depth in irradiated silicon as a function of depth for different energies of the incident ions: 1 – 245; 2 – 350; 3 – 500 keV.

The distribution curves of implanted ions for different energies as for arsenic and indium have distinct maxima, and their concentration is irregularly distributed across the depth of silicon.

As calculations show the distribution of energy losses for the ionization and excitation in the depth of the material represents quite sharply decreasing functions. The distributions of knocked-on atoms in depth at a certain extent repeat the distribution of the implanted impurity. Comparing the calculations of the distributions of implanted particles for arsenic (245 keV) and indium (350 keV) in silicon, we can observe a good agreement between them. The small difference of calculations and experimental data [5] for arsenic is connected with the incomplete survey on the temperature environment effect.

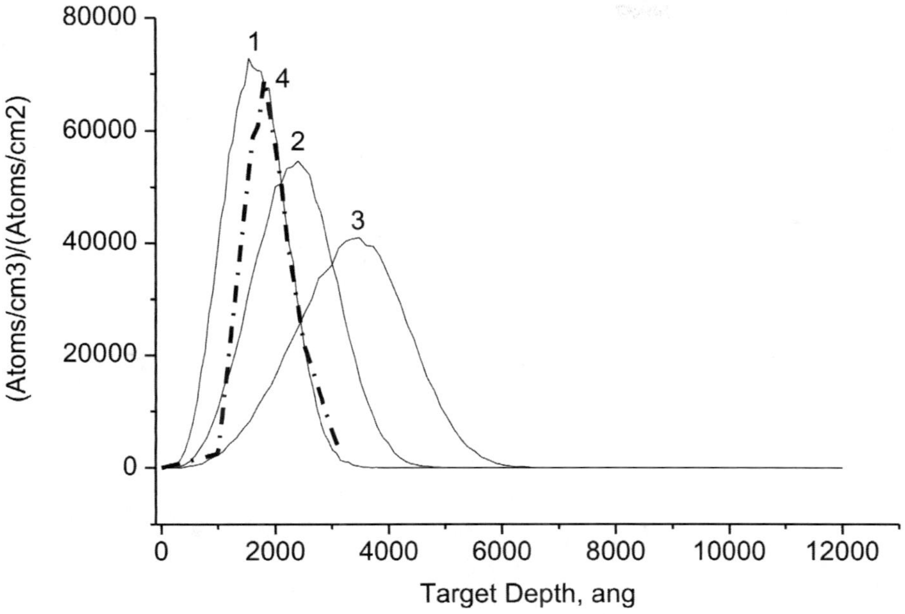

1 – 245 keV; 2 – 350 keV; 3 – 500 keV; 4 – experiment (245 keV)

Figure 1. The distribution of the implanted As in depth of silicon at the irradiation by ion beams of different energies

The calculations of cascade-probability functions were performed in double precision over the entire interval of observation depths. The correlation coefficient (Table 1) between σ calculated by the formula (2) and using the Rutherford formula is more than 0.99, that testifies about a good approximation (2). The calculation results show that the CPF depending on h and n have the following behavior: they increase, reaching a maximum and then they decrease. The ratio dependence of the displacement of left and right boundaries of the domain results CPF on the penetration depth for arsenic (245keV) and indium (350 (keV)) in silicon are shown in Tables 2 and 3. It follows from them that the real region of existence of the CPF on various parameters displacement of boundaries (C1, C2, C3) may differ from h/λ to 30%. Although all the curves at first glance have the same behavior, the areas of result findings are different. All the CPF dependencies are located in a very narrow range of depths and number of interactions. Depending on n the CPF with increasing the depth of observation is displaced to the right, the function value at the maximum point is decreased, the distance between two adjacent curves increases. With the increase of atomic weight of the incident particle and the target the CPF behavior is similar. Depending on h the CPF with increasing number of interactions is also shifted to the right, the values of functions at the maximum point decrease, with increasing atomic number of the incident particle and the target the distance between two adjacent curves, the real area a results finding and the values of the maxima are significantly decreased (Figures 5 and 6). The obtained results can significantly improve understanding of radiative processes in materials and simplify the calculations of other parameters and characteristics of the particles and defects.

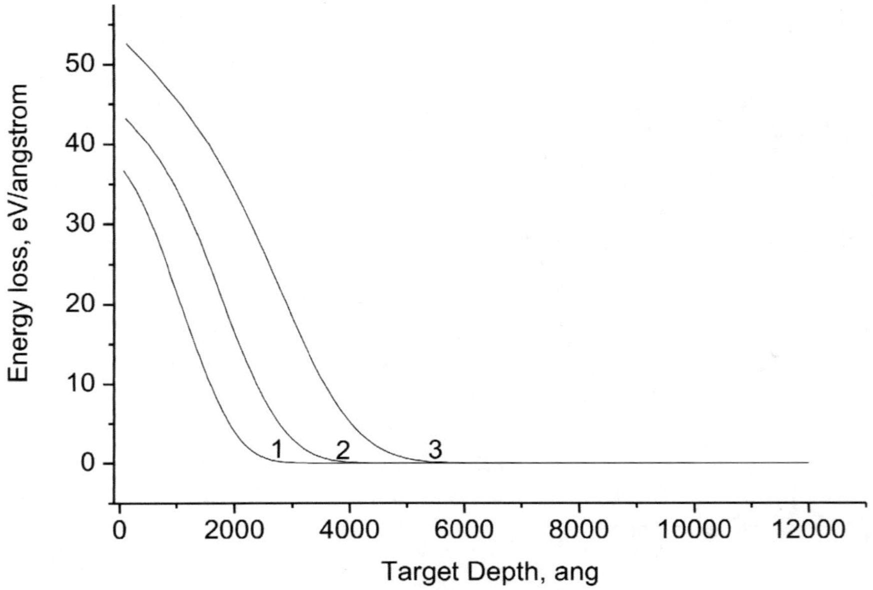

1 – 245 keV; 2 – 350 keV; 3 – 500 keV

Figure 2. The distribution of the energy loss in depth when irradiated silicon by As ions of the different energies

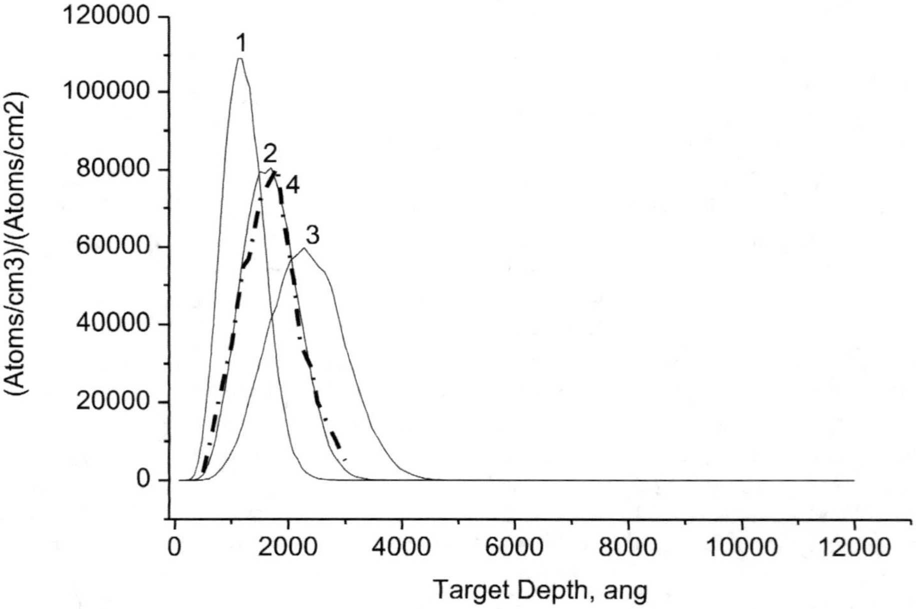

1 – 245 keV; 2 – 350 keV; 3 – 500 keV; 4 – 350 keV; experiment

Figure 3. The distribution of the implanted *In* in depth of silicon when bombarded by the streams of the particles of the different energies

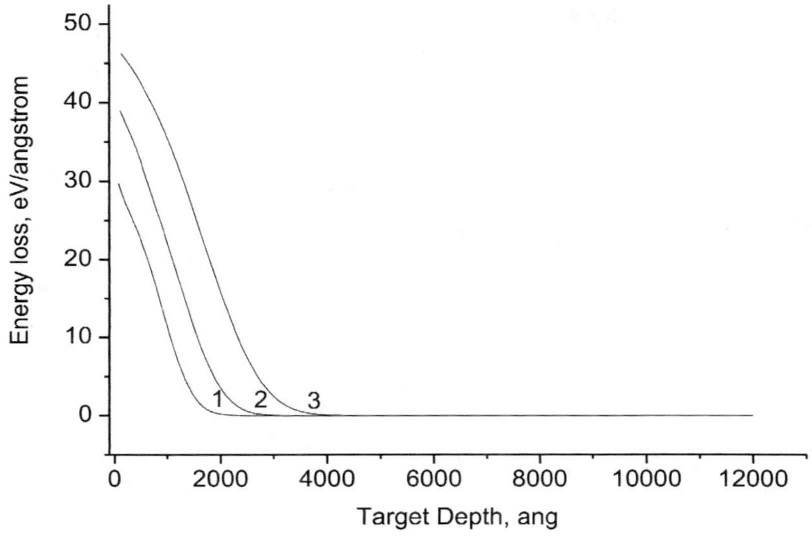

1 – 245 keV; 2 – 350 keV; 3 – 500 keV

Figure 4. The distribution of the energy loss in depth when irradiated silicon by *In* ions of the different energies

Table 1. The approximate parameters for arsenic and indium in silicon

E_0, *keV*	$\sigma_0 *10^{10}$	*a*	E_0	*k*	η	chi^2
245 arsenic	0.39188	0.23283	0.74915	1311.42461	0.99619	14.85065
350 indium	0.7804	0.15684	0.66457	1108.63961	0.99372	225.53162

Table 2. The ratio dependence of the displacement of left and right boundaries of the domain results CPF on the penetration depth for arsenic in silicon at E_0 = 245KeV

$h*10^4$, cm	h/λ, cm	C_1, %	C_2, %	N_h	C_3, %
1	116611	- 9.1	12.5	500	3.4
2	306622	- 20.24	22.15	1000	1.91
3	651245	- 28.625	29.58	2200	0.955
4	1421513	- 29.36	29.65	6500	0.29

Table 3. The ratio dependence of the displacement of left and right boundaries of the domain results CPF on the penetration depth for indium in silicon at E_0 = 350 keV

$h*10^4$, cm	C_1, %	C_2, %	N_h	C_3, %	h/λ, cm
1	-8.60	10.5	900	1.9	409946
1.5	-13.75	15.1	1300	1.35	689775
2	-18.68	19.7	1900	1.02	1044469
2.5	-23.25	24.05	2400	0.8	1506130
3	-27.12	27.685	3000	0.565	2128270
3.5	-29.65	30.035	4500	0.385	3007239
4	29.925	30.14	8000	0.215	4335823

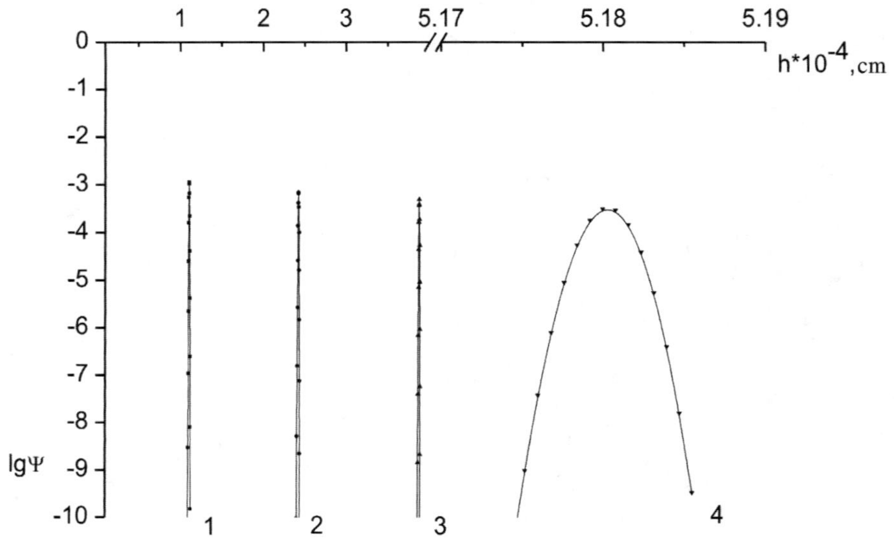

n = 116611, 306622, 651245, 1421513 (1-4)

Figure 5. The dependence of the CPF for arsenic in silicon on h at E_0 = 245 keV

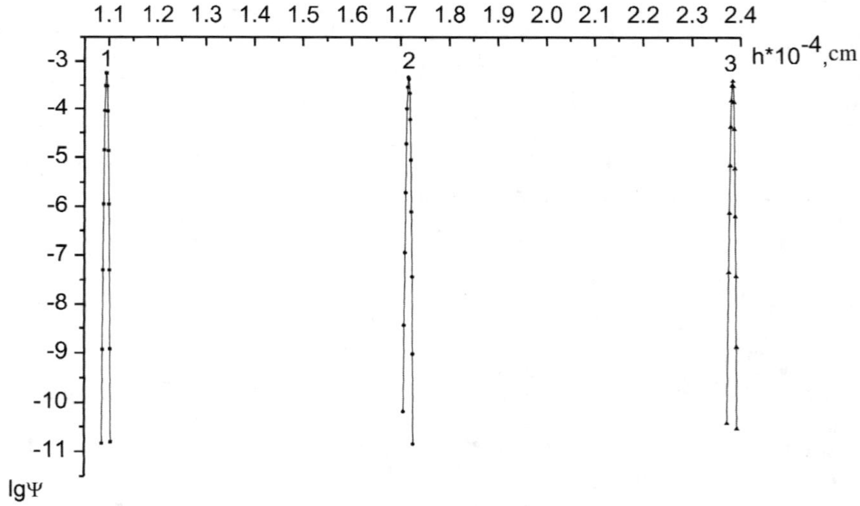

n = 409946; 689775; 1044469 (1-3)

Figure 6. The dependence of the CPF for indium in silicon on *h*, with E = 350 keV

3. Conclusions

1. The computer simulation of the implanted arsenic and indium ions distributions in depth with energies E = 245, 350 and 500 keV, as well as energy loss in the silicon. It was found that the curves for different energies for arsenic and indium in silicon have distinct maxima, and their concentration is irregularly distributed in depth. With the energy increasing the curves are shifted to the right and the value of their maxima decreases. Comparison of the calculations for arsenic (245 keV) and indium (350 keV) gives a good agreement with experiment.

2. The calculations show that the distribution of energy losses for the ionization and excitation in the depth of the material (arsenic and indium in silicon) represents quite sharply decreasing functions for all the calculated energies (245, 350, 500 keV).
3. The computer modeling of cascade-probability functions depending on the depth of penetration of arsenic and indium in silicon was carried out. As it follows from the calculations, all CPF dependences are located in a very narrow range of depths and number of interactions. Depending on n the CPF are shifted to the right relatively h/λ with increasing the depth of observation, the value of the maximum point is decreased. With increasing atomic weight of the incident particle and the target the CPF behavior is similar. Depending on the depth the CPF (relative to h/λ) with increasing number of interactions is also shifted to the right, the values of functions at the maximum point and the real area of the result finding is significantly decreased.
4. The analysis of the characteristics of ion was executed depending on the depth of penetration and the number of interactions. It is found that the approximation coefficient of the cross section (Eq. 2) is quite good (with a correlation coefficient of more than 0.99) describes a similar value calculated using the Rutherford formula and energy losses listed in Tables Kumahova- Komarova. The percentage displacement of the left and right boundaries of the domain results of the CPF depending on the penetration depth for arsenic and indium ions in silicon (C1, C2, C3) may differ (relative h/λ) by 30%.

Acknowledgement
This work was supported by the grants of the Ministry of Education and Science of the Republic of Kazakhstan.

References

[1] Deans J, Viniyard J 1960 *Radiation effects in solids* (Moscow: FL) p 243.
[2] Kupchishin A A, Kupchishin A I, Shmygaleva T A 2007 *A computer modeling of radiation-physical problems* (Almaty: "Kazak universiteti") p 432.
[3] Burenkov A F, Komarov F F, Kumahov M A, Temkin M M 1980 *Table parameters of the spatial distribution of ion-implanted impurities* (Minsk: Izd. BSU nam. Lenin) p 352.
[4] Jorabayev A B, Kupchishin A A, Kupchishin A I, Smygalev E V, Shmygaleva T A 2013 *Bulletin. Chemistry series.* **42 (71)** 112-117.
[5] Komarov F.F., Milchanin O V, Kupchishin A I, Togambaeva A K, Shmygaleva T A 2008 *News of higher education institutions. Series physics* **11 (3)** 31-37.

Study of the fluorescence blinking behavior of single F_2 color centers in LiF crystal

S V Boichenko[1], K Koenig[1], S A Zilov[1,2*], V P Dresvianskiy[1], A P Voitovich[3] A L Rakevich[1], A V Kuznetsov[1], A V Bartul[1], E F Martynovich[1,2]
[1]Irkutsk Branch of Institute of Laser Physics of Siberian Branch of Russian Academy of Sciences, Irkutsk, Russia
[2]Irkutsk State University, Irkutsk, Russia.
[3]Institute of Physics, National Academy of Sciences of Belarus, Minsk, Belarus

E-mail: zilov@ilph.irk.ru

Abstract. Using confocal fluorescence microscopy technique, we observed experimentally the luminescence of single F_2 color centers in LiF crystal. It is disclosed that the fluorescence shows blinking behavior. It is shown that this phenomenon is caused by the F_2 center reorientation occurring during the experiment. The ratio of luminescence intensities of differently oriented centers is assessed theoretically for two different experiment configurations. The calculated ratios are in fine agreement with experimental result.

1. Introduction

Color centers in alkali-halide crystals are interesting for quantum optics, nano-optics, laser physics and other applications. They have long been studied and thoroughly investigated. There are a lot of reviews and monographs on this subject [1, 2]. However, most fluorescent studies in the 20th century were dedicated to the study of color centers on the ensemble level (that is fluorescence of a large number of centers was analyzed). With the development of single-molecule fluorescence microscopy [3], it became possible to examine individual quantum systems. This allows to get rid of averaging of properties over the ensemble, and thus to obtain new more detailed information. Apparently, the paper [4] was the first work in which a single color centers in crystals of lithium fluoride (F_2 centers) were used as light sources for scanning optical microscopy of high resolution. In [5] single color centers of the same type in the same crystal were studied by confocal scanning fluorescence microscopy for the first time.

In this paper, we make an attempt to explore the reorientation of single F_2 centers in LiF crystals with laser scanning confocal fluorescence microscopy.

2. Experiment results and discussion

γ-irradiated LiF crystal was used as an experimental sample. It was gradient annealed for studying areas with different density of color centers. In particular, the sample contained areas with a concentration low enough to observe single color centers. Observation of fluorescence of single F_2 centers in LiF was carried out using fluorescent confocal laser scanning microscope MicroTime 200 (PicoQuant, GmbH). Fluorescence was excited with pulse semiconductor laser with wavelength 470 nm. Repetition rate was 5 and 10 MHz, pulse duration was 60 ps. Excitation laser light had linear polarization. Laser beam was perpendicular to crystal surface. Vector of polarization was oriented along the two crystallographic axes by rotation of the crystal: [100] axis (orientation "0°") or [110] (orientation "45°").

Content from this work may be used under the terms of the Creative Commons Attribution 3.0 licence. Any further distribution of this work must maintain attribution to the author(s) and the title of the work, journal citation and DOI.
Published under licence by IOP Publishing Ltd

Journal of Physics: Conference Series **552** (2014) 012048 doi:10.1088/1742-6596/552/1/012048

A scanned fluorescent image of 8×8 μm area in LiF sample is shown at Fig. 1. Excitation wavelength was 470 nm. Light spots at Fig. 1 are single F_2 color centers (it will be shown below). After recording of the fluorescent image the excitation laser was aimed at one of the light spots for recording of dependence of fluorescence intensity on time (fluorescence quantum trajectory). Examples of fluorescence quantum trajectory are represented at Fig. 2. At the first quantum trajectory (left part of Fig. 2) the intensity of fluorescence decreases to the background level. Then the intensity is restored to its initial level. At the second quantum trajectory (right part of Fig. 2) the intensity in the beginning falls to the background level, then rises to a level higher than initial level, then again decreases to the background level. Abrupt and complete disappearance of fluorescence ("step" on the quantum trajectory) is proof that we observe a single color center.

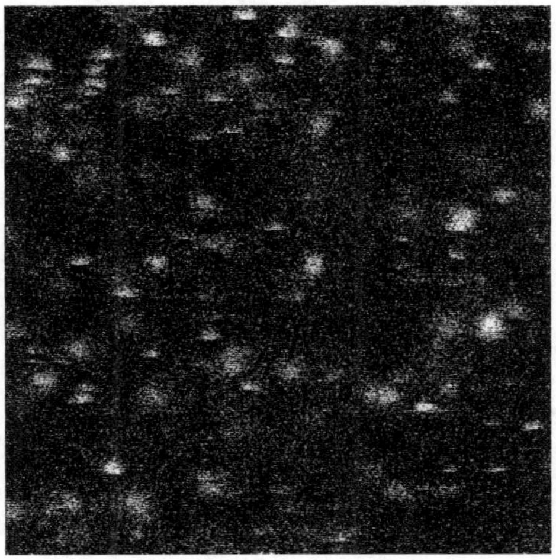

Fig. 1. Fluorescence of single F_2 color centers in the LiF crystal. Area of scanning is 8×8 μm. Wavelength of excitation is 470 nm.

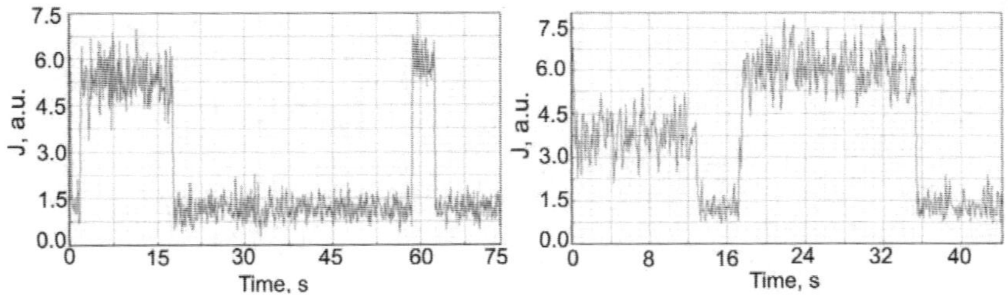

Fig. 2. Quantum trajectories of the fluorescence of single F_2 color centers.

As is known, F_2 center consists of two adjacent anion vacancies placed along C_2 axis in LiF lattice with a pair of trapped electrons. Absorption at 440 nm and fluorescence at 670 nm of F_2 center are described with linear oscillator oriented along one of six possible axes C_2 [6]. Reorientation of a center means a jump of one of the anion vacancies to the next site of the crystal lattice. It is impossible for F_2 centers in the ground state at room temperature since these centers are stable and persist for years. However, if it is assumed that the two-photon absorption center has metastable (possibly triplet) mobile state, such reorientation under laser excitation can be possible. Recently reorientation diffusion of F_2 color centers in LiF crystals was studied by analyzing the correlation of polarization components of the fluorescence [7]. Fluorescence of ensemble of F_2 centers in a sample was detected in these recent works. Accuracy of measurement allowed obtaining very rough estimates of the diffusion parameters and the probability of reorientation only. Observation of the fluorescence of single F_2 centers allows a more clear and precise study of this phenomenon.

Based on the model of a linear oscillator, it is possible to calculate the ratio of the intensities of the fluorescence of a single F_2 center for its six possible orientations of the crystal, for "0°" and "45°" experimental configurations respectively (calculation is illustrated at Fig. 3). We obtained the following ratio of the intensities of the luminescence: $I_{1,2}$:$I_{3,6}$:$I_{4,5}$ =1:2:0 for the "0°" configuration, I_6:$I_{1,2,4,5}$:I_3 =8:1:0 for the "45°" configuration. Accordingly to it, if reorientation of a center actually occurs in experiment, then fluorescent signals with intensity ratios 1:1 и 1:2 should be observed in quantum trajectories for the «0^0» configuration and intensity ratios 1:1 и 1:8 for the «45^0» configuration. The quantum trajectories at Fig. 2 were recorded for the "0°" configuration. The first trajectory has the ratio of intensity of fluorescent signals 1:1.15 and the ratio at the second trajectory is 1:1.86. These ratios correspond to theoretical considerations with account of experimental inaccuracy. Thus, it can be stated that under laser excitation an F_2 center "disappears" for some time (stops fluorescence) and in some cases (statistically) reappears in a different orientation. In order to check this claim we processed the nine different quantum trajectories for the «0^0» configuration. Statistical processing gave the ratio 1:2.2 with a relative error of 20%. This ratio is in agreement with theoretical value. A similar study for the "45^{0}" configuration gave a mean value of the intensity ratio 1:7.5 with a relative error about 60%. The reason of so high value of error is low intensity $I_{1,2,4,5}$ (Fig. 3) at quantum trajectories (it is comparable to noise level). Thus, this result is not precise but it does not contradict to the expected value of the intensity ratio. The investigation suggests that the experimental values of the ratio of the intensities agree with the calculated values. It supports the hypothesis on the reorientation of the color center in the excited state.

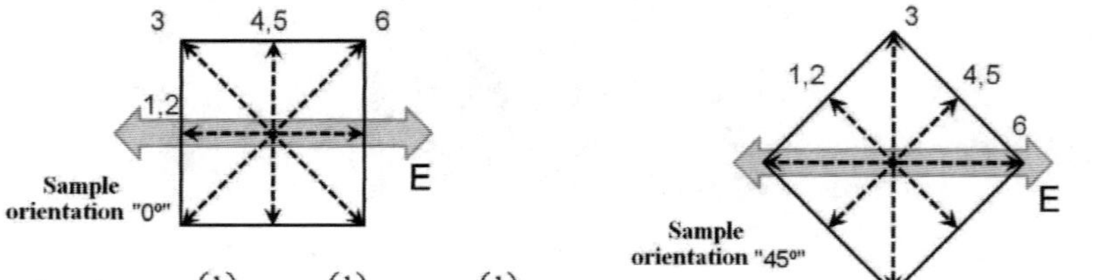

$$P_{1,2}^{abs} = C\cos^2\alpha = C\left(\frac{1}{2}\right);\quad P_{1,2}^{emit} = C\left(\frac{1}{2}\right);\quad I_{1,2} = C\left(\frac{1}{4}\right);$$
$$P_{3,6}^{abs} = C\left(\frac{1}{2}\right);\quad P_{3,6}^{emit} = C(1);\quad I_{3,6} = C\left(\frac{1}{2}\right);$$
$$P_{4,5}^{abs} = 0\quad I_{4,5} = 0;$$

$$P_{1,2,4,5}^{abs} = C\left(\frac{1}{4}\right);\quad P_{1,2,4,5}^{emit} = C\left(\frac{1}{2}\right);\quad I_{1,2,4,5} = C\left(\frac{1}{8}\right);$$
$$P_{6}^{abs} = C(1);\quad P_{6}^{emit} = C(1);\quad I_6 = C(1);$$
$$P_{3}^{abs} = 0\quad I_3 = 0;$$

Fig. 3. The illustration of calculation of intensity ratio of color centers with different orientation in two experimental configurations.

3. Conclusion

Thus, we have shown that under the action of the laser excitation with a wavelength of 470 nm a single F_2 color center goes to non-fluorescent metastable mobile state (most likely in result of two-photon absorption) in which reorientation is possible.

Acknowledgments

This work was carried out in the Program SB RAS –NASB (the project №13) and the RFBR project № 14-02-90033.

References

[1] Hayes W, Stoneham AM 1985 *Defect and Defect Processes in Nonmetallic Solids* (Wiley).
[2] Seitz F 1946 *Rev. Mod. Phys.* **18** 384
[3] Osad'ko IS 2006 *Phys.-Usp.* **49** 19
[4] Sekackii SK, Letokhov VS 1996 *JETP Letters* **63** 311
[5] Martynovich EF, Kuznetsov AV, Rakevich AL 2010 *Nanophysics and nanoelectronics. Proceedings of the XIV International Symposium 15-19 March 2010* (Nizhny Novgorod, IPM RAS) [in Russian].

[6] Feofilov PP 1959 *Polarized luminescence of atoms, molecules and crystals* (Moscow, State Publishing company of Physical and Mathematical Literature) [in Russian].
[7] Zilov SA, Baskakov DB, Martynovich EF, Rakevich AL 2011 *Izvestia vysshikh uchebnykh zavedeniy. Fizika* **2/2** 69 [in Russian].
[8] Schwille P, Haustein E 2004 *Fluorescence Correlation Spectroscopy: An Introduction to its Concepts and Applications* (Max-Planck-Institute for Biophysical Chemistry).

Formation of neutron fields for radiation technologies

V A Varlachev, E G Emets, S I Kuznetsov*, A M Bogdan and N V Varlacheva
National Research Tomsk Polytechnic University, TPU, Tomsk, Russia

E-mail: smit@tpu.ru

Abstract. The article considers the problem of forming a neutron field in the experimental channels of the water-cooled research reactors. Using the software package MCU5 we calculated a neutron field in different moderators which the experimental channel passes through. It is shown that the most appropriate moderator for neutron transmutation doping is the beryllium. The optimal location of the channel in the moderator was determined.

1. Introduction

In experiments conducted with the use of research nuclear reactors, selection of a moderator is often a decisive factor for the formation of a neutron field with the required parameters. In particular, we considered the problem of neutron transmutation doping (NTD) of silicon in water-cooled research reactors, for example, in a typical research reactor (IRT). Using the software package (MCU5TPU), we calculated a neutron field in various moderators. The experimental channel for NTD of silicon passed through these moderators. Calculations were carried out to compare the capabilities of various moderators to form the neutron field for NTD of silicon. Fig. 1 shows the "geometry" of calculations.

2. Description of reactor's construction

The core [1] consists of 20 fuel assemblies and a beryllium "trap" of thermal neutrons. The core 2 is surrounded by two reflectors 1 and 3 on four sides. The reflector 1 is made of beryllium; the reflector 3 is made of either beryllium or material of the moderator. The moderator 5 adjoins the reflector 3. The height (along the axis z) of the core, reflectors and moderator is equal to 60 (cm). In the horizontal section, the thickness of the reflector 1 is 7 (cm), the core 2 has the following dimensions: 42 (cm) (along the axis x) and 28 (cm) (along the axis y). The thickness of the reflector 3 (along the axis y) is either 7 (cm) or 14 (cm). The total thickness of the reflector 3 and the moderator 5 (along the axis y) is 63 (cm), and the width is 56 (cm) (along the axis x). All this is surrounded by water from a reactor pool.

Content from this work may be used under the terms of the Creative Commons Attribution 3.0 licence. Any further distribution of this work must maintain attribution to the author(s) and the title of the work, journal citation and DOI.
Published under licence by IOP Publishing Ltd

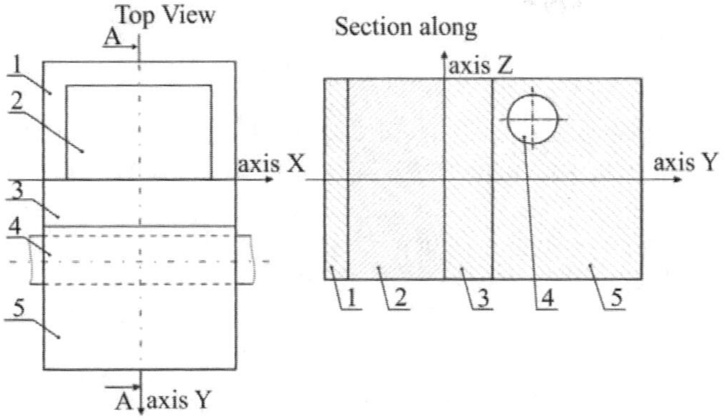

Figure 1. Scheme of a core with a reflector and a moderator:
1 – reflector from Be; 2 – core; 3 – reflector; 4 – HEC-4; 5 – moderator

In some calculations, the experimental channel is made of steel 4 with a wall thickness of 4 (mm) passed through the moderator 5. The internal diameter of the channel is 15 (cm). The channel centerline is at a distance of 12.5 (cm) from the top of the core. We considered the moderators from beryllium, heavy water and graphite. In the calculations for the core we took into account the constructional materials and coolant (water) at a distance of 12.5 (cm) from the top of the core.

Fig. 2 shows the flux distribution of thermal neutrons versus the distance to the core (along the axis y) (experimental channel is not available). We calculated the average flux density of thermal neutrons in layers of an information prism. The thickness of each layer is equal to 2.3 (cm). The total thickness of such information prism is equal to the total thickness of the reflector and moderator (63 (cm)), the height is 15 (cm), and the width is equal to the width of the moderator (56 (cm)).

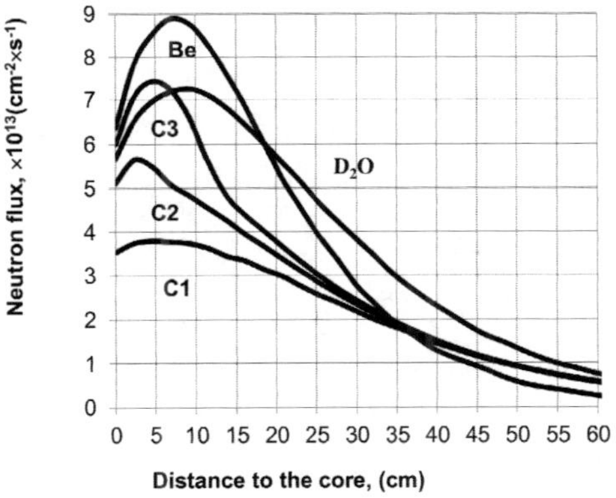

Figure 2. Flux density of thermal neutrons in different moderators

The prism axis, parallel to the side of the core, as well as a channel axis that will be considered in the further calculations, is at a distance of 12.5 (cm) from the top of the core.

The point '0' on the X-axis corresponds to the boundary of "core 2 – reflector 3 with a moderator 5" (Fig. 1). For all distributions the material of the reflector 1 (Fig. 2) was beryllium. We used the following materials as a reflector 3 and a moderator 5: distribution C1 – graphite – graphite; distribution C2 – beryllium (7 cm) – graphite; distribution C3 – beryllium (14 cm) – graphite; distribution D2O – heavy water – heavy water; distribution Be – beryllium – beryllium.

Beryllium and heavy water are more favorable moderators in comparison with other moderators in terms of the maximum flux density of thermal neutrons in the channel. However, beryllium is easy to operate and readily available for us than heavy water. Therefore, it was selected as a reflector 3 and a moderator 5 (Fig. 2).

The distribution of thermal neutron flux density along the X-axis in homogeneous reflectors of beryllium, heavy water, graphite, perpendicular to the faces of the core, has a maximum at a distance ($\tau^{1/2}$ – age thermal neutrons) from the reactor core and reflectors of ordinary water – at a distance, equal to the diffusion length of thermal neutrons [3]. If this maximum will be in the reactor channel, even due to the rotation radial inhomogeneity of doping persists. Therefore, when silicon is irradiated in the tangent channel passing through the reactor's reflector, the reflector thickness between the core and the channel must be greater than the specified distance (in our opinion, 1.2-1.4 times). The other two axes distribution of neutrons in the first approximation is the cosine. The peak in the distribution of one of them (the Y-axis), perpendicular to the channel axis, should not be in the reactor channel for the same reason. The distribution of neutrons along the channel (Z-axis) affects only the axial uniformity of doping of silicon ingots. And in a translation container with bars on the channel across the irradiation zone axial doping uniformity is achieved for any distribution of the neutron channel. These considerations were taken into account in the development of the irradiation device for NTD technology at the research reactor of IRT type in Tomsk which power is 6 MW.

The results of calculations for the neutron field in the experimental channel are shown in Table 1 (L is the distance from the edge of the channel to the core; φaverage is the average neutron flux density for the length of 164 (cm); δ is the fraction of thermal neutrons in the spectrum; D is the radial flux density distribution of thermal neutrons for the diameter of 14 (cm); Q is the spectral factor (the ratio of the thermal neutron flux density to the integrated neutron flux density with an energy above 3 (MeV)). Based on these calculations, we determined the location of the channel L = 15 (cm) (the distance between the canal centerline and the fuel assembly is 22.5 (cm)). Table 1 shows the results of calculations.

Table 1. The results of calculations for the neutron field in the experimental channel

L, (cm)	$\varphi_{average}$, 10^{12}(cm^{-2}s^{-1})	δ, (%)	*D*, (%)	*Q*
2	3.87	28.3	3.59	8.15
4	4.53	35.6	4.09	17.1
6	4.72	41.4	2.26	31.4
8	4.84	48	3.45	52.5
10	5.00	55	2.66	97.9
12	4.82	60	1.10	156
14	4.66	66	1.56	245
16	4.29	72	1.54	426
18	4.11	76	2.19	599
20	3.78	81	2.03	879
22	3.45	84	2.63	1453
24	3.08	87	2.18	1622
26	2.74	90	3.09	3442

For creating a new facility for radiation technology, it was necessary to determine the parameters of the irradiation zone. To do this, experimental researches of the characteristics of the neutron field in the HEC-4 were carried out. Activation detector method was used in measurements of the neutron field. The relative measurement technique was used to determine the distribution of thermal neutron flux. As activation detectors the copper foil and copper wire were used, which were fixed on a special device and using a rod were fed into the reactor channel. The plotted reference points on the bar allowed to set the device in the reactor channel with an accuracy better than 3 mm, i.e. each activation detector was set at a predetermined point of the irradiation zone with an accuracy of ± 2 mm, and their mutual arrangement – with an accuracy of about 1 mm. The irradiation of the activation detector was performed at the reactor power of 20 kW for 30 minutes.

After 24-hour incubation, portions of interest were excised from the irradiated foil and wire. Then they were weighed and dissolved in nitric acid. Activity measurements were carried out in the solutions of the same geometry using the installation of γ-γmatches. In this case the relative measurement error did not exceed 3 % of relative one. The measurement results are shown in Fig. 3.

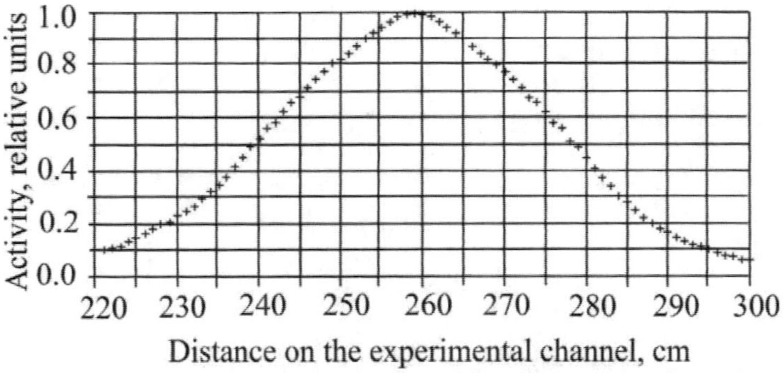

Figure 2. Distribution of the thermal neutron flux through the channel HEC-4.

3. Conclusion

Using the complex of neutron doping of silicon, we received the following parameters of the neutron field: the fraction of thermal neutrons in the spectrum is 73 %; the spectral factor (the ratio of the thermal neutron flux density to the integrated neutron flux density with an energy above 3 (MeV)) is 106; the effective temperature of thermal neutrons is 337 K; the maximum thermal neutron flux density in the channel with a irradiator is $1.5 \cdot 10^{13}$ (cm^{-2} s^{-1}); the average thermal neutron flux density (along the length of the reciprocating movement of the container) is $4.1 \cdot 10^{12}$($cm^{-2}s^{-1}$); radial irregularity of neutron fluence during rotation of the container is less than 1% . All this parameters allowed us to dope silicon with a world-class quality. Furthermore, the obtained results can be used to create other radiation technologies.

Acknowledgment

This work was supported by the grant of the Ministry of Education and Science of Russian Federation (RFMEFI59114X0001).

References

[1] Gluhov G G and Didenko A N 1988 *Atomic Energy* **64 5** 366-370 (in Russian)
[2] Varlachev V A, Solodovnikov E S 2010 *Russian Physics Journal* **10/2** 310-312. (in Russian)
[3] Bat G A 1985 *Research Nuclear Reactors* (Moscow: Atomizdat) (in Russian)
[4] Kulakov V M, Ladygin E A and Shahovtsov V I 1980 *Effect of Iionizing Radiation on Electronic Products* (Moscow: Soviet Radio)

Journal of Physics: Conference Series **552** (2014) 012050 doi:10.1088/1742-6596/552/1/012050

Electric pulse breakdown and rock fracture in a coupled environment of increased pressure and temperature

V F Vazhov, S Yu Datskvich, M Yu Zhurkov, V M Muratov
Institute of High Technology Physics, Tomsk Polytechnic University

Email: vazhov@tpu.ru

Abstract: This paper represents the research results of electric breakdown of drilling fluid Versa Pro and rocks, as well as rock fracture caused by pulsed electric discharge in a coupled environment of increased pressure and temperature. It proves that rock destruction including rock boring by means of electric discharge is possible at a depth of down to 3500 m.

1. Introduction

Conventional mechanical methods of rock destruction (drilling) have reached their engineering potential. Nowadays it is necessary to develop new, more efficient, ways and methods of rock destruction. An electric discharge technology of rock fracture discovered and developed in the Tomsk Polytechnic University, Russia, meets most requirements specified to new ways of rock destruction [1]. Physical phenomenon of the technology lies in exceeding the electric strength of liquid dielectrics over the electric strengths of solid dielectrics and rocks when applying an electric pulse voltage for 1 microsecond and less [1,2]. When placing electrodes on a rock surface and applying the high voltage to one of the electrodes, an electric discharge channel penetrates into solid dielectrics. Here it is crucially important that liquid dielectric covers the rock surface, i.e. a rock block should be under the liquid. Breakdown of a solid dielectric (rocks) is a ground for the electric discharge technology. This technology fundamentally differs from other methods of rock fracture by electric field. When a breakdown has occurred, the energy, that has been stored in the pulse voltage generator, releases into a discharge channel in the solid dielectric for a period of 1-10 microseconds. This results in blast-like processes which cause the rock fracture mainly due to tension forces. A working tool that fractures the rock is plasma of a discharge channel; the plasma restores again and again, from pulse to pulse, as pulses come to working electrodes from the high-voltage generator and such an electrode system allows drilling for hundreds meters without replacing the system. The very essence of the technology lies in possibility of getting higher, compared to conventional methods, performance in rock destruction in combination with lower energy consumption.

The electric discharge technology has nowadays found its application in drilling, crushing and cutting off rocks. At present a method of electric discharge boring is of most-in-demand. The increase in drilling depth causes the increase in hydrostatic pressure and temperature on the bottomhole. There are very few references to the cooperative effect of increased pressure and temperature on the breakdown voltage and rock fracture performance at a pulse voltage with a pulse rise time of 1 microsecond and less, available.

Physical and mechanical and electro-physical properties of rocks in the interior of the Earth are significantly determined by a joint effect of increased pressure and temperature occurring at great depths that surely influence the electric discharge rock fracture characteristics. At the electric

Content from this work may be used under the terms of the Creative Commons Attribution 3.0 licence. Any further distribution of this work must maintain attribution to the author(s) and the title of the work, journal citation and DOI.
Published under licence by IOP Publishing Ltd

Journal of Physics: Conference Series **552** (2014) 012050 doi:10.1088/1742-6596/552/1/012050

discharge rock fracture, one observes reduction of the fracture zone in dielectrics and reduction of rock destruction performance when the hydrostatic pressure increases up to 15 MPa [3,4].

2. Methodology

The research has been aimed at determining the feasibility of electric discharge boring of rocks at depths of down to 3500 m. In order to carry out the research work, there has been developed a test bench allowing to create a pressure of up to 35 MPa, temperature of up to 120 C^0 and to apply the pulse voltage of up to 400 kV to rock sample blocks. We used granite, limestone and sandstones as rock samples and drilling fluid Versa Pro developed by the company Schlumberger as a working medium. A two-electrode system (electrode-electrode arrangement) with an interelectrode gap of 10-30 mm was placed on a rock sample surface and 5 pulses were applied to the rock surface. Before being tested the rock blocks were kept in a drilling fluid for an hour to be impregnated. The depth of rock impregnation with drilling fluid exceeded the depth of penetration of a discharge channel into rock.

3. Results and discussion

An average breakdown strength E_{br} of the rocks mentioned varies as a function of pressure P and temperature T increasing and having their maximum at P=5-10MPa and T=35C^0 that is conditioned by the fluid electric strength when the pressure is increasing (Fig.1) [5]. Further increase of P and T causes decrease in E_{br} of the rocks, like of the liquid. Such a decrease is caused by the increase in the temperature as being a factor which, to a large extent, effects a dielectric liquid [6], i.e. under these conditions the electric strength of the rock impregnated with the liquid is mainly determined by the liquid. And the probability of penetration of a discharge channel into the rock is 100% that allows one to conclude that the rock drilling at the depth of down to 3500 m is possible.

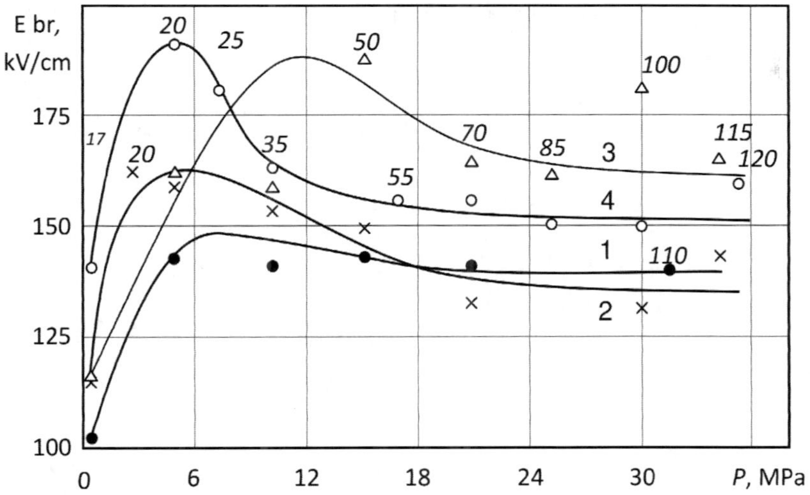

Figure 1. The function of rock electric strength and pressure and temperature:
1 – sandstone, 2 –limestone, 3 – granite, 4 – drilling fluid (figures next to dots are temperature values)

The rock destruction occurs in a blast-like way due to the energy released in the discharge channel. The joint increase in both the pressure of up to 35 MPa and the temperature of up to 120C^0 reduces the fracture performance by 5-11 times, particularly, the most significant reduction of the performance occurs in the range of 0.1 – 5.0 MPa and up to 30C^0 that is mainly caused by the pressure increase. (Fig.2). A similar phenomenon is also typical of mechanical rock destruction including mechanical drilling. For example, the rate of penetration by both diamond and roller drilling technologies is reduced by 12-16 times when drilling the rocks of drillability index VI-IX. The electric discharge technology allows improving the fracture performance and, hence, the rate of penetration by increasing the electrode gap.

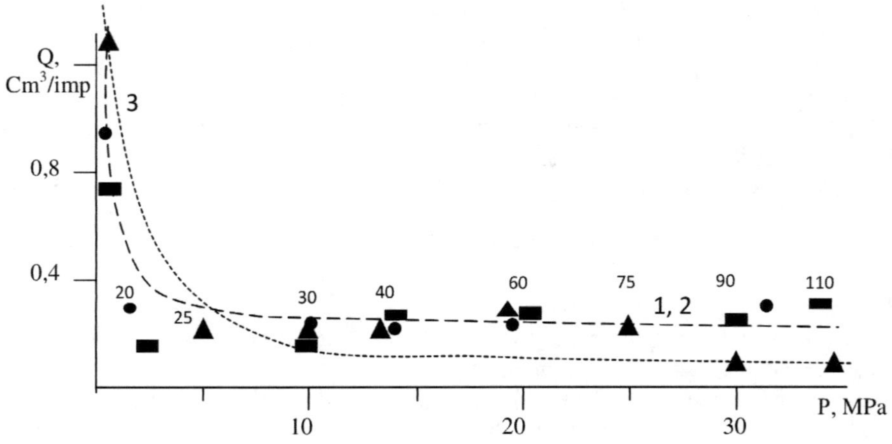

Figure 2. The function of rock destruction performance (1-3) and pressure and temperature: 1- sandstone; 2 - limestone; 3 – granite; figures next to dots are temperature values.

We have carried out test experiments on effecting the coupled environment of increased pressure and temperature on destruction performance and energy consumption as a function of the electrode gap *S* = 10-30 mm.

The table presents figures averaged for 3-5 tests for each gap distance. One can see that the bigger the gap *S*, the higher the destruction performance and the less energy consumption is.

Table 1. Fracture characteristics of the granite rock

S, mm	P, MPa	T, ^{0}C	Q, cm3 /pulse
10	25	85	0.02
20	25	85	0.11
30	25	84	0.31

As for deep drilling, here big-diameter boreholes are mainly used; it is possible for the electric discharge boring technology to increase the gap *S* of up to 30 mm and more.

4. Conclusion

The research results prove that rocks are possible to be fractured by the electric discharge technology at a depth of 3500 m and the electric strength of drilling fluid complies with requirements of electric discharge boring technology.

To our mind the technology of electric discharge boring has significant benefits and advantages compared to conventional mechanical methods of drilling. Few advantages are as follows:

1) Rather high destruction performance when having larger gaps S;

2) The drill bit does not turn ;

3) Significant depth (hundreds – down to thousands meters deep) of drilling without replacing the drill bit;

References

[1] Syomkin B V, Usov A F, Kurets V I. 1995 Basics of electric discharge destruction of materials *SPB; Nauka.* p 276.

[2] Vorob'yov G A, Chepikov A T, Vazhov V F 1992 Dynamics of electro burst in solids: I. Power characteristics of electro burst. *J. Iizvestiya vuzov. Fizika.* **12**. pp 110-113.

[3] Kalyatsky I I, Krivko V V. 1996. Izvestiya TPI. **149**. pp 163-168

[4] Kalyatsky I I, Krivko V V. 1964. *Proceedings of Interuniversity Conference "Breakdown and destruction of dielectrics and semi-conducting materials"* (M-L. Energiya) pp 249-251

[5] Under the editorship of Val'ter A F. 1932. Dielectrics Physics (GTII) p. 560

[6] Val'ter A F. 1933. Breakdown of solid dielectrics (GTII)

AUTHOR INDEX

Aduev, S B 177
Akhmetshin, R 79
Aktaev, N E 184
Anan'Eva, M V 206
Artemenko, S 120
Artyomov, A P 58
Artyomov, K P 194
Astrakchantsev, N V 36
Avgustinovich, V 120
Bakulin, A V 230
Baranova, A A 189
Barengolts, S A 125
Bartul, A V 268
Beniyash, A 18
Bogdan, A M 272
Boichenko, S V 268
Bolotov, A V 101
Borduleva, A O 52
Borisov, D P 1
Borodin, E N 6
Budko, A A 143
Bukharkin, A A 64
Bykov, N M 166
Chaikovsky, S A 58, 68, 83, 108, 125, 148
Chinkov, E P 251
Chistyakova, E K 199
Chulkov, E V 1
Dabagov, S B 194
Datsko, I M 68, 108
Datskvich, S Yu 276
Devyatkov, V N 73
Dik, A V 194
Dresvianskiy, V P 268
Dushkin, I V 31
Eltchaninov, A A 172
Emets, E G 272
Fedunin, A V 58
Filkov, A I 143
Frants, O B 101
Frolov, E N 194
Gao, Y 24
Gaydaychuk, A V 256
Geyman, V G 101
Grishkov, A A 95
Ionov, I V 52
Isaev, Yu N 143
Ishchenko, S A V 199
Ivashov, R V 24
Kabyshev, A V 13, 256
Kaikanov, M 130
Kalenskii, A V 206
Kalmykov, P I 246
Kaminsky, V 120
Kamrikova, A A 212
Karipbaev, Z 225
Kasyanov, V S 24, 101
Khailov, I 136
Khamzakhan, G 83
Khokhlov, K O 189
Kholodnaya, G 89, 130, 136
Koenig, K 268
Konusov, F V 13, 256
Korepanov, V I 212
Kornilov, S 18
Korolev, Y D 24, 101
Korotaev, A D 1
Kortov, V 218
Kortov, V S 235, 241
Koryashov, I A 64
Koval, N N 73
Kovalchuk, A N 52
Kovalchuk, B M 172
Kozyrev, A V 46
Krasnikov, V S 6
Krasov, V I 36
Kriger, V G 206
Kulkova, S E 230
Kupchishin, A A 261
Kupchishin, A I 261
Kurkan, I K 95, 166, 172
Kuskova, N I 108
Kuznetsov, A V 268
Kuznetsov, S I 272
Kuznetsov, V M 1
Kuznetsov, V V 148
Kuznetsova, N S 154, 159
Labetskaya, N A 58, 68, 108
Landl, N V 24, 101
Lavrinovich, I V 114
Lebedev, N V 36
Linnik, S A 256
Lisitsyn, V 225
Lisitsyna, L 225
Lopatin, V V 46, 64, 154, 159
Lysyk, V V 246
Makarova, A F 42
Manko, N G 199
Martemyanov, S M 64
Martynovich, E F 268
Mayer, A E 6
Mayer, P N 6
Muratov, V M 276
Murray, N 18
Nekhoroshev, V O 101
Nikiforov, S V 235
Nikitenkov, A N 31
Nikitenkov, N N 31
Nikitin, A 177
Novikov, S 120
Nurmukhametov, D 177

AUTHOR INDEX

Oleshko, V 79
Oleshko, V I 246
Oreshkin, E V 125
Oreshkin, V I 58, 68, 108, 114, 125, 148
Osipov, V 120
Ovchinnikov, V..79
Paperny, V L 36
Pavlov, S K 13
Pegel, I V..95
Petikar, P V 212
Petrovykh, K A...241
Ponomarev, D 89, 130, 136
Rabotkin, S V 52
Rakevich, A L 268
Rakhmatullin, I A 42
Ratakhin, N A...68, 148
Razin, A 79
Remnev, A G 46
Remnev, G 89, 130, 136
Remnev, G E...13, 256
Rempe, N 18
Rempel, A A 241
Rousskikh, A G 58, 148
Rud, A D 108
Sazonov, R 89, 130, 136
Shalimov, L N...199
Shao, T 24
Shemyakin, I A 24
Shestakov, G V 199
Shipilova, A V...52
Shmygalev, E V 261
Shmygaleva, T A 261
Shonokhova, A A..199
Shtan'Ko, V F 251
Shtykov, A N...199
Shulgin, B V..199
Sivkov, A A 42
Skripin, A..79
Solovyev, A A 52
Spiridonova, T I 230
Stepanov, S A 251
Stepchenko, A S..166
Sun, Y H..24
Sypchenko, V S...31
Tlebaev, K B 261
Tsipilev, V 79
Tsipilev, V P 246
Tuch, E V 230
Tyurin, Yu I...31
Uemura, K 46
Valiev, D 225
Varlachev, V A 272
Varlacheva, N V 272
Vasileva, O V 143
Vazhov, V F 276
Vikanov, A 89
Vilkhivskaya, O V 31
Voitenko, N V 154
Voitovich, A P 268
Vokhmintsev, A S 199
Weinstein, I A 199
Yakovlev, A...79
Yudin, A S 154, 159
Yushkov, Yu 120
Zarko, V 79
Zarko, V E 246
Zherlitsyn, A A 172
Zhigalin, A S 58, 148
Zhurkov, M Yu..276
Zilov, S A...268
Zvekov, A 177
Zykov, I Yu 206

Institute of Physics
Dirac House, Temple Back
Bristol BS1 6BE UK

CURRAN ASSOCIATES INC
proceedings.com